INDUSTRIAL ORGANIZATION

ADVANCES IN APPLIED MICROECONOMICS

Series Editor: Michael R. Baye

ADVANCES IN APPLIED MICROECONOMICS VOLUME 9

INDUSTRIAL ORGANIZATION

EDITED BY

MICHAEL R. BAYE

Bert Elwert Professor of Business Economics & Public Policy,
Kelley School of Business, Indiana University, USA

2000

JAI
An Imprint of Elsevier Science

Amsterdam – London – New York – Oxford – Paris – Shannon – Tokyo

ELSEVIER SCIENCE Inc.
655 Avenue of the Americas
New York, NY 10010, USA

First edition 2000

Library of Congress Cataloging in Publication Data
A catalog record from the Library of Congress has been applied for.

ISBN: 0-7623-0687-4

∞ The paper used in this publication meets the requirements of ANSI/NISO Z39.48-1992 (Permanence of Paper).
Printed in The Netherlands.

CONTENTS

LIST OF CONTRIBUTORS

Maria Arbatskaya	Department of Economics, Emory University, Atlanta, GA 30322-2240, USA
Roger Blair	University of Florida, Professor of Economics, Gainesville, Florida, USA
Keith Crocker	University of Michigan Business School, Ann Arbor, MI 48104, USA
Raymond J. Deneckere	Department of Managerial Economics and Decision Sciences, Northwestern University, Evanston, IL 60201, USA
Neil Doherty	Department of Insurance and Risk Management, University of Pennslyvania, Philadelphia, PA 19104, USA
James M. Fesmire	University of Tampa, Gunter Professor of Economics, Tampa, Florida, USA
Heidrun Hoppe	Institut fuer Allokation und Wettbewerb, von-Melle-Park 5, D-20146 Hamburg, Germany
Dan Kovenock	Purdue University, Department of Economics, West Lafayette, IN 47907, USA
Scott E. Masten	The University of Michigan School of Business Administration, Ann Arbor, MI 481099-1234, USA
John W. Maxwell	Indiana University, Kelley School of Business Economics and Public Policy, Bloomington, IN 47405, USA
Guido Merzoni	Faculty of Political Sciences & ISEIS, Università Cattolica del Sacro Cuore, 20123 Milano, Italy

Mark J. Roberts The Pennsylvania State University & NBER, 513 Kern Building, University Park, PA 16802, USA

Richard E. Romano University of Florida, Department of Economics, Gainesville, FL 32611, USA

Kamal Saggi Southern Methodist University, Department of Economics, Dallas, TX 75275-0496, USA

Yong Yeop Sohn Chonnam National University, College of Business Administration, Buk-ku, Kwangju, 500-757, Korea

Dale O. Stahl II University of Texas, Department of Economics, Austin, TX 78712, USA

Mark W. Stegeman VPI & SU, Department of Economics, Blacksburg, VA 24061, USA

Dylan Supina University of Vienna, Department of Economics, A-1201 Wien, Austria

Nikolaos Vettas Duke University, Fuqua School of Business, Durham, NC 27708-0120, USA

EDITOR'S NOTE

Guidelines for submitting papers to *Advances in Applied Microeconomics* and information about planned future volumes may be obtained via the internet on my Homepage:

http://php.indiana.edu/ ~ mbaye

Michael R. Baye
Bloomington, IN
June 2000

OUTPUT PRICE AND MARKUP DISPERSION IN MICRO DATA: THE ROLES OF PRODUCER HETEROGENEITY AND NOISE

Mark J. Roberts and Dylan Supina

ABSTRACT

This chapter provides empirical evidence on the extent of producer heterogeneity in the output market by analyzing output price and price-marginal cost markups at the plant level for thirteen homogeneous manufactured goods. It relies on micro data from the U.S. Census of Manufactures *over the 1963–1987 period. The amount of price heterogeneity varies substantially across products. Over time, plant transition patterns indicate more persistence in the pricing of individual plants than would be generated by purely random movements. High-price and low-price plants remain in the same part of the price distribution with high frequency, suggesting that underlying time-invariant structural factors contribute to the price dispersion. For all but two products, large producers have lower output prices. Marginal cost and the markups are estimated for each plant. The markup remains unchanged or increases with plant size for all but four of the products and declining marginal costs play an important role in generating this pattern. The lower production costs for large producers are, at least partially, passed on to purchasers as lower output prices. Plants with the highest and lowest markups tend to*

Industrial Organization, Volume 9, pages 1–36.

ISBN: 0-7623-0687-4

remain so over time, although overall the persistence in markups is less than for output price, suggesting a larger role for idiosyncratic shocks in generating markup variation.

I. INTRODUCTION

Two common features of producer micro data are that, regardless of the dimension examined, producers often differ as much within the same industry as they do across industries and these differences can survive over long periods of time. Besides well-recognized variation in size and capital intensity, recent empirical studies have documented extensive heterogeneity in producer profit rates, failure probabilities, growth rates, the adoption of advanced technologies, R&D expenditures, productivity levels and growth rates, and labor compensation.[1] Coincident with this accumulation of empirical evidence has been the development of theoretical models of firm and industry dynamics capable of explaining some of the empirical patterns. The key components that provide these models with their power to explain the data include: a source of firm-level heterogeneity, often modeled as differences in production efficiency, idiosyncratic shocks that lead to different firm-level learning, investment or productivity paths, and sunk entry costs that slow or prevent the sorting out of inefficient from efficient producers.[2]

While the combination of firm-level heterogeneity in costs, idiosyncratic shocks to firm productivities, and a process of market selection appear to be very important in explaining the dynamic patterns of firm growth and failure, what is missing from the literature is systematic evidence on how these forces affect the output market; specifically, the output prices and markups charged by producers and their evolution over time. This gap in the empirical record can be traced to the difficulty of measuring output prices for individual producers, particularly for representative cross-sections or at more than one point in time. This evidence is also directly relevant to the long-standing question in industrial organization of whether the superior profit performance of some firms reflects more efficient production, the ability to charge higher markups, or a combination of the two (see Demsetz (1973), Peltzman (1977), Mueller (1986), and Schmalensee (1989)). Direct assessment of this issue requires micro data on prices, costs, and markups.

The goal of this chapter is to provide empirical evidence on the extent of producer heterogeneity in the output market, specifically, the heterogeneity in output prices and markups. We document both the cross-sectional and time-series patterns in prices and markups at the plant level for thirteen products relying on data drawn from the U.S. Census of Manufactures over the 1963–1987 period. We focus on homogeneous manufactured products in order to remove, as much as possible, price differences due to heterogeneity in the output itself. The data source is unusual in several dimensions. It provides wide coverage of the cross-section of plants in an industry and is not limited to just the large or long-lived producers. It allows us to construct important cost variables, particularly the price of material inputs, at the plant level so that plant-level markups control for cost heterogeneity among producers. Finally, we can trace individual manufacturing plants over time and document the persistence of price, cost, and markup patterns.

A number of earlier studies including Stigler (1961), Stigler & Kindahl (1970), and Pratt, Wise & Zeckhauser (1979) have documented that the price of a good can vary substantially across different buyers or sellers. Output prices can vary across manufacturing plants in an industry for many reasons including, unobserved product heterogeneity or quality variation across producers, high search costs by consumers, or differences in production costs or competition across local geographic markets. It can also arise in micro data as a result of reporting or measurement errors.[3] Just as the extent of price dispersion is likely to vary across products, no single explanation will be relevant for all the products we study. Instead, the explanation will depend on the particular details of the technology, demand, market competition, and random shocks that pertain to each product. To organize the evidence for the broad range of products we study, we focus on distinguishing price and markup differences at the plant level that persist over time from transitory differences that do not. To the extent that observed price and markup differences persist, it is likely they arise from differences across producers in technology, product quality, demand conditions, market competition, or other structural factors that are constant or change slowly over time. In contrast, if uncorrelated random shocks or measurement errors are the main sources of producer heterogeneity then price and markup differences are much more likely to be transitory.

The remainder of this chapter is organized as follows. The second section discusses some unique characteristics of the *Census of Manufactures* establishment data that are relevant for output price and markup measurement. Section III summarizes the patterns of output price variation across producers, their persistence over time, and the correlation between plant size and prices. Section IV summarizes the estimation of plant-level marginal cost and Section V presents the cross-sectional and time-series patterns in plant markups.

II. OUTPUT PRICE MEASUREMENT WITH CENSUS ESTABLISHMENT DATA

In order to make meaningful comparisons of output prices across producers it is important to begin with physically identical goods that are produced by a number of plants. The data are drawn from the U.S. *Census of Manufactures*, which is a survey of all manufacturing establishments conducted in the years 1963, 1967, 1972, 1977, 1982, and 1987. As part of the census, each surveyed establishment reports the value of shipments of each seven-digit SIC product manufactured in the plant as well as the physical quantity produced for a subset of homogeneous products with well-defined units of measure. This allows us to construct a plant-specific average price, as the ratio of the annual value of shipments to the annual quantity of output shipped, for very disaggregated products in each census year.[4]

In order to compare markups across producers it is necessary to control for cost variation. Since raw materials can often account for more than 50% of production costs in a manufacturing plant it is important to control for variation in material prices across establishments. The *Census of Manufactures* also collects data on the plant's expenditure on material inputs and the physical quantity of inputs purchased for detailed raw material categories. This allows us to construct a plant-specific average price of materials for important inputs. The products we have chosen to study all have one or two well-defined material inputs that account for a significant fraction of total material expenditures in the plant and for which expenditure and quantity data are collected.

This study will examine output price dispersion across manufacturing plants for 13 products: white pan bread, roasted coffee, cotton sheeting and allied fabrics, 100% spun polyester blends with cotton, finished wool apparel fabrics, hardwood oak flooring, hardwood plywood, softwood plywood, newsprint, corrugated boxes, motor gasoline, ready-mixed concrete, and tin-plate steel cans.[5] Each of these is a seven-digit SIC product with the exception of coffee, wool fabrics, and corrugated boxes which are aggregates over several seven-digit categories.[6] Each of the thirteen products is fairly homogeneous and has been chosen to eliminate, as much as possible, price differences that result from output heterogeneity among producers in the seven-digit product category.[7] With this data we can analyze the cross-sectional distribution of plant-level output prices, shifts of the price distribution over time, and the movement of individual producers through the different yearly distributions.

III. PRICING PATTERNS AT THE PLANT LEVEL

Cross-Section Variation in Output Prices

In this section we summarize the heterogeneity in output prices for each product. Table 1 reports three summary statistics of the output price distributions in each year. The first is the trimmed mean, which deletes the lowest 5% and highest 5% of observations in each year in order to remove the effect of outliers in the data. The second measure is the coefficient of variation, the ratio of the standard deviation to the mean multiplied by 100. The final summary statistic is a measure of dispersion that is likely to be more robust to measurement errors and outliers in the data. It is constructed as the difference between the 90th and 10th percentile of the distribution divided by the median. The final column of the table reports the average value of the two dispersion measures across all available years and the products are ranked in the table from highest to lowest dispersion based on the average value of the robust dispersion measure.[8]

The first pattern evident from Table 1 is that each of the products is characterized by cross-sectional dispersion, but the magnitude of the dispersion varies substantially across products. At the extremes, the robust dispersion measure averages 2.34 for hardwood plywood and 0.146 for newsprint. The second pattern is that the amount of dispersion for a given product is similar across the different years. For example, the robust dispersion measure for hardwood plywood falls between 2.213 and 2.479 for the different census years, while newsprint varies from 0.126 to 0.161. As a result, the ranking of products from high to low dispersion in Table 1 would be virtually the same in every year and any shifts in ranking that did occur would be due to small movements in the summary statistic.[9]

The coefficient of variation and robust dispersion measure can easily be compared across industries and across time but they do not provide a complete picture of the price distribution, particularly the higher moments of the distribution or whether there exist different modes. In order to summarize the yearly price distributions more fully, without placing much structure on the data, we construct kernel density estimates for each product in each year.[10] In order to remove the effect of general price-level changes so the separate cross-section price distributions can be compared, each plant's price in year t is normalized by the geometric mean price in year t. Specifically, we estimate the density of $\ln P^*_{it} = \ln(P_{it}) - n^{-1}\ln(\Pi_i(P_{it}))$ where n is the number of plants in year t and the term after the minus sign is the log of the geometric mean price over all plants in year t.[11]

Table 1. Summary Measures of Output Price Distribution by Year

Trimmed Mean
(Coefficient of Variation)
[Robust Dispersion Measure]

Product	1963	1967	1972	1977	1982	1987	Mean
Hardwood Plywood	0.255	0.292	0.347	0.548	0.757	0.893	
($/sq. ft. (surface measure))	(80.84)	(79.97)	(78.89)	(90.19)	(70.70)	(65.60)	(77.70)
	[2.213]	[2.270]	[2.242]	[2.427]	[2.479]	[2.434]	[2.344]
100% Spun Polyester Blend		0.555	0.501	0.846	0.779		
with Cotton ($/lin. yd.)	—	(58.06)	(53.51)	(44.01)	(59.89)	—	(53.87)
		[1.414]	[1.041]	[1.108]	[1.868]		[1.358]
Cotton Sheeting & Allied	0.300	0.328	0.453	0.804	0.867		
Fabrics ($/lin. yd.)	(52.33)	(42.50)	(52.50)	(78.11)	(54.59)	—	(56.01)
	[1.052]	[1.091]	[1.156]	[1.721]	[1.728]		[1.350]
Finished Wool	2.249	2.293	2.507	4.910	6.721	7.525	
Apparel Fabrics ($/lin. yd.)	(34.09)	(36.02)	(37.63)	(34.03)	(18.80)	(33.83)	(32.40)
	[0.931]	[1.017]	[0.909]	[0.986]	[0.671]	[1.191]	[0.951]
Tinplate Steel Cans	0.017	0.019	0.024				
($1000/base box of steel)	(26.20)	(30.99)	(24.96)	—	—	—	(27.38)
	[0.663]	[0.667]	[0.629]				[0.653]
White Pan Bread	0.164	0.178	0.198	0.297	0.408	0.475	
($/lb.)	(27.36)	(26.44)	(25.99)	(28.02)	(49.63)	(33.79)	(31.87)
	[0.475]	[0.571]	[0.493]	[0.656]	[0.790]	[0.850]	[0.639]
Corrugated Shipping	0.227	0.238	0.256	0.391	0.538	0.621	
Containers ($1000/short ton)	(21.79)	(23.87)	(36.94)	(27.73)	(39.61)	(37.19)	(31.19)
	[0.517]	[0.563]	[0.613]	[0.579]	[0.458]	[0.505]	[0.539]
Roasted Coffee	0.622	0.702	0.846	2.955	2.290	2.601	
($/lb.)	(16.41)	(15.01)	(14.35)	(16.08)	(17.49)	(25.121)	(17.41)
	[0.409]	[0.392]	[0.365]	[0.418]	[0.394]	[0.539]	[0.420]

Table 1. Continued

Product	1963	1967	1972	1977	1982	1987	Mean
Hardwood Oak Flooring	0.147	0.155	0.204	0.419			
($/board ft.)	(13.84)	(34.48)	(15.86)	(85.67)	—	—	(34.50)
	[0.312]	[0.394]	[0.282]	[0.685]			[0.412]
Ready-Mixed Concrete	0.014	0.015	0.018	0.028	0.043	0.046	
($/cubic yd.)	(20.30)	(13.22)	(15.25)	(13.59)	(17.38)	(37.20)	(19.49)
	[0.328]	[0.318]	[0.370]	[0.321]	[0.355]	[0.425]	[0.353]
Interior Softwood Plywood	0.058	0.057	0.088	0.152			
($/sq. ft. 3/8' basis)	(22.20)	(14.66)	(17.11)	(75.33)	—	—	(32.33)
	[0.316]	[0.371]	[0.389]	[0.202]			[0.320]
Motor Gasoline	4.830	5.108	5.337	15.606	39.591	22.873	
($/barrel)	(11.50)	(11.62)	(11.75)	(6.36)	(6.22)	(8.17)	(9.27)
	[0.307]	[0.304]	[0.261]	[0.159]	[0.133]	[0.167]	[0.222]
Newsprint	0.122	0.126	0.147	0.280			
($1000/short ton)	(7.41)	(8.16)	(5.34)	(4.48)	—	—	(6.35)
	[0.126]	[0.157]	[0.161]	[0.140]			[0.146]

The density of $\ln P^*_{it}$ in year t is estimated as:

$$\hat{d}_h = \Sigma_i \, K_h(\ln P^* - \ln P^*_{it})$$

where $\ln P^*$ is an ordered vector of normalized output prices, K_h is the kernel function, and the summation is taken over all plants in year t. For K_h we use the Epanechnikov kernel:

$$K_h(\ln P^* - \ln P^*_{it}) = \frac{3}{4^h}\left(1 - \left(\frac{\ln P^* - \ln P^*_{it}}{h}\right)^2\right) I(|\ln P^* - \ln P^*_{it}| \leqslant h)$$

The function $I(\cdot)$ is an indicator function that takes the value one if $\ln P^*$ and $\ln P^*_{it}$ are within an exogenously specified bandwidth h of each other, and zero otherwise. This implies that observations outside of the bandwidth are given a weight of zero while those within the bandwidth are given a weight that varies inversely with their distance from $\ln P^*_{it}$.

Figures 1a–1f graph the estimated densities for $\ln P^*_{it}$ for each year for hardwood plywood, cotton sheeting, bread, corrugated boxes, ready-mixed concrete, and gasoline. These products were chosen to span the set from high to low dispersion and because they had sufficient observations to support the non-parametric estimation.[12] The two products with the highest dispersion, hardwood plywood (panel a) and cotton sheeting (panel b), also show the most variation in the densities across years. Even in these cases several similarities are present. The similarity in the overall range of prices across years is clearly evident. Also, a second higher-price mode at approximately 1 (approximately twice the mean price) is evident in each year for hardwood plywood. This suggests that some largely time-invariant characteristic such as product quality may be an important contributor to the high level of price dispersion observed for this product.

The remaining products in panels (c) through (f) of Fig. 1, show price densities that are more similar across the different years.[13] In particular, the skewness of the densities for bread and corrugated boxes is present in every year, indicating the consistent presence of some high-price producers. Both of these products are sold in small geographic markets and the skewness could arise if there are some local markets with consistently higher prices. Concrete and gasoline are more symmetric distributions and, as seen in Table 1, have the lowest overall levels of dispersion among these six products.

Overall, Table 1 and Fig. 1 indicate clearly that, for all of the products, there is cross-sectional variation in plant output prices at every point in time. There are many possible explanations for this including: variation in quality that is not captured by differences in the product definition, differences in production

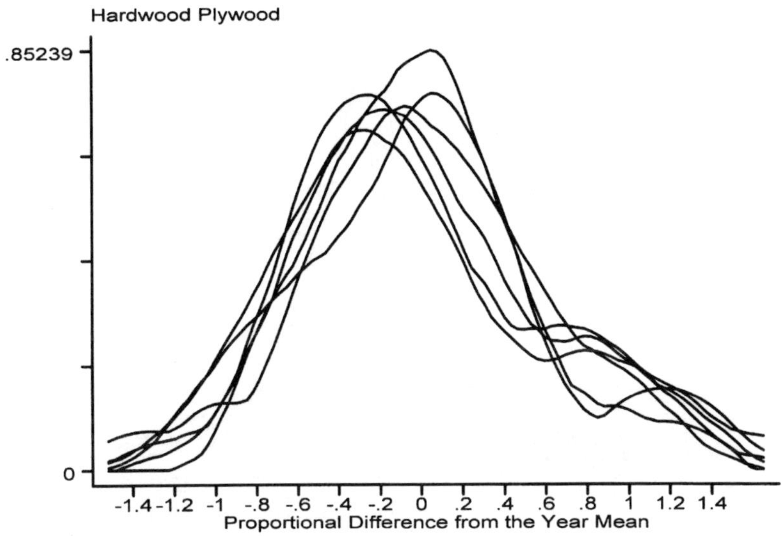

Fig. 1a. Density of the Output Price Distribution by year.

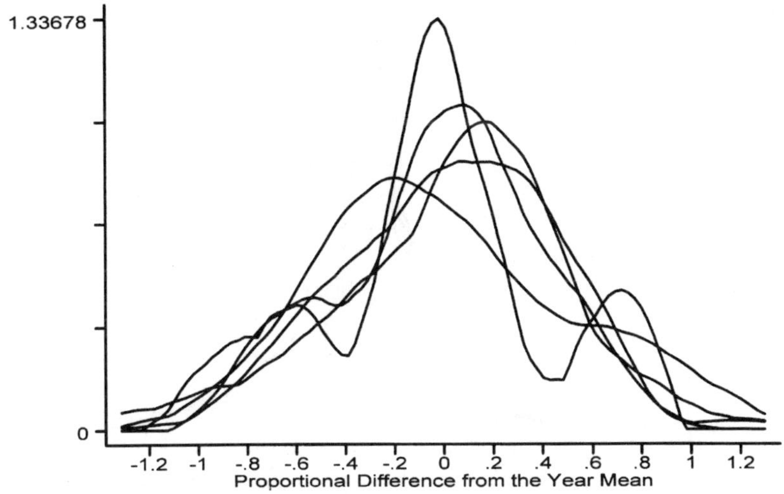

Fig. 1b. Density of the Output Price Distribution by year.

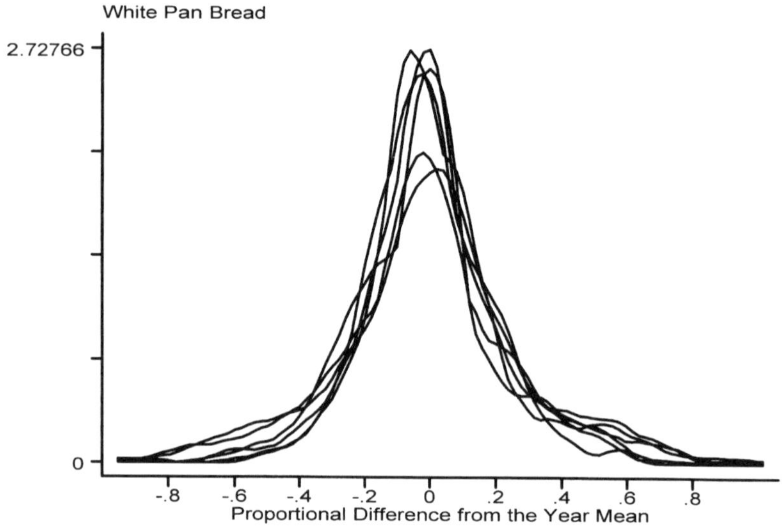

Fig. 1c. Density of the Output Price Distribution by year.

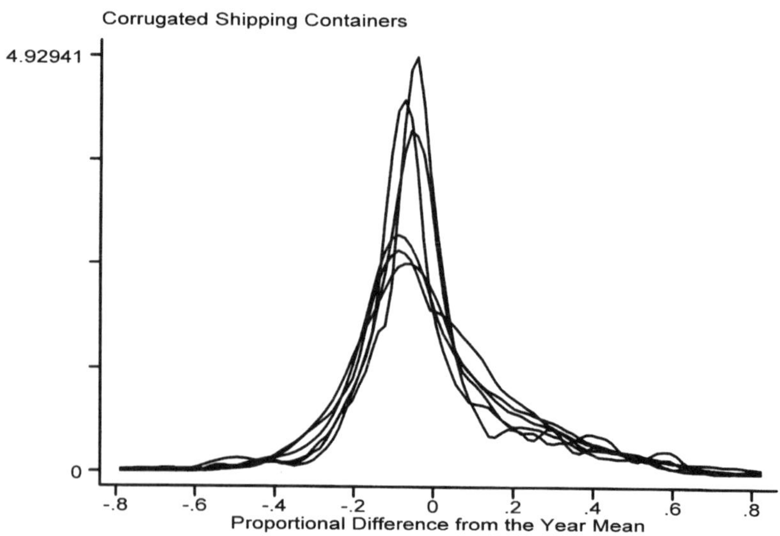

Fig. 1d. Density of the Output Price Distribution by year.

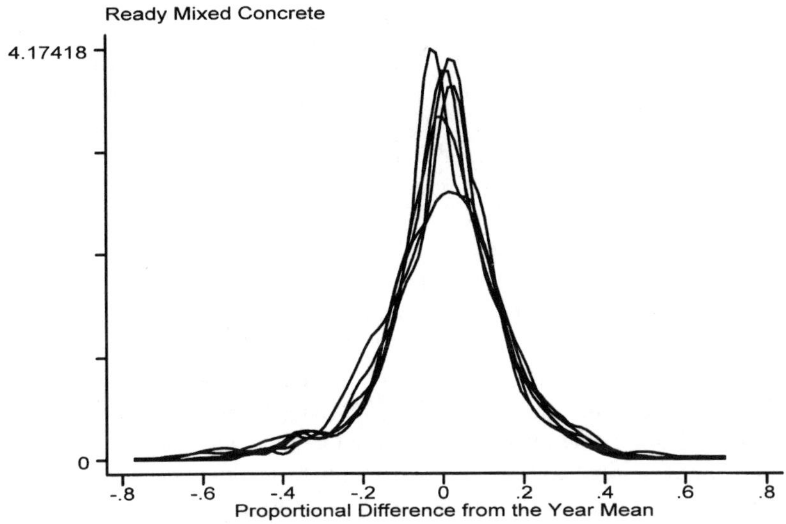

Fig. 1e. Density of the Output Price Distribution by year.

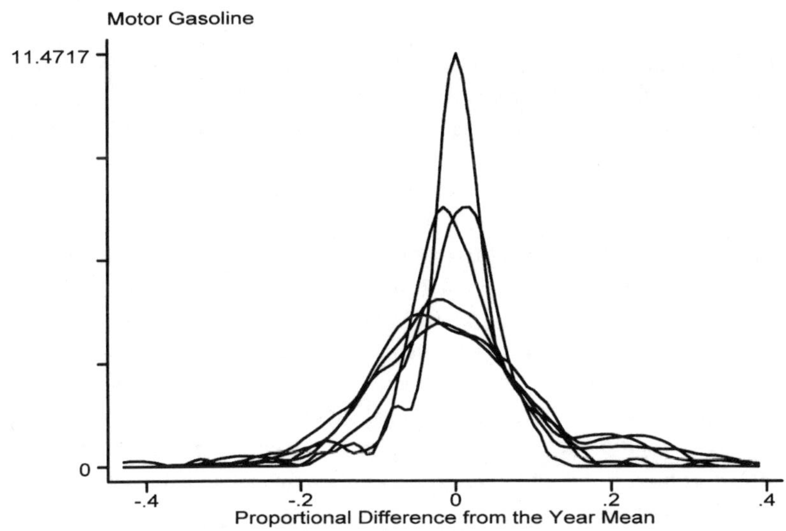

Fig. 1f. Density of the Output Price Distribution by year.

costs among plants that are partly or fully passed on to consumers, high search costs by consumers which can generate price dispersion even with identical costs across producers, differences in the competitiveness of output markets across producers, explicit adjustment costs associated with changing prices that result in producers responding differently to cost or demand shocks, or measurement or reporting errors in the value or quantity of shipments data. As a first step toward understanding the source of the cross-sectional dispersion, we will exploit the multiple time periods in our data and measure to what extent the price differences reflect permanent time-invariant differences among plants versus year-specific idiosyncratic differences.

In order to categorize the likely sources of price variation, assume each plant produces a unique differentiated product, facing a demand elasticity for its product of η_{it} and marginal cost MC_{it}. The plant chooses price to maximize profit by equating marginal revenue and marginal cost. Both the demand elasticity and marginal cost can vary across plants and time because of market-level time effects (α_t), plant-specific factors that do not change over time (γ_i), and idiosyncratic shocks that are specific to a plant and time period (ε_{it}). The log of the plant's output price can be written as:

$$\ln P_{it} = -\ln(1 - 1/_{it}) + \ln MC_{it} = \alpha_t + \gamma_i + \varepsilon_{it}.$$

The market level factors reflect such things as changes in costs due to common input price changes or shifts in market demand for the product. The γ_i will reflect factors that vary across plants but are fixed or change very slowly over time. This can include differences in product quality among producers, persistent differences in productive efficiency, or differences in the competitiveness of the local market in which the plant operates. The idiosyncratic shocks will reflect all time-varying factors that alter a plant's marginal cost or demand elasticity as well as measurement or reporting errors in the observed prices. In the case where plants are producing an identical product and selling it in a single, perfectly competitive output market then observed prices should be explained only by market factors that pertain in that year (α_t) and measurement or aggregation errors (ε_{it}) that result in plant-level price variation. There should be no role for permanent between-plant differences in prices arising from γ_i and all cross-sectional variation in prices should reflect measurement and reporting errors. We will assess the importance of between-plant price variation versus within-plant price variation over time with both a variance decomposition and evidence on the persistence of plant-level pricing patterns over time.

The first two columns of Table 2 report the amount of between- and within-plant variation in $\ln P^*_{it}$.[14] Several patterns are evident. First, moving down the

Table 2. Between and Within-Plant Variation in Price and Output (Output price expressed as deviation from year mean)

Product	Between-Plant Variance ln P*	Within-Plant Variance ln P*	Between-Plant Variance ln Q	Within-Plant Variance ln Q
Hardwood Plywood	0.294	0.149	1.155	0.325
100% Spun Polyester Blends with Cotton	0.141	0.125	1.493	1.228
Cotton Sheeting & Allied Fabrics	0.162	0.119	1.422	0.889
White Pan Bread	0.044	0.045	2.510	0.343
Corrugated Shipping Containers	0.031	0.032	1.389	0.287
Roasted Coffee	0.017	0.026	2.440	0.220
Hardwood Oak Flooring	0.020	0.049	0.871	0.730
Ready-Mixed Concrete	0.015	0.020	0.856	0.280
Softwood Plywood	0.012	0.021	1.366	0.768
Motor Gasoline	0.004	0.006	2.745	0.261
Newsprint	0.002	0.002	0.887	0.344

table from highest to lowest dispersion products, both variances decline substantially indicating smaller differences in the average price across plants and less time-series variation in prices for each plant. Second, the between-plant variance declines much more rapidly than the within-plant variance as you move from high to low dispersion products. For the three highest dispersion products, hardwood plywood, polyester, and cotton fabric, the between-plant variance is larger than the within-plant variance, indicating that permanent price-level differences across plants are the major source of cross-section price variation. This is consistent with large, permanent differences in demand elasticities or marginal costs across plants. For the remaining products, time-series variation in each plant's price contributes as much (bread, corrugated boxes, and newsprint) or more (coffee, oak flooring, concrete, softwood plywood, and gasoline) to total variation as the between-plant price differences. For these products there is a potentially larger role for time-varying demand or cost conditions at the plant level, as well as noise, to play a role in generating price dispersion.

Table 2 indicates that the cross-sectional dispersion observed in Table 1 and Fig. 1 arises from a different combination of permanent between-plant differences and idiosyncratic factors for each product. These two sources have different implications for the persistence of the plant-level patterns over time.

If price heterogeneity arises from structural factors that are slow to change over time, such as the quality of a plant's output or efficiency advantages that the plant may have, then a plant's location in the price distribution should tend to persist over time. In contrast, if the price variability arises solely from plant-level shocks that are not serially correlated over time, such as idiosyncratic cost or demand shocks or pure measurement errors, then there should be no persistence in the rankings of plants over time. In the next subsection we summarize the time-series persistence as one way of categorizing the importance of structural and random factors in generating price dispersion.

Persistence in Pricing Patterns

We explore the persistence in the output price distributions by constructing transition matrices that summarize the movement of plants among different quartiles of the price distribution over time. These transition rates are less likely to be affected by small measurement or reporting errors than the variances reported in Table 2. Table 3 disaggregates plants by the quartile of the price distribution in which they were located in one census year and then reports the proportion that move to each of the other quartiles in the following census year.[15] To simplify the table we have aggregated the second and third quartiles of the price distribution into a single category. The first row of the table indicates that, of all hardwood plywood producers in the highest price quartile in census year t, 83.5% remained in the highest price quartile 5 years later, 14.1% had dropped to the middle two quartiles of the price distribution and only 2.4% had fallen to the lowest quartile. Of plants in the middle two quartiles in year t, 71.9% remained there in year $t+1$. Also, 63.2% of the low-price plants in year t remained there in $t+1$.

As a basis for comparison, if all price variation within a year is due to idiosyncratic shocks, so that $\gamma_i = 0$, then the probability of being in any quartile in year $t+1$ is independent of the location in year t and the first column of Table 3 should be 0.25, the second column 0.5, and the final column 0.25. The same pattern should be observed in columns 4–6 and columns 7–9. We test that each sample proportion is significantly different than 0.25 or 0.5, as relevant, and denote cases where the hypothesis is not rejected.

Focusing first on the transition rates of high-price producers in columns 1–3, two patterns are evident. First, the probability of remaining a high-price producer is greater than would be indicated by the random movement of prices. The probability varies from a low of 0.398 for gasoline to a high of 0.835 for hardwood plywood. For all of the products the transition rate is significantly different than 0.25. Second, for many of the thirteen products, the probability

Table 3. Transition Rates Across Output Price Quartiles

Product	Plants in the highest price quartile in year t			Plants in the middle two price quartiles in year t			Plants in the lowest price quartile in year t		
	Highest quartile in $t+1$	Middle two quartiles in $t+1$	Lowest quartile in $t+1$	Highest quartile in $t+1$	Middle two quartiles in $t+1$	Lowest quartile in $t+1$	Highest quartile in $t+1$	Middle two quartiles in $t+1$	Lowest quartile in $t+1$
Hardwood plywood	0.835	0.141	0.024	0.110	0.719	0.171	0.039	0.329	0.632
100% Spun polyester blends with cotton	0.426	0.519[a]	0.055	0.164	0.746	0.090[b]	0.135	0.308	0.558
Cotton sheeting and allied fabrics	0.538	0.404[a]	0.058	0.173	0.655	0.173	0.063	0.266	0.672
Finished wool apparel fabrics	0.529	0.352[a]	0.118[b]	0.206[b]	0.647[a]	0.147[b]	0.052	0.368[a]	0.579
Tinplate steel cans	0.610	0.293	0.098	0.176	0.631	0.193	0.046	0.402[a]	0.552
White pan bread	0.501	0.374	0.125	0.148	0.637	0.216	0.083	0.382	0.535
Corrugated shipping containers	0.501	0.423	0.075	0.187	0.619	0.194	0.073	0.396	0.531
Roasted coffee	0.429	0.438[a]	0.133	0.174	0.601	0.225[b]	0.138	0.459[a]	0.404
Hardwood oak flooring	0.521	0.375[a]	0.104	0.182[b]	0.625	0.193[b]	0.130	0.457[a]	0.413
Ready-mixed concrete	0.505	0.407	0.088	0.196	0.593	0.212	0.099	0.426	0.475
Interior softwood plywood	0.483	0.449[a]	0.067	0.181	0.559[a]	0.260[b]	0.110	0.425[a]	0.466
Motor gasoline	0.398	0.393	0.209[b]	0.220[b]	0.573	0.207	0.146	0.466[a]	0.388
Newsprint	0.538	0.385[a]	0.077	0.193[b]	0.710	0.097	0.000	0.267	0.733

[a] do not reject that the proportion equals 0.5 at the 0.05 significance level; [b] do not reject that the proportion equals .25 at the .05 significance level

of moving into the middle two quartiles does not differ significantly from 0.5, while the probability of moving into the low price category is substantially less than 0.25. This implies that the producers that move out of the high-price category predominantly move into the middle of the price distribution but that too few move into the low price quartile for the transitions to be driven solely by random factors.[16] While plants clearly move through the price distribution over time, the persistence of plants in the high-price part of the distribution suggests that structural, rather than purely idiosyncratic, forces are at work in determining pricing patterns.

The transition rates for plants with prices in the middle two quartiles of the price distribution tell a similar story. The probability of remaining in the middle-price category exceeds 0.5 for all products, and the difference is significant for eleven of the products, indicating more persistence than would be generated by random price movements. For plants that move out of the middle-price category, the probability of moving to lower prices is greater than the probability of moving to higher prices for 8 of the 13 products. Five of the products have a higher probability of plants moving up in the price distribution and for one product the probabilities are equal. Among the plants in the lowest quartile of the price distribution in year t, between 38.8% (gasoline) and 73.3% (newsprint) remain in the lowest price quartile five years later. All of these transition rates are significantly different than 0.25. Transitions out of the low-price category are also heavily weighted toward the middle, rather than the upper tail, of the price distribution.

Overall, the transition rates indicate more persistence in the ranking of plant prices than is consistent with purely idiosyncratic shocks generating price dispersion and suggest that structural factors that change slowly over time are responsible for some of the price heterogeneity that is observed. In particular, the probability of remaining in the highest or lowest quartile of the price distribution is too large, relative to random chance, and the probability of moving to the other tail of the distribution is too small.

When comparing the probabilities of remaining in the same quartile of the price distribution (cols. 2, 5, and 8) across products, one additional pattern is evident. The degree of persistence is highest for the products with the largest overall levels of price dispersion such as hardwood plywood, the three fabric products, and tin cans. In general, the lower dispersion products such as coffee, oak flooring, concrete, softwood plywood, and gasoline have less persistence and transition rates that more closely approach random noise.[17] This suggests that products with larger price dispersion are not ones with larger idiosyncratic shocks but rather ones where serially-correlated structural factors are more important. The reduction in total dispersion as we move down the rows in

Tables 1 and 2 results from a reduction in the relative importance of permanent to idiosyncratic variation in the price data.

Output Prices and Producer Size

Theoretical models of industry evolution often generate differences in producer size with a combination of permanent differences in cost or productivity between producers and idiosyncratic demand or cost shocks over time. The importance of permanent between-plant size differences versus within-plant idiosyncratic variation in size for the products in this study is summarized in the last two columns of Table 2. For every product the between-plant size variation is larger than the total within-plant variation, in some cases by a factor of ten, indicating an important role for permanent, or at least slowly evolving, size differences. If output prices are correlated with plant size or productivity, then the heterogeneity and persistence in output prices documented in the two sections may be another manifestation of the underlying process that generates persistent differences in size.

There are several reasons to suspect that plant size will be systematically related to output price. If larger plants serve larger, more competitive markets then their prices may be lower as a result of the increased competition. If larger plants have lower production costs, either because they are more efficient or can exploit scale economies, these may be passed on to purchasers (Demsetz (1973) and Peltzman (1977)). If larger plants produce very homogeneous, standardized products while smaller plants manufacture specialty products and serve niche markets then costs and prices may vary with size (Caves & Porter (1977), Porter (1980)).

To summarize the output price-size relationship we regress the log of the plant's normalized output price $\ln P^*_{it}$ on dummy variables for the size quartile in which the plant is located in year t and conduct simple parametric tests of the equality of mean prices across quartiles of the plant size distribution.[18] The regression results and tests for equality of means are reported in Table 4. In these regressions β_1 is the intercept and represents the mean price in the smallest size quartile. The other coefficients represent deviations from this for the three larger quartiles. The estimates of β_1 show that, with the exception of gasoline, the smallest producers have prices above the mean. The significant negative estimates for the other coefficients indicate a statistically significant decline in the mean price across the size distribution.[19]

The hypothesis that the mean price is constant across the four quartiles ($\beta_2=\beta_3=\beta_4=0$) is tested and the F-statistic is reported in column 5 of the Table. The hypothesis is rejected for all products except tin cans, gasoline, and

Table 4. Output Price Differences by Size Quartile (standard errors in parentheses) Dependent Variable: ln P^*_{it}

Product	β_1	β_2	β_3	β_4	$\beta_2=\beta_3=\beta_4=0$	$\beta_2=\beta_3=\beta_4$	$\beta_3=\beta_4$	Obs.	R^2
Hardwood plywood	0.522**	–0.415**	–0.733**	–0.936**	60.95**	31.49**	10.88**	491	0.33
	(0.055)	(0.078)	(0.074)	(0.066)					
100% Spun polyester blends with cotton	0.200**	–0.083	–0.258**	–0.459**	15.43**	19.20**	13.31**	367	0.15
	(0.056)	(0.072)	(0.067)	(0.069)					
Cotton sheeting and allied fabrics	0.213**	–0.150**	–0.324**	–0.379**	8.26**	6.88**	0.65	337	0.09
	(0.056)	(0.071)	(0.073)	(0.076)					
Finished wool apparel fabrics	0.281**	–0.338**	–0.343**	–0.436**	7.40**	1.26	2.04	94	0.25
	(0.069)	(0.090)	(0.082)	(0.085)					
Tinplate steel cans	0.037	–0.037	–0.061**	–0.049*	1.32	0.40	0.24	712	0.01
	(0.025)	(0.032)	(0.029)	(0.030)					
White pan bread	0.091**	–0.088**	–0.136**	–0.139**	35.21**	11.90**	0.06	2807	0.05
	(0.011)	(0.014)	(0.013)	(0.013)					
Corrugated shipping containers	0.136**	–0.119**	–0.185**	–0.240**	242.09**	138.51**	83.28**	4571	0.18
	(0.009)	(0.011)	(0.010)	(0.010)					
Roasted coffee	0.047**	–0.021	–0.079**	–0.089**	8.28**	8.51**	0.31	762	0.04
	(0.013)	(0.018)	(0.019)	(0.019)					
Hardwood oak flooring	0.107**	–0.122**	–0.131**	–0.175**	3.98**	1.57	1.74	280	0.05
	(0.040)	(0.044)	(0.046)	(0.048)					
Ready-mixed concrete	0.046**	–0.028**	–0.059**	–0.096**	122.05**	119.92**	71.50**	11360	0.04
	(0.003)	(0.004)	(0.004)	(0.005)					
Interior softwood plywood	0.059**	–0.032	–0.091**	–0.113**	15.69**	17.17**	2.59	418	0.13
	(0.029)	(0.031)	(0.031)	(0.030)					
Motor gasoline	–0.004	0.003	0.008	0.005	0.23	0.21	0.16	964	0.001
	(0.006)	(0.009)	(0.009)	(0.009)					
Newsprint	0.024	–0.038	–0.034	–0.025	0.81	0.39	0.26	67	0.05
	(0.024)	(0.026)	(0.027)	(0.026)					

** Significant at the =0.05 level; * Significant at the =0.1 level

newsprint. The latter two are the products with the lowest overall price dispersion.[20] The remaining 10 products all show significant price declines with increases in plant size. We next test the hypothesis that the price differences arise only for the smallest plants in the industry by testing that plants in the three largest quartiles have average prices that are equal ($\beta_2 = \beta_3 = \beta_4$). This hypothesis is rejected for 8 of the 10 products. Finally, the hypothesis that plants in the largest two quartiles have equal prices ($\beta_3 = \beta_4$), is rejected for four of the products, hardwood plywood, cotton sheeting, corrugated boxes, and concrete. The results indicate that these four products have systematic price declines that continue across the whole size distribution. Overall in the cross-section, output prices decline with plant size for ten of the thirteen products. Two of the products with no significant size-price correlation, gasoline and newsprint, are the two for which there is very little overall dispersion. These findings further reinforce the conclusion of the previous section that the observed cross-sectional price variation is not simply random noise. Differences in output prices across plants persist over time and the level of a plant's output price is negatively correlated with the plant's size in the cross-section.[21] In the next section we examine whether production costs vary systematically across plants as one possible explanation for the cross-sectional pattern of output prices and its persistence over time.

IV. PLANT-LEVEL MARGINAL COST

An important component of any explanation of price heterogeneity must be heterogeneity in cost, yet cost data is absent from most empirical studies of price dispersion. As part of this research project we have estimated cost function models using the plant-level panel data and constructed estimates of marginal cost that vary by plant and year reflecting differences in factor prices and the level and mix of plant output.[22] Several unique characteristics of the census establishment data are important for modeling production costs. First, virtually all plants produce more than one output. Simply ignoring the output of the plant that is not in the product category we study would lead us to systematically underestimate total plant output and, depending on whether the omitted output was positively or negatively correlated with the output we are interested in, will lead to upward or downward bias in scale parameters.[23] To account for this we aggregate the shipments of all other products manufactured in the plant into a single secondary output, deflate this with an industry-level price index, and include it as an additional control in our multi product cost models. Second, we are able to construct plant-specific prices for materials, labor, and energy and utilize flexible functional forms that allow estimated marginal cost to vary with these factor prices. Third, we have relatively poor information on the plant's capital use. Data for the book value of capital stocks is available by plant in most years but no information on plant service prices or expenditures on capital input is collected. Because of this, cost is defined as the plant's expenditure on materials, labor, and energy inputs.[24] Because of the missing capital data we will systematically underestimate the level of a plant's marginal cost because the expansion in capital costs with increases in output will not be captured. Perhaps more importantly, since large plants are generally more capital intensive the bias in our estimate of marginal cost will increase with plant size, leading us to potentially underestimate marginal cost and overestimate markups more for large plants.

The basic 'long-run' cost model is $C(p_l, p_m, p_e, Q, R, D_t, U)$ where the first three arguments are the prices of labor, material, and energy, Q is the output level of the product we are studying, and R is the output of all other products manufactured in the plant. Cost is defined as the expenditure on labor, materials, and energy.[25] Year effects are controlled for with a set of time dummies D_t that distinguish census years. These are interacted with the factor prices to recognize that the factor demand equations may also shift over time. A single-unit/multi-unit ownership dummy U is also included to control for possible differences in capital prices, or differences in the way that central office expenditures are allocated.[26]

The empirical model uses the translog functional form and estimates the cost function together with cost share equations for materials and energy. Estimates of a plant's marginal cost for output Q are constructed as:

$$MC_{it} = \frac{C_{it}}{Q_{it}} \frac{\partial \ln C_{it}}{\partial \ln Q_{it}} = \frac{C_{it}}{Q_{it}} (\beta_Q + \gamma_{QQ} \ln Q_{it} + \gamma_{QR} \ln R_{it} + \Sigma_j \gamma_{jQ} \ln P_{jit} + \gamma_{QU} U_{it})$$

where the Greek letters are the cost function parameters that are estimated. Estimated marginal cost will vary across plants with differences in factor prices, output levels, and output mix.

Two problems that can arise in production models estimated with micro data are unobserved heterogeneity and measurement error in output. Both can lead to bias in output-related parameters in the cost function and thus bias in estimates of returns to scale or marginal cost (see Griliches (1986), Griliches & Hausman (1986), and Mairesse (1990) for discussion). In general, measurement error in output, which can arise from improper deflation (which is not a problem in our data set) or from deviations between observed and planned output caused by random demand or input supply shocks, will result in a downward bias in output coefficients and an upward bias in estimates of scale economies. Similarly, unobserved efficiency differences, which will tend to dominate in cross-section data, will generally lead to upward biased estimates of returns to scale, because more efficient producers will be larger and have lower average cost. If we overestimate the degree of scale economies we are likely to underestimate the marginal cost of large producers and thus overestimate their markups.

We estimate the cost model using the seemingly unrelated regression (SUR) estimator. In an attempt to control for measurement error arising from deviations between planned and actual output we use the plant's beginning-of-year capital stock as an instrumental variable for output. Like the SUR estimator, this IV estimator relies primarily on cross-sectional variation in the data. Second, to eliminate plant-specific, time-invariant heterogeneity in the cost function we estimate the cost function in first-differences. Griliches (1986) and Griliches & Hausman (1986) show that, while removing one problem, difference estimators can exaggerate the bias from measurement error. The use of longer time differences, in our case five-year differences, will reduce but not eliminate this problem. To attempt to correct for both sources of bias we also apply the instrumental variable estimator to the time-differenced data using the plant's capital stock as an instrument for the growth in output.

Among the different estimators the SUR and IV estimators that exploit the cross-section variation in the data provide the most plausible estimates.[27] In general the SUR estimates show slight increasing returns to scale across most of the plant size distribution, while the IV estimates indicate constant returns to scale for plants in the largest one or two quartiles of the size distribution. This is consistent with slight measurement error problems in the cross-sectional output data. In the remainder of this chapter we will rely on the SUR estimates of the cost function. The main way this will affect our results is to load unobserved heterogeneity effects into the scale-related cost parameters. Given that our interest is in constructing point estimates of marginal cost for each plant that recognize the plants' scale of operation and factor prices, and not separating heterogeneity and scale parameters, this should not be a serious problem.

Table 5 provides summary measures of the variation in estimated marginal cost across the plant size distribution. The format of the Table is identical to Table 4. The dependent variable is $\ln MC^*_{it} = \ln(MC_{it}) - n^{-1}\ln(\Pi_{ii}(MC_{it}))$ which is the log of plant-level marginal cost normalized by the log of the geometric mean of marginal cost in the same year. This is regressed on a set of dummy variables for the quartiles of the size distribution as a way of summarizing the cross-sectional marginal cost variation with plant size.

The marginal cost variation observed in Table 5 is straightforward to summarize for most of the products because it parallels the patterns observed

Table 5. Marginal Cost Differences by Size Quartile (standard errors in parentheses) Dependent Variable: ln MC^*_{it}

Product	β_1	β_2	β_3	β_4	$\beta_2=\beta_3=\beta_4=0$	$\beta_2=\beta_3=\beta_4$	$\beta_3=\beta_4$	Obs.	R^2
Hardwood plywood	0.201**	–0.326**	–0.259**	–0.214**	3.51**	0.89	0.21	272	0.05
	(0.064)	(0.092)	(0.106)	(0.081)					
100% Spun polyester blends with cotton	0.387**	–0.354	–0.525**	–0.665**	26.54**	13.12**	7.61**	268	0.29
	(0.066)	(0.082)	(0.075)	(0.076)					
Cotton sheeting and allied fabrics	0.274**	–0.237**	–0.413**	–0.448**	9.15**	6.04**	0.23	297	0.11
	(0.065)	(0.077)	(0.080)	(0.086)					
Tinplate steel cans	–0.034	0.022	0.035	0.081**	1.56	1.92	2.19	669	0.01
	(0.035)	(0.044)	(0.042)	(0.041)					
White pan bread	0.135**	–0.131**	–0.182**	–0.231**	61.59**	25.13**	11.85**	2638	0.09
	(0.011)	(0.015)	(0.015)	(0.016)					
Corrugated shipping containers	0.026**	–0.028**	–0.026**	–0.052**	6.33**	6.11**	10.56**	4393	0.01
	(0.012)	(0.014)	(0.013)	(0.013)					
Roasted coffee	0.113**	–0.101**	–0.127**	–0.222**	13.61**	10.68**	12.48**	637	0.08
	(0.022)	(0.028)	(0.028)	(0.030)					
Hardwood oak flooring	0.231**	–0.195**	–0.352**	–0.374**	6.22**	3.73**	0.07	200	0.11
	(0.063)	(0.078)	(0.086)	(0.084)					
Ready-mixed concrete	0.009	–0.030**	–0.022*	0.015	6.45**	12.29**	16.02**	6045	0.01
	(0.010)	(0.012)	(0.012)	(0.012)					
Interior softwood plywood	0.384**	–0.357	–0.511**	–0.662**	48.92**	33.72**	20.33**	167	0.55
	(0.057)	(0.062)	(0.060)	(0.063)					
Motor gasoline	–0.053**	0.050**	0.075**	0.084**	7.74**	3.79**	0.54	918	0.03
	(0.016)	(0.018)	(0.018)	(0.018)					
Newsprint	0.098**	–0.041	–0.158	–0.201**	3.26**	3.09*	0.40	47	0.24
	(0.044)	(0.066)	(0.085)	(0.060)					

** Significant at the $\alpha = 0.05$ level; * Significant at the $\alpha = 0.1$ level

for output price.[28] For most of the products, particularly the ones with high price dispersion, marginal cost falls with increases in plant size. Of the first ten products listed in the Table, there is a significant decline in marginal cost for eight of them. The same products also had significant declines in output price. Tin cans does not have a significant pattern in marginal cost across the size distribution, but this also parallels the finding of a weak correlation between output price and size in Table 4. Concrete has a slight decline in marginal cost for plants in the middle of the size distribution, but a much more systematic decline in output price. The final two products, gasoline and newsprint, have no systematic output price pattern but both have systematic variation in marginal cost. Overall, for most of the twelve products and for all of the products with the largest price declines, marginal costs are lower for large producers.

This cross-sectional decline in marginal cost reflects a number of factors. Scale effects are important for three of the products with large cost declines, hardwood plywood, polyester fabric, and cotton sheeting. The scale elasticity, evaluated at the means of the data, equals 1.78, 1.46, and 1.52 for the three products, respectively.[29] For a second group of products, bread, oak flooring, concrete, softwood plywood, and newsprint, the scale elasticity at the means of the data varies from 1.13 to 1.20, thus contributing to a decline in marginal cost with plant size. For the final group of products, cans, corrugated boxes, coffee, and gasoline, the scale elasticity lies between 1.02 and 1.04, indicating that increases in plant size have little effect on marginal cost. As discussed above, the scale elasticity will capture both true scale economies as well as the effect of unobserved heterogeneity that results in shifts of the intercept of the cost function across plants. For our purpose, which is to account for across-plant differences in marginal cost due to all sources, this is fine. A second way that scale differences affect our marginal cost estimates is through economies of scope. For virtually all of the products we find significant economies of scope which implies that increases in the plants' other outputs will lower the marginal cost of the product we are focusing on. Since larger plants generally produce more of both outputs Q and R this will contribute to a decline in the marginal cost of output Q as plant size increases. Factor price differences also contribute to heterogeneity in plant marginal costs. Material prices decline with plant size for many of the products, particularly corrugated boxes, coffee, concrete, oak flooring, and softwood plywood, which will lower the marginal cost of larger producers. Countering this is the fact that larger plants almost always pay higher wages. Taken together, these factors contribute to substantial cross-sectional variation in marginal cost, as evidenced by the coefficients in Table 5, with the net effect being lower marginal costs for large plants in all but three of the products. This suggests that the lower output prices charged by large plants may reflect underlying cost advantages and this issue is explored in the next section.

V. PLANT-LEVEL MARKUPS

In this section we combine the measured output prices and estimated marginal costs into a plant markup and summarize its cross-sectional and time-series patterns. The plant's markup is defined as $\ln(P_{it}/MC_{it})$, which can be interpreted as $-\ln(1 - 1/\eta_{it})$ where η_{it} is plant i's demand elasticity in year t. Since both the output price and marginal cost decline with plant size for most products, the cross-sectional variation in the markup will reflect which variable declines more rapidly. Table 6 summarizes the difference in average markups across the quartiles of the plant size distribution. The format of the Table is identical to Tables 4 and 5 and the markup is expressed as a deviation from the geometric mean in the relevant year.

There are three reasons we might expect markups to rise with plant size. If larger plants have lower demand elasticities because, for example, there are fewer alternative products available to purchasers, then markups should rise with plant size. If plants have different marginal costs and act as Cournot competitors in a homogeneous-good output market then markups will increase with market share.[30] Finally, bias in marginal cost estimates resulting from our inability to measure capital expenditures will generate an increase in the markup with plant size if large plants are more capital intensive.[31] There is at least one reason to expect markups to decline with plant size. If larger plants operate in more populated geographic markets they are likely to face more competitors than small geographically-isolated producers and should thus face higher demand elasticities for their product and have lower markups.

The markup-size relationship varies substantially across the 12 products. For six of the products, polyester blend fabrics, bread, coffee, oak flooring, softwood plywood, and newsprint, there is a significant increase in the markup with plant size and the increase is often monotonic across the size class dummies. Two of the 12 products, cotton sheeting and gasoline, have no significant change in markups with plant size. The final four products, hardwood plywood, cans, corrugated boxes and concrete, have markups that decline significantly as plant size increases.

Hardwood plywood is the most unusual case, with a markup that declines by 70% as we move from the smallest to largest quartile of plants. This is also the product with the most price dispersion (Table 1) and the highest ratio of between-plant to within-plant price variation (Table 2). It is likely that, among our set of products, this has the most substantial variation in the quality of

Table 6. Markup Differences by Size Quartile
(standard errors in parentheses)
Dependent Variable: ln (P_{it}/MC_{it})*

Product	β_1	β_2	β_3	β_4	$\beta_2=\beta_3=\beta_4=0$	$\beta_2=\beta_3=\beta_4$	$\beta_3=\beta_4$	Obs.	R^2
Hardwood plywood	0.392**	–0.306**	–0.551**	–0.700**	34.07**	20.59**	3.58*	272	0.34
	(0.054)	(0.073)	(0.088)	(0.064)					
100% Spun polyester blends with cotton	–1.62**	0.212**	0.197**	0.237**	3.72**	0.43	0.85	268	0.05
	(0.070)	(0.078)	(0.076)	(0.076)					
Cotton sheeting and allied fabrics	0.014	0.029	–0.027	–0.056	0.82	1.55	0.32	297	0.01
	(0.070)	(0.079)	(0.081)	(0.077)					
Tinplate steel cans	0.041	–0.032	–0.054	–0.078**	2.24*	1.76	1.01	669	0.01
	(0.028)	(0.034)	(0.034)	(0.032)					
White pan bread	–0.057**	0.054**	0.069**	0.108**	10.69**	6.84**	6.42**	2638	0.02
	(0.012)	(0.016)	(0.016)	(0.018)					
Corrugated shipping containers	0.112**	–0.096**	–0.164**	–0.189**	149.85**	66.99**	15.55**	4393	0.12
	(0.009)	(0.012)	(0.010)	(0.010)					
Roasted coffee	–0.050**	0.045	0.041	0.115**	3.73**	4.56*	7.65**	637	0.02
	(0.023)	(0.029)	(0.029)	(0.031)					
Hardwood oak flooring	–0.150**	0.102	0.257**	0.239**	3.37**	3.23*	0.07	200	0.06
	(0.067)	(0.080)	(0.084)	(0.084)					
Ready-mixed concrete	0.027**	0.013	–0.022*	–0.099**	44.66**	78.23**	77.83**	6045	0.03
	(0.010)	(0.012)	(0.011)	(0.011)					
Interior softwood plywood	–0.303**	0.289**	0.397**	0.522**	29.63**	25.93**	14.42**	167	0.42
	(0.067)	(0.071)	(0.071)	(0.072)					
Motor gasoline	0.019	–0.023	–0.029	–0.022	0.72	0.15	0.24	918	0.00
	(0.015)	(0.019)	(0.018)	(0.019)					
Newsprint	–0.086**	0.016	0.140*	0.187**	3.53**	4.52*	0.39	46	0.25
	(0.042)	(0.059)	(0.076)	(0.058)					

** Significant at the $\alpha = 0.05$ level; * Significant at the $\alpha = 0.1$ level

output across plants and the price and markup decline is consistent with the large plants specializing in standardized commodity products, which are close substitutes, while smaller plants serve specialized niche markets. The other three products with declining margins, cans, boxes, and concrete, share the common feature that they are sold in relatively small geographic markets. Thus the cross-sectional markup variation in Table 6 can reflect differences in local markup competition. The decline in both price and markup across the size distribution is consistent with larger plants serving more densely populated, and more competitive geographic markets. We note, however, that the other product sold in small markets, bread, does not show evidence of this competitive effect possibly because there is more product differentiation among bread producers.

The two products with no size related variation in markups, cotton sheeting and gasoline, have the lowest level of markups but very different underlying price and cost patterns. Like hardwood plywood, cotton fabric has a high degree of price dispersion and a high ratio of between-plant to within-plant price variation, suggesting that unobserved quality variation among producers is playing a significant role. The average price of cotton fabric falls by 38% from the smallest to largest quartile of plants, but this is closely matched by a decline in marginal cost, again consistent with quality variation, resulting in markups that are small and fairly constant across the range of plant sizes. In the other case, gasoline, there is no systematic price variation and a slight increase in marginal cost, resulting from higher crude oil prices, for larger plants. Markup differences among different size plants appear random.

For the six products with increasing markups, the common thread is that all of them have substantial declines in marginal cost. The coefficient β_4 in Table 5 indicates that, on average, the marginal cost for the largest quartile of producers is between 20 and 66% lower than the cost for the smallest quartile of producers. With the exception of newsprint, for which there is little output price variation, these lower costs are partially passed through to the output market in the form of lower prices.[32] This pattern is consistent with large producers that face more inelastic demand than their smaller rivals. Their lower costs are reflected partially in lower output prices and partially in higher markups. Two of the products, bread and coffee, are ones where advertising and product differentiation play a role in consumer markets and could lead to more inelastic demand for the output of the large, national producers. The four remaining products are generally sold as intermediate materials to other manufacturing firms and it is not clear without more detailed study of the using industries what to expect about the pattern of demand elasticities faced by large versus small producers.

Differences in demand elasticities and markups across plants can result from idiosyncratic shocks or from more permanent across-plant differences in product quality or other structural factors. We quantify the relative importance of permanent plant-specific factors versus transitory fluctuations in the markup by examining transition rates through the separate markup distributions over time. These are summarized in Table 7, which has the same format as Table 3.[33] The first row of the Table shows that, of the 25% of the plants with the highest markups in year *t*, 0.484 of them remained in the highest markup quartile five years later, 0.387 had moved to the middle two quartiles of the markup distribution, and 0.129 had moved to the lowest quartile. The Table also reports test results for the hypothesis that the transition rates are equal to those that would be generated by purely random movements in markups.

Two patterns are evident. First, as we observed with the price distribution, there is too much persistence in the tails of the markup distribution for the variation to be driven purely by idiosyncratic shocks. For all but one product, softwood plywood, the probability of remaining in the highest markup quartile or remaining in the lowest markup quartile is significantly different than the 0.25 that would be observed with random movements, and frequently exceeds 0.50. There is also no obvious difference in the persistence patterns for the products that have lower markups for the large plants (hardwood plywood, cans, corrugated boxes, and concrete). Second, when compared with the transition rates for output prices in Table 3, the transitions for plants in the middle of the markup distribution are much more frequently consistent with random movements in markups. That is, we reject the hypothesis of random movements less often with the markups than with the output prices. For five products, polyester blends, cotton sheeting, coffee, oak flooring, and softwood plywood, we cannot reject the null hypothesis of random movements for plants that begin in the middle two quartiles of the markup distribution.

The implication of these patterns is that random shocks or noise in the data are a more important source of markup variation than they are of output price variation but that persistent structural factors are still important in accounting for differences among plants in the high and low markup tails of the distribution. Given our earlier findings on the decline in prices and marginal costs across the size distribution and the importance of permanent between-plant differences in size, the markup transitions are not surprising. If the between-firm variation in prices arises largely from between-firm differences in costs then the markup effectively nets out much of the between-firm variation in these two components and leaves a larger role for idiosyncratic shocks as the source of markup variation. While the larger idiosyncratic variation makes movements among quartiles of the markup distribution more likely, there is still

Table 7. Transition Rates Across Quartiles of the Markup Distribution

Product	Plants in the highest price quartile in year t			Plants in the middle two price quartiles in year t			Plants in the lowest price quartile in year t		
	Highest quartile in $t+1$	Middle two quartiles in $t+1$	Lowest quartile in $t+1$	Highest quartile in $t+1$	Middle two quartiles in $t+1$	Lowest quartile in $t+1$	Highest quartile in $t+1$	Middle two quartiles in $t+1$	Lowest quartile in $t+1$
Hardwood plywood	0.484	0.387[a]	0.129	0.111	0.778	0.111	0.071	0.214	0.714
100% Spun polyester blends with cotton	0.469	0.438[a]	0.094	0.270[b]	0.556[a]	0.175[b]	0.154[b]	0.308	0.538
Cotton sheeting and allied fabrics	0.467	0.400[a]	0.133[b]	0.174[b]	0.581[a]	0.244[b]	0.022	0.457[a]	0.522
Tinplate steel cans	0.563	0.350	0.088	0.152	0.591	0.258[b]	0.083	0.524[a]	0.393
White pan bread	0.541	0.363	0.096	0.164	0.631	0.204	0.098	0.376	0.526
Corrugated shipping containers	0.491	0.410	0.099	0.133	0.611	0.255[b]	0.075	0.487[a]	0.438
Roasted coffee	0.505	0.398	0.097	0.250[b]	0.543[a]	0.207[b]	0.094	0.406[a]	0.500
Hardwood oak flooring	0.476	0.429[a]	0.095	0.213[b]	0.596[a]	0.191[b]	0.037	0.333[a]	0.630
Ready-mixed concrete	0.396	0.476[a]	0.129	0.184	0.614	0.202	0.096	0.362	0.542
Interior softwood plywood	0.471[b]	0.353[a]	0.176[b]	0.125	0.563[a]	0.313[b]	0.154[b]	0.461[a]	0.385[b]
Motor gasoline	0.456	0.443[a]	0.101	0.205	0.573	0.223[b]	0.110	0.433[a]	0.457

[a] do not reject that the proportion equals 0.5 at the 0.05 significance level; [b] do not reject that the proportion equals 0.25 at the 0.05 significance level

a permanent component that is identifiable for the highest and lowest markup plants.

VI. SUMMARY AND CONCLUSIONS

In this chapter we exploit data on output prices at the plant-level to study the dispersion of prices, their persistence over time, correlation with producer size, and correlation with production costs for thirteen homogeneous manufactured products. The data reveal that the amount of micro-level price dispersion varies substantially across products but is relatively constant over time for any specific product. Patterns of movement through the price distribution over time reveal that there is more persistence in the pricing of individual plants than would be generated by purely random movements. In particular, high-price plants tend to remain high price and low-price plants remain low price with high frequency, suggesting that underlying time-invariant structural factors contribute to the output price dispersion. For all products except gasoline and newsprint, the two products with the least overall price dispersion in our group, large producers have lower output prices.

To explore if price differentials are cost based we estimate cost function models, construct plant-specific estimates of marginal cost, and examine time-series and cross-sectional patterns of plant-level price-cost markups. The markup remains unchanged or increases with plant size for all but four of the products and declining marginal costs play an important role in generating the markup patterns. The plants with the highest and lowest markups for each product tend to remain in the same location in the cross-sectional distributions in different years, although the degree of persistence for plants in the middle of the distribution is not as strong as it is with the output price. This suggests that idiosyncratic factors are more important in generating markup variation than they are in generating output price variation.

The evidence we offer highlights the heterogeneity in output market performance across different producers. While no manufactured product is perfectly homogeneous across sellers, we have chosen products to come as close as possible to that norm. Even in these cases, there exists substantial and persistent variation in output prices and markups across plants for most of the products we study. This emphasizes that underlying producer heterogeneity in demand elasticities and production costs are likely to be important determinants of micro level price and markup patterns even in relatively homogeneous goods markets.[34]

There are, however, differences in the price, cost, and markup relationships across products and we cannot point to a single structural explanation that is consistent with the observations for all. Rather, this diversity suggests that the underlying structural features of the industry differ substantially across products. For example, in corrugated boxes differences in the competitiveness of large and small geographic markets combined with small differences in production costs are consistent with the lower prices and margins for large plants. In contrast, the substantially lower marginal costs of large coffee producers, combined with brand name effects supported through advertising, may be responsible for the slightly lower prices, but significantly higher markups of the large producers. Finally, the small price dispersion, high transition rates, and low correlation of prices, costs, and markups with plant size for the gasoline refiners suggest a process in which random shocks and other sources of noise are the dominant source of output market heterogeneity.

Overall, our empirical findings support the view articulated by Demsetz (1973) that producers differ in their efficiency and that this has implications for how we view the competitiveness of market outcomes. In addition, we provide evidence that these efficiency differences are at least partially reflected in output prices. Large producers not only have lower costs in many product markets, they also have lower output prices. The results suggest the value of further theoretical and empirical work that recognizes the existence of within-industry producer heterogeneity and can link the production and demand sides of the firm's decisions.

NOTES

1. Studies that rely on plant or firm-level panel data sets include: Mueller (1986) on profit differences, Dunne, Roberts & Samuelson (1989) on growth and failure rates, Davis, Haltiwanger & Schuh (1996) on growth rates, Lichtenberg & Siegel (1991) and Klette (1996) on R&D investment, Dunne (1994) on technology adoption, Bailey, Hulten & Campbell (1992), Griliches & Regev (1995), Olley & Pakes (1996), and Tybout (1996) on productivity differences, and Davis & Haltiwanger (1991) on wages.

2. Dynamic models of entry and exit with differential efficiency among producers include Jovanovic (1982), Lambson (1991), Hopenhayn (1992), and Ericson & Pakes (1995).

3. Several recent empirical studies use micro data sets to examine one or more sources of price dispersion. Borenstein & Rose (1994) examine ticket prices on specific airline routes and use differences in demand conditions among the routes to explain differences in the degree of price dispersion. Bresnahan & Reiss (1991) and Dunne & Roberts (1992) use variation in the number of competitors across small geographic markets to explain price differences among tire retailers and bread manufacturers, respectively. Wiggins & Maness (1993) use time-series variation in the number of manufacturers to explain price differences for pharmaceutical products. Shepard (1991) exploits differences in the ability of single and multi-product firms to price discriminate

to explain differences in the average price of full and self-serve gasoline among retail service stations in a single geographic market. Abbott (1989) develops a test to distinguish real price dispersion from pure measurement error in the micro data and applies it to output prices for the U.S. cement industry. Van Hoomissen (1988) and Lach & Tsiddon (1992) study the effect of inflation on price dispersion. Pratt, Wise & Zeckhauser (1979) and Dahlby & West (1986) use models of costly consumer search to explain price dispersion.

4. This annual average price is a quantity-weighted average of the price of each unit of output sold over the year. Even if all plants sell at exactly the same price at any point in time, we can observe cross-sectional variation in our plant output prices if the common market price fluctuates over the year and plants do not sell identical shares of output at each of the different prices. This type of aggregation bias, however, should not create correlation in a plant's price across census years.

5. The number of products for which quantity data is collected and the number of producers covered have both fallen over time and, since we want to study changes in a producers' price over time, this also affected the set of products we study.

6. Coffee and wool fabrics are each an aggregate of two seven-digit products. Corrugated boxes is an aggregate over 9 to 11 seven-digit categories that are defined by the end use of the product.

7. Data is available from each of the six censuses between 1963 and 1987 for all products except: cotton sheeting, where 1987 is not available, polyester blends, where 1963 and 1987 are not available, softwood plywood, hardwood oak flooring, and newsprint, where 1982 and 1987 are not available, and tin cans, where data is not available after 1972. The missing years result from redefinition of the seven-digit product codes which made it impossible to construct a consistent definition over all census years. In general, products are aggregated in the later census years. In each case we have constructed consistent product definitions based on the original survey forms from each year and this allows us to avoid problems introduced by large-scale changes in the SIC codes in 1972.

8. We will maintain this ranking of products from highest to lowest average dispersion in all remaining Tables in this paper.

9. The coefficient of variation for softwood plywood is unusually large in one year, 1977, because of a small number of high-price outliers. The average coefficient of variation over 1963–1972 is 17.99 which is in line with the product ranking based on the robust dispersion measure.

10. Deaton (1989) provides an introduction to kernel density and regression estimators.

11. This is identical to normalizing by the arithmetic mean of the log prices in each year. An alternative normalization is to subtract the log of the arithmetic mean of the prices. In this data the arithmetic and geometric means of the prices are virtually identical and the choice of normalization has no effect on the results. Our preference for the geometric mean arises because it is less sensitive to outliers than the arithmetic mean.

12. Four of the products bread, corrugated boxes, concrete, and gasoline have at least 150 observations in each year. The other two products, hardwood plywood and cotton sheeting, have fewer producers and price observations. They each have at least 60 price observations in each year but this may account for the fact that the density estimates are more variable over time for these two products.

13. Some of this stability may simply reflect the fact that the sample sizes in every year are larger for these four product than for the products in panels (a) and (b).

14. This variable nets out the geometric mean price in each year so common market-level factors α_t are not included as a source of price variability. Two products, wool fabrics and tin cans, are not included because of the small number of observations or the short time-series available.

15. A plant must be in operation in two adjacent census years to be included in these calculations.

16. Due to a small number of observations, the probabilities for wool apparel fabrics are not precisely estimated.

17. The one prominent exception to this conclusion is newsprint. This product has the lowest overall price dispersion but the highest level of price persistence at the plant level. It suggests that price dispersion arises from small persistent cost, demand, or product quality differences at the plant level and there are very few or very small idiosyncratic shocks.

18. The quartiles are constructed from the distribution of real output levels observed across all years. The size distributions are very stable across census years. The regressions are estimated with weighted least squares. The observations within each quartile are used to estimate the standard deviation within the quartile and these are used as quartile-specific weights in the regressions.

19. The use of quartile regressions is robust to measurement errors in output. Simple regressions of plant price on output are subject to two sources of bias. The first is the standard error-in-variables bias that results from mismeasurement of an explanatory variable and will bias the coefficient on output toward zero. The second bias arises from the fact that price is constructed as the ratio of value of shipments to output so that measurement errors in output are transmitted to price and introduce a spurious negative correlation between price and output. If the true output coefficient is negative then the two biases work in the opposite direction. The magnitude of the biases will decrease as the ratio of the variance of the true output level to the measured output level increases. Because the true cross-sectional size variation is so large in the manufacturing data, and the size differences are so persistent over time, it is unlikely that output measurement errors play a large role in generating the cross-sectional size-price relationship which is summarized in Table 4. Roberts & Supina (1996) provide pictures of the nonparametric regressions of price on output, which clearly illustrate the negative cross-sectional correlation, for six of the products studied here.

20. Tin cans is unusual because, while there are a large number of plants, they are virtually all owned by a small number of firms. Can plants tend to be constructed near the location of the food processing or beverage plant that will purchase their output. If all cost and price heterogeneity arises at the firm, rather than plant, level and if the size of can plants largely reflects the demand of the near-by users, then there may be persistent price heterogeneity yet no correlation between a plant's size and price.

21. This finding also has implications for the measurement of productivity and scale economies with micro data (see Abbott (1992) and Klette & Griliches (1996)). In most applications real output is constructed by deflating the value of firm or plant shipments by a common industry-level price index. Our finding that output prices are lower for large plants implies that the use of a common price deflator will underestimate the output and productivity of large producers relative to small. This problem may be

mitigated somewhat in industry-level studies if the output deflator accurately reflects the prices of the large producers since they will account for the bulk of industry output and input use.

22. The complete details of the cost function models are described in Supina (1994) and here we will simply summarize some important aspects of the model and estimation methods.

23. In general the omitted and included output will be positively correlated because large plants tend to have higher levels of both outputs. These will lead to an upward bias in the coefficient on output in cost function regressions and thus upward biased estimates of marginal cost.

24. Given the large sample sizes and number of products we examine in this project, it would be an enormous task to construct plant-level service prices for capital that adequately reflected plant-level variation in the opportunity cost of funds, tax variables, and other components of the service price. We will include an ownership-type variable in the cost function model to control for possible differences in capital costs between single and multi-plant producers. Also, because investment flow data is generally only collected for large plants on an annual basis (as part of the *Annual Survey of Manufactures*) it is not possible to construct perpetual-inventory capital stocks for most of the plants in the samples. We do utilize the available information on the book value of the plant's capital in several ways. In the cost model it is used as an instrumental variable for the plant's output and, in the next section, as an additional variable to explain markup differences across plants.

25. If larger plants have more capital intensive technologies, then our inability to measure capital expenditures means that our underestimate of long-run marginal will increase with plant size. This is true assuming capital prices are identical for small and large plants. In practice the price of capital is likely to decline with plant size because large plants are often owned by corporations or large firms who can borrow at lower interest rates. If this is true then the measurement bias could increase or decrease with plant size.

26. We also estimate 'short-run' cost models that include the book value of the plant's capital stock as an additional argument, $C(p_l, p_m, p_e, Q, R, Dt, U, K)$. This approach is justified if plants choose labor, materials, and energy inputs to minimize expenditure on these three variable inputs subject to a fixed capital stock. In this case the total cost we measure in the data is the theoretically correct one. Estimates of the short-run model were unsatisfactory because in micro data sets that are dominated by cross-sectional variation among producers, such as the *Census of Manufactures* data, capital acts as a scale parameter like output and reflects long-run differences in plant size, rather than the short-run substitution between fixed and variable inputs. As a result estimates of the cost elasticity with respect to capital were generally positive, reflecting the fact that plants with more capital are larger and have higher variable costs, rather than negative as it would be if it reflected substitution effects.

27. The returns to scale estimates from the time-differenced data are implausibly high, which is consistent with differencing exaggerating the measurement error problem in output, and attempts to instrument the growth in plant output are of little value, probably because the instruments are fairly poor. In addition to capital we used lagged values of output, as suggested by Griliches & Hausman (1986), as additional

instruments but the results were very similar. As expected, the use of instrumental variables also led to substantial increases in standard errors of the coefficients.

28. Wool apparel fabric is not included because the data was too incomplete to estimate the cost model.

29. The scale elasticity is defined as $SE = (\partial \ln C/\partial \ln Q + \partial \ln C/\ln R)^{-1}$. Both of the cost elasticities vary with the plant's output mix and factor prices.

30. If all plants are selling an identical product at the same market price, as assumed in the Cournot model, then all the observed cross-sectional variation in output price would have to arise from random measurement errors. This is inconsistent with the persistence patterns documented in Tables 2 and 3 for most of the products.

31. To control for variation in capital use, we also estimate the markup regression including the log or the plant's capital stock and its square as explanatory variables. This controls for the fact that the marginal cost estimates do not capture difference in the capital intensity of the plants and the markup we estimate must, in the long run, cover the plant's capital costs. The capital stock data is not available for 1963 and 1967 and this results in reductions is the sample sizes available for the markup regressions. The effect of including capital in the regressions is discussed below.

32. When we also include the plant's capital stock as an additional control variable that may account for rising margins, the markups for these six products rise more substantially than those reported in Table 6. The increased markup for large plants that we observe does not appear to be due to more capital inputs in larger plants.

33. Newsprint is dropped from this Table because there were insufficient observations to precisely estimate the transition rates.

34. Beginning with Stigler (1961), models with buyer search costs have been used to generate price dispersion for identical products. Reinganum (1979) and Carlson & McAfee (1983) develop models in which marginal cost differerences among sellers are reflected in the price distribution. Fishman & Robb (1995) use a combination of search costs, serially-correlated firm cost shocks, and repeat purchases by consumers to generate persistent price and profit differences among producers. At present we do not have any evidence on whether search costs are important or likely to differ across our products and so are not able to draw any conclusions about how relevant these models are for the patterns we observe.

ACKNOWLEDGMENTS

The authors thank Mike Baye, Keith Crocker, Tim Dunne, Paul Geroski, Zvi Griliches, Adam Jaffe, Robert McGuckin, Pinelopi Goldberg, Dennis Mueller, Peter Reiss, Rafael Rob, John Sutton and the staff of the Center for Economic Studies of the U.S. Census Bureau for helpful comments and discussions on earlier drafts of this chapter. This research was conducted at the Center for Economic Studies while the authors were participants in the Census Bureau Research Fellows Program. Any opinions, finding, or conclusions expressed here are those of the authors and do not in any way reflect the views of the U.S. Census Bureau.

REFERENCES

Abbott, T. A. (1989). *Producer Price Dispersion and the Analysis of Production*, unpublished Ph.D. dissertation, Harvard University.

Abbott, T. A. (1992). Producer Price Dispersion, Real Output, & the Analysis of Production. *Journal of Productivity Analysis*.

Baily, M., C. Hulten, & Campbell, D. (1992). Productivity Dynamics in Manufacturing Plants. *Brookings Papers on Economic Activity: Microeconomics*, 187–227.

Borenstein, S., & Rose, N. (1994). Competitive Price Discrimination in the U.S. Airline Industry. *Journal of Political Economy, 102*, 653–683.

Bresnahan, T. F., & Reiss, P. C. (1991). Entry and Competition in Concentrated Markets. *Journal of Political Economy, 99*, 977–1009.

Carlson, J. A., & McAfee, R. P. (1983). Discrete Equilibrium Price Dispersion. *Journal of Political Economy, 91*, 480–493.

Caves, R. E., & Porter, M. E. (1977). From Entry Barriers to Mobility Barriers. *The Quarterly Journal of Economics, 91*, 241–261.

Dahlby, B., & West, D. S. (1986). Price Dispersion in an Automobile Insurance Market. *Journal of Political Economy, 94*, 418–438.

Davis, S. J., & Haltiwanger, J. (1991). Wage Dispersion between and within U.S. Manufacturing Plants, 1963–86. In: M. N. Bailey & C. Winston (Eds), *Brookings Papers on Economic Activity: Microeconomics*, 115–180.

Davis, S. J., Haltiwanger, J. & Schuh, S. (1996). *Job Creation and Destruction*. Cambridge: MIT Press.

Deaton, A. (1989). Rice Prices and Income Distributions in Thailand: A Non-Parametric Analysis. *Economic Journal, 99*, 1–37.

Demsetz, H. (1973). Industry Structure, Market Rivalry, & Public Policy. *Journal of Law and Economics, 6*, 1–10.

Dunne, T. (1994). Plant Age and Technology Use in U.S. Manufacturing Industries. *The Rand Journal of Economics, 25*, 488–499.

Dunne, T., & Roberts, M. J. (1992). Costs, Demand, & Imperfect Competition as Determinants of Plant-Level Output Prices. In: D. B. Audretsch & J. J. Siegfried (Eds), *Empirical Studies in Industrial Organization*. Kluwer Academic Publishers.

Dunne, T., Roberts, M. J. & Samuelson, L. (1989). The Growth and Failure of U.S. Manufacturing Plants. *The Quarterly Journal of Economics, 104*, 671–698.

Ericson, R., & Pakes, A. (1995). Markov-Perfect Industry Dynamics: A Framework for Empirical Work. *Review of Economic Studies, 62*, 53–82.

Fishman, A., & Rob, R. (1995). The Durability of Information, Market Efficiency and the Size of Firms. *International Economic Review, 36*, 19–36.

Griliches, Z. (1986). Economic Data Issues. In: M. Intrilligator & Z. Griliches (Eds), *Handbook of Econometrics*. Amsterdam: North -Holland.

Griliches, Z., & Hausman, J. A. (1986). Errors in Variables in Panel Data. *Journal of Econometrics, 31*, 93–118.

Griliches, Z., & Regev, H. (1995). Firm Productivity in Israeli Industry: 1979–1988. *Journal of Econometrics, 65*, 175–203..

Hopenhayn, H. (1992). Entry and Exit in Long Run Equilibria. *Econometrica, 60*, 1127–1150.

Jovanovic, B. (1982). Selection and the Evolution of Industry. *Econometrica, 50*, 649–670.

Klette, T. J. (1996). R&D, Scope Economies, & Plant Performance. *The Rand Journal of Economics*, *27*, 502–522.

Klette, T. J., & Griliches, Z. (1996). The Inconsistency of Common Scale Estimators when Output Prices are Unobserved and Endogenous. *Journal of Applied Econometrics*, *11*, 343–361.

Lach, S., & Tsiddon, D. (1992). The Behavior of Prices and Inflation: An Empirical Analysis of Disaggregated Price Data. *Journal of Political Economy*, *100*, 349–389.

Lambson, V. E. (1991). Industry Evolution with Sunk Costs and Uncertain Market Conditions. *International Journal of Industrial Organization*, *9*, 171–196.

Lichtenberg, F., & Siegel, D. (1991). The Impact of R&D Investment on Productivity – New Evidence Using Linked R&D-LRD Data. *Economic Inquiry*, *29*, 203–229.

Mairesse, J. (1990). Time-Series and Cross-Sectional Estimates on Panel Data: Why are They Different and Why Should They be Equal?. In: J. Hartog, G. Ridder, & J. Theeuwes (Eds), *Panel Data and Labor Market Studies*. Elsevier Science Publishers B.V. (North-Holland).

Mueller, D. C. (1986). *Profits in the Long Run*. Cambridge: Cambridge University Press.

Olley, G. S., & Pakes, A. (1996). The Dynamics of Productivity in the Telecommunications Equipment Industry. *Econometrica*, *64*, 1263–1297.

Peltzman, S. (1977). The Gains and Losses from Industrial Concentration. *Journal of Law and Economics*, *20*, 229–263.

Porter, M. E. (1980). *Competitive Strategy*. New York: The Free Press.

Pratt, J. W., Wise, D. A. & Zeckhauser, R. (1979). Price Differences in Almost Competitive Markets. *Quarterly Journal of Economics*, *93*, 189–211.

Reinganum, J. F. (1979). A Simple Model of Equilibrium Price Dispersion. *Journal of Political Economy*, *87*, 851–858.

Roberts, M. J., & Supina, D. (1996). Output Price, Markups, & Producer Size. *European Economic Review Papers and Proceedings*, *40*, 909–921.

Schmalensee, R. (1989). Inter-Industry Studies of Structure and Performance. In: R. Schmalensee & R. Willig (Eds), *The Handbook of Industrial Organization*. Amsterdam: North-Holland.

Shepard, A. (1991). Price Discrimination and Retail Configuration. *Journal of Political Economy*, *99*, 30–53.

Stigler, G. (1961). The Economics of Information. *Journal of Political Economy*, *64*, 213–225.

Stigler, G. J., & Kindahl, J. K. (1970. *The Behavior of Industrial Prices*. New York: Columbia University Press.

Supina, D. (1994). *Price and Markup Dispersion Among U.S. Manufacturing Plants*, unpublished Ph. D.dissertation, The Pennsylvania State University.

Tybout, J. R. (1996). Heterogeneity and Productivity Growth: Assessing the Evidence. In: M. Roberts & J. Tybout (Eds), *Industrial Evolution in Developing Countries: Micro Patterns of Turnover, Productivity, & Market Structure*. Oxford: Oxford University Press.

Van Hoomissen, T. (1988). Price Dispersion and Inflation: Evidence from Israel. *Journal of Political Economy*, *96*, 1303–1314.

Wiggins, S. N., & Maness, R. (1993). *Price Competition in Pharmaceutical Markets*, mimeo.

QUOTAS AND TARIFFS WITH ENDOGENOUS CONDUCT

Raymond J. Deneckere, Dan Kovenock and Yong Yeop Sohn

ABSTRACT

This chapter utilizes the results of Deneckere & Kovenock (1988, 1989, 1992, 1996) on price setting games with capacity constraints and different unit costs up to capacity to analyze the effects of quotas and tariffs in a model in which a domestic market for a homogeneous product is supplied by a duopoly consisting of a domestic and a foreign firm. A model of the timing of price setting is constructed in which the existence of price leadership, as well as the identity of the leader, depends upon the vector of unit costs and capacities (k_1, k_2, c_1, c_2). *With firm 1 the foreign firm and firm 2 the domestic firm, the levying of a tariff raises the foreign unit cost of production up to capacity to* $c_1^t = c_1 + t$, *while the imposition of a 'binding' quota reduces capacity from* k_1 *to* k_1^q *(the level of the quota). The effects of quotas and tariffs on the equilibrium in the game of timing are examined starting from an initial vector in which costs are identical,* $c_1 = c_2 = c$. *It is shown that, due to the endogeneity of price leadership, trade restrictions can have surprising effects. In addition to the traditional view that quotas hurt the foreign firm and help the domestic firm, and the results of Harris (1985) and Krishna (1989) that quotas may help both firms, we show that with endogenous timing a quota can make the domestic firm worse off and the foreign firm better off (by altering the identity of the price leader). However, a quota will always (weakly)*

Industrial Organization, Volume 9, pages 37–68.

ISBN: 0-7623-0687-4

increase price. In contrast, a tariff always (weakly) hurts the foreign firm and (weakly) helps the domestic firm but may, by affecting the leadership role, lower price. The question of the equivalence of quotas and tariffs is also examined. In contrast to the result of Deneckere & Kovenock (1989) for the simultaneous price setting game, we show that with the endogenous timing of price-setting there are certain initial conditions (k_1, k_2, c_1, c_2) for which the prices and quantities generated by a 'binding' quota can be duplicated by a tariff (and vice versa). It is possible, however, that a quota that reduces the foreign capacity slightly is equivalent only to a very severe tariff. We conclude by showing how the model allows for simple welfare comparisons in environments in which protection is likely.

I. INTRODUCTION

It is now well-known that in an imperfectly competitive market open to international trade quotas and tariffs may have substantially different effects. As first demonstrated by Bhagwati (1965) for a domestic monopolist facing a competitive fringe of foreign suppliers, for any given quota there may not exist a tariff which is equivalent. Since Bhagwati's pioneering work, and especially since the game theoretic revolution in the analysis of imperfect competition, there has been a keen interest among economists in evaluating the effects of tariffs and quotas across a large number of plausible models of imperfectly competitive market behavior. While an outside observer might find the abundance of models troubling, the fact that different models might best describe the behavior of firms in different industries makes these extensions worthwhile.

This chapter attempts to contribute to our understanding of the effects of quotas and tariffs by analyzing a model in which the conduct of firms in the market arises endogenously and is affected by the trade restrictions imposed. Our focus is on price leadership, an institution which is prevalent in many markets.[1] More specifically, we use the results of Deneckere & Kovenock (1988, 1989, 1992, 1996) on price setting games with capacity constraints and different unit costs up to capacity to analyze the effects of quotas and tariffs in a domestic market for a homogeneous product that is supplied by a duopoly consisting of a domestic and a foreign firm. We construct a model of the timing of price setting in which both leader-follower and simultaneous price setting equilibria are allowed, and in which the existence of price leadership, as well as the identity of the leader, depends upon the vector of unit costs and capacities (k_1, k_2, c_1, c_2). With firm 1 the foreign firm and firm 2 the domestic firm, levying a tariff raises the foreign unit cost of production up to capacity to $c_1^t = c_1 + t$, while imposing a 'binding' quota reduces capacity from k_1 to k_1^q (the level of the quota). We examine the effects of quotas and tariffs on the equilibrium in the game of timing starting from an initial vector in which costs are identical, $c_1 = c_2 = c$.[2] We show that, due to the endogeneity of price leadership and the switching of leadership regimes that might result, trade restrictions can have surprising effects. In addition to the traditional view that quotas hurt the foreign firm and help the domestic firm, and the results of Harris (1985) and Krishna (1989) that quotas may help both firms, we show that with endogenous timing of price setting a quota can make the domestic firm worse off and the foreign firm better off (by altering the identity of the price leader). However, a quota will always (weakly) increase price. In contrast, a tariff always (weakly) reduces the foreign firm's profit and (weakly) increases the profit of the domestic firm. However, a tariff may lower the equilibrium price when it causes the price leadership role to switch from the domestic to the foreign firm.

The equivalence of quotas and tariffs is also examined. In contrast to the result of Deneckere & Kovenock (1989) for the simultaneous move price setting game, we show that with the endogenous timing of price setting there are certain initial conditions (k_1, k_2, c_1, c_2) and binding quotas k_1^q for which equivalent tariffs exist. However, it is possible that a quota that reduces the foreign capacity slightly is equivalent only to a very severe tariff.

While there are many papers dealing with the effects of tariffs and quotas in an imperfectly competitive environment, to our knowledge the papers most closely related to ours are those of Dean & Gangopadhyay (1992), Harris (1985), Ito & Ono (1982) and Krishna (1989). Harris analyzes the effects of a voluntary export restraint (VER) and tariff in a duopoly model in which the foreign and domestic goods are imperfect substitutes. He assumes that the firms initially have unlimited capacity and are simultaneous price setters but that the VER forces the domestic firm to become a price leader. Krishna, in a similar model,[3] assumes simultaneous price setting both before and after the imposition of a VER. While consistency of treatment is, in our opinion, a desirable approach, the model presented here provides limited support for the Harris assumption in the context in which it was made. In particular, although endogenous timing does not always yield the domestic leadership outcome assumed by Harris, domestic leadership does indeed arise under the homogeneous good analogues of the assumptions Harris makes regarding capacities and the size of the VER. Hence, one contribution of this chapter might be viewed as providing a game theoretic justification of the Harris leadership assumption. Ito and Ono analyze quotas and tariffs in a model with both exogenous and endogenous price leadership. However, they identify price leadership with behavioral assumptions that do not have to hold in the equilibrium of a Stackelberg price setting game – specifically, that the follower always sells its supply at the price quoted by the leader. This renders the follower a price taker – an assumption that we find unappealing. The implications of this assumption regarding the effects of quotas and tariffs are important as is shown below. Finally, Dean & Gangopadhyay (1992) find results similar to some of those obtained in Deneckere & Kovenock (1988, 1992) and construct a scenario in which leadership is obtained with preplay communication. However, their paper focuses on price setting in an environment in which the imposition of the quota is uncertain and in which prices cannot adjust after the quota is imposed. We find this an unappealing assumption as it implies that prices adjust less quickly than governments. Their model also does not accommodate the analysis of tariffs.

In Section II we present the basic model of price setting. Equilibria with exogenously specified roles are described in Section III, and Section IV examines the determination of endogenous leadership. Section V looks at the effects of quotas (or VER's) on the endogenous timing game and Section VI follows with the effects of tariffs. In Section VII we examine the (non)-equivalence of quotas and tariffs. Section VIII closes with an example that shows how the methods developed allow for simple welfare comparisons in economic environments in which protection is likely.

II. THE MODEL

Consider a market shielded from entry in which two firms produce a homogeneous good. Aggregate demand for the firms' output as a function of price is $d(p): R_+ \rightarrow R_+$. We assume that $d(p)$ satisfies the following assumptions:

A1. There exists a $p_0 > 0$ such that $d(p) = 0$ for every $p \geqslant p_0$ and $d(p) > 0$ for $p < p_0$. Furthermore, $d(p)$ is twice continuously differentiable, strictly decreasing and concave on $[0, p_0)$.

Each firm i $(i = 1, 2)$ has a fixed capacity k_i, and produces the good at a constant unit cost $0 \leqq c_i < p_0$ up to k_i. Since a firm may not be able to serve the entire market demand at its announced price it is necessary to specify a rule by which consumers are rationed. Following Levitan & Shubik (1972), Kreps & Scheinkman (1983), and Osborne & Pitchik (1986), we assume that rationing is efficient. That is, if $p_i < p_j$ and $d(p_i) > k_i$, the demand facing firm j is $\max(0, d(p_j) - k_i)$. As noted by Dixon (1987a, b) such a contingent demand would obtain if there were a single consumer in the market and income effects of low priced purchases were compensated. Alternatively, with many identical consumers it can be obtained from a rationing rule in which each consumer obtains an equal share of the output available when there is excess demand. Finally, and perhaps most naturally, it could be obtained in a model in which consumers have inelastic demand for one unit of the good, and consumers with the highest reservation prices obtain the good from the low priced firm, either directly or through resale of the product by consumers with low reservation prices.[4]

Given these assumptions, the profit of firm i when it sets price p_i and firm j sets price p_j is

$$\Pi_i(p_i, p_j \mid k_i, k_j) = \begin{cases} L_i(p_i \equiv (p_i - c_i)\min(k_i, d(p_i)) & \text{if } p_i < p_j \\ T_i(p_i) \equiv (p_i - c_i)\min(k_i, \min(0, d(p_i) - I^i k_j)) & \text{if } p_i = p_j \\ H_i(p_i) \equiv (p_i - c_i)\min(k_i, \max(0, d(p_i - k_j)) & \text{if } p_i > p_j \end{cases} \quad (2.1)$$

where I^i is an indicator which, in a price leader game, takes on the value 1 if i is a leader and 0 if i is a follower. In a simultaneous move game I^i takes on the value 1 if $c_j < c_i$ or $c_1 = c_2$ and $i = 2$, and takes on the value 0 if $c_i < c_j$ or $c_1 = c_2$ and $i = 1$. Thus in leader-follower games we will assume that in the case of a tie in prices the second firm to set price sells its capacity first. This assumption is made for the purely technical reason of avoiding the need to have a follower undercut the leader by an arbitrarily small amount. In the simultaneous move game we will assume that in a case of a tie the low cost firm sells its capacity first.[5] Note that in (2.1) $L_i(p_i)$ refers to the profit from being the low priced seller at p_i and $H_i(p_i)$ the profit from being the high priced seller at p_i. $T_i(p_i)$ is the profit from tying at p_i. The functions L_i and H_i are illustrated in Fig. 1.

In all of the price setting games which follow we assume that firm i chooses a price in the set $S_i = [c_i, p_0]$. Note that this rules out strategies in which firm i prices below its unit cost.[6] Let P_i^L be the set of prices which maximize $L_i(p_i)$ and P_i^H be the set of maximizers of $H_i(p_i)$. Given (A1) there exists a unique $p_i^L \in P_i^L$; $L_i(p_i)$ is continuous and strictly increasing in p_i for $p_i < p_i^L$. Let $H_i^* \equiv H_i(P_i^H)$. If H_i^* is nonzero (A1) guarantees a unique element p_i^H in P_i^H; $H_i(p_i)$ is continuous and strictly increasing in p_i for $p_i < p_i^H$. If $H_i^* = 0$ we define $p_i^H = c_i$. Let

$$\underline{p}_i = \min\{p_i : L_i(p_i) = H_i^*\} \quad i = 1, 2$$

Observe that $\underline{p}_i \leq p_i^H < p_i^L$, $i = 1, 2$. In the analysis that follows we shall sometimes write p_i^L, p_i^H, and $\underline{p}_i$ as functions of (k_1, k_2, c_1, c_2) to indicate the dependence of these prices on capacities and unit costs. It will also prove useful to refer to the inverse demand function $P(q)$, defined by: $P(q) = d^{-1}(q)$ for $0 < q \leq d(0)$, $P(0) = p_0$, and $P(q) = 0$ for $q > d(0)$. Finally, we will often use the expression $r_i(k \mid c_i) \equiv \text{argmax}_x (P(x + k) - c_i)x$, which is firm i's Cournot best

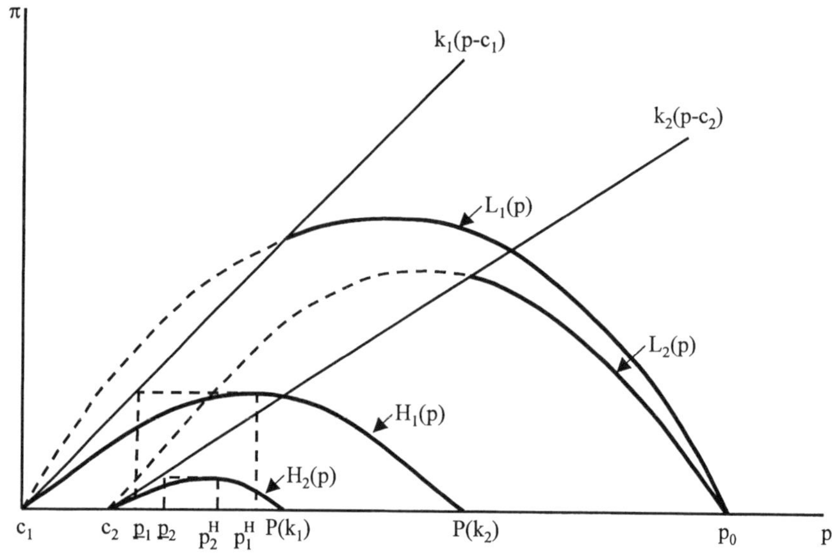

Fig. 1. The functions L_1 and H_1 with the critical prices.

response when its rival's output is k and firm i has a constant unit cost of production c_i.

III. EQUILIBRIA WITH EXOGENOUSLY SPECIFIED ROLES

We now summarize the properties of the simultaneous move price setting game and the exogenously specified price leader games. To analyze the former the strategy space of firm i ($i = 1, 2$) needs to be extended to the set of mixed strategies over S_i, while firm i's payoff function becomes its expected profit as a function of the mixed strategies played. The following proposition is a summary of the detailed characterization of the Nash equilibria of the simultaneous move price setting game provided in Deneckere & Kovenock (1989, 1996):

Proposition 1. *For any vector of capacities and costs* (k_1, k_2, c_1, c_2) *the simultaneous move price setting game has a Nash equilibrium in mixed strategies.*

(*a*) *If* $k_i \leqslant r_i(k_j | c_i)$, $i = 1, 2$, $j \neq i$, *the unique equilibrium is in pure strategies. Both firms charge* $p = P(k_1 + k_2)$ *and profits are given by* $\Pi_i^S = k_i(P(k_1 + k_2) - c_i)$.

(*b*) *If* $c_j \geqslant P(k_i)$ *and* $c_j \geqslant \underline{p}_i$, *for some i,* $j \neq i$ *then there exists an equilibrium in pure strategies (if* $c_i \neq c_j$ *there will be equilibria in which firm j mixes as well). Firm i charges* $p_i = \min(p_i^L, c_j)$ *and firm j places all probability at* c_j *(or possibly above* c_j *if* $c_i \neq c_j$*).*
Equilibrium profits are

$$\Pi_i^S = [\min(p_i^L, c_j) - c_i]d(\min(p_i^L, c_j)) \quad \textit{and} \quad \Pi_j^S = H_j^* = 0.$$

(*c*) *If neither the conditions of (a) nor the conditions of (b) apply, there is a unique equilibrium which is in nondegenerate mixed strategies. If* $\underline{p}_i \geqslant \underline{p}_j$, $i \neq j$, *then in equilibrium* $\Pi_k^S = L_k(\underline{p}_i)$, $k = 1, 2$.

An illustration of the regions of capacity for which (a), (b), and (c) hold is provided in Fig. 2 for the case where $c_1 = c_2$ and in Fig. 3 for the case where $c_1 < c_2$ and $r_1(0 | c_1) < d(c_2)$. The proposition can be interpreted as follows: Case (a) corresponds to the region of capacities below both firms' Cournot reaction functions for the given unit costs. Over this region perfect competition, price competition and quantity competition all coincide and price clears all capacity off the market. In case (b) the standard Bertrand result for constant unit costs arises; if these unit costs differ the low cost firm prices the high cost firm out of the market. In region (c), the Edgeworth cycle case, no pure strategy equilibrium in prices exists. The characterization of the equilibrium profits and distributions depends crucially on the ranking of $\underline{p}_1$ and $\underline{p}_2$. This ranking in turn depends on the vector (k_1, k_2, c_1, c_2). As shown in the figures, within region (c), when $c_1 = c_2$, $\underline{p}_1 \geqslant \underline{p}_2$ if and only if $k_1 \geqslant k_2$; when $c_1 < c_2$, $\underline{p}_2 > \underline{p}_1$ if $k_2 \geqslant k_1$. Note also that $\underline{p}_i$ is a continuous function of the vector (k_1, k_2, c_1, c_2).

Consider now the games in which one of the firms is an exogenously determined price leader. Without loss of generality we shall assume that $c_2 \geqslant c_1$. When the capacities of the two firms are in region (a) of Proposition 1, that is, when both firms sell all capacity in the simultaneous move equilibrium, the leader-follower equilibria coincide with the simultaneous move equilibrium. When the capacities are in region (b) of Proposition 1, both leader-follower games have a continuum of equilibria, in which firm 2 generally earns zero profit if $c_2 > c_1$. Finally, in the range of capacities where no pure strategy simultaneous move equilibrium exists, i.e. region (c) of Proposition 1, the nature of the leader-follower equilibria depend on the ranking of $\underline{p}_1$ and $\underline{p}_2$. If $\underline{p}_i > \underline{p}_j$, in $i \in \{1, 2\}$, $j \neq i$, and firm i leads, it sets p_i^H and firm j follows by matching. If the inequality holds and firm j leads, it sets $p_j = \underline{p}_i$, and firm i follows with $p_i = p_i^H$.[7]

The logic underlying the derivation of the equilibrium in region (c) is straightforward and instructive. If firm i leads it must charge a price less than or equal to $\underline{p}_j$ to induce firm j to set a higher price as follower. For prices above $\underline{p}_j$, firm j would prefer to undercut (if $p_i > p_j^L$) or match, and sell all its capacity. Hence, when firm i leads it must decide which it prefers: it can set a price $p_i \leqslant \underline{p}_j$ and be the low-priced firm, or set its preferred price given that it is going to be undercut, p_i^H. If $\underline{p}_i > \underline{p}_j$, then since $L_i(p)$ is increasing over the relevant interval, $L_i(\underline{p}_j) < L_i(\underline{p}_i) = H_i^*$. Therefore, firm i will set p_i^H, and firm j will match and sell its capacity. If $\underline{p}_i < \underline{p}_j$, $L_i(\underline{p}_j) > L_i(\underline{p}_i) = H_i^*$ and firm i prefers to set $\underline{p}_j$ as leader. Firm j follows with p_j^H.

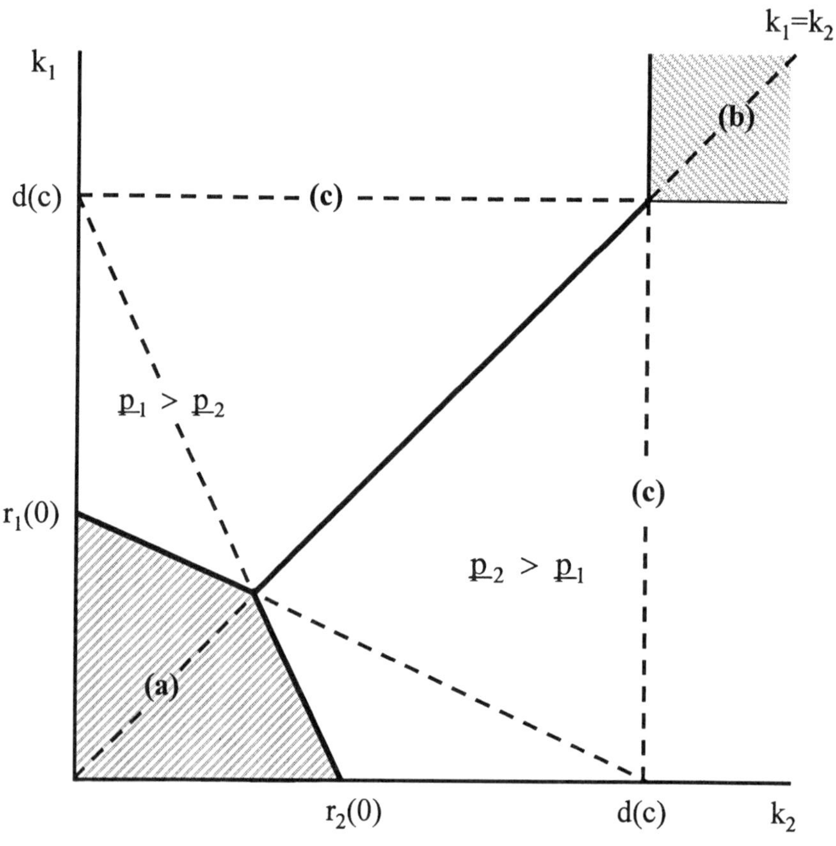

Fig. 2. The ranking of $\underline{p}_1$ and $\underline{p}_2$ and the regions of Proposition 1 for $c_1 = c_2 = c$.

Proposition 2 summarizes the characterization of the set of equilibria when firm 2 is an exogenously specified leader. The characterization for a low cost leader is provided in Proposition 3.

Proposition 2. *For any quadruple* (k_1, k_2, c_1, c_2) *with* $c_2 \geqslant c_1$, *the price setting game with firm 2 as a price leader has the following equilibria:*

(a) If $k_i \leqslant r_i(k_j \mid c_i)$, $i = 1, 2$, $j \neq i$, *then both firms set* $p = P(k_1 + k_2)$ *and profits are given by* $\Pi_i = k_i(P(k_1 + k_2) - c_i)$.

(b) If $c_2 \geqslant P(k_1)$ *and* $c_2 > \underline{p}_1$ *the set of (pure strategy) equilibrium prices is given by* $\{(p_1,p_2) \mid p_1^L \geqslant p_2 \geqslant c_2,\ p_1 = p_2\} \cup \{(p_1,p_2) \mid p_2 \geqslant \max(c_2,p_1^L),\ p_1 = p_1^L\}$. *If* $c_2 \geqslant P(k_1)$ *and* $c_1 < c_2 = \underline{p}_1$, *then in addition to these equilibria,* $p_2 = c_2$ *and* $p_1 = p_1^H$ *is also an equilibrium price pair. If* $c_2 \geqslant P(k_1)$ *and* $c_1 = c_2 = \underline{p}_1$ *then* $\{(p_1,p_2) \mid p_2 = c_2, p_1 \geqslant c_1\}$ *is also in the set of equilibria. In all of these equilibria the profit of firm 2,* Π_2^L, *is equal to zero. The equilibrium profit of firm 1,* Π_1^F, *depends upon the particular equilibrium chosen.*[8]

(c) If the conditions in (a) and (b) do not hold, and $\underline{p}_2 \geqslant \underline{p}_1$ *then there exists an equilibrium in which* $p_2 = p_2^H$ *and* $p_1 = p_2$. *Profits in this equilibrium are*

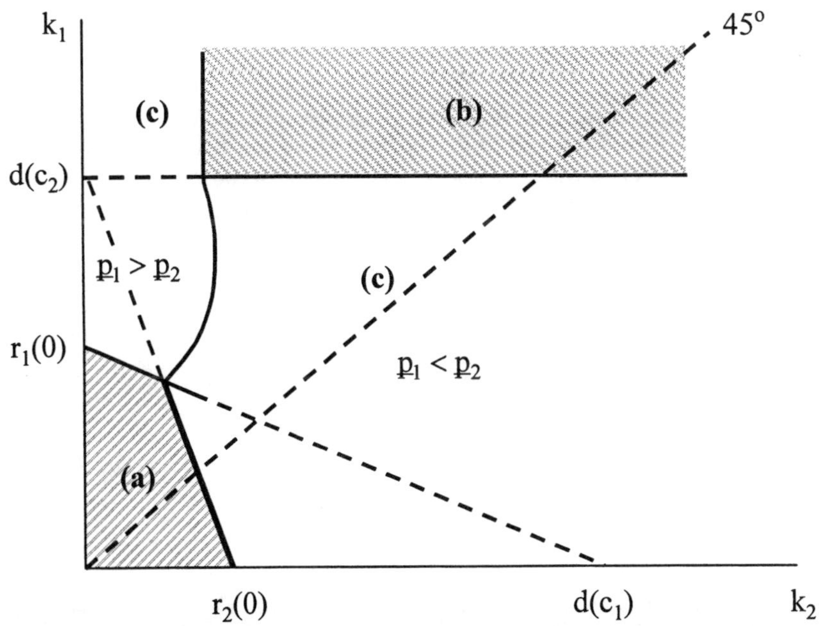

Fig. 3. The ranking of $\underline{p}_1$ and $\underline{p}_2$ and the regions of Proposition 1 for $c_1 < c_2$.

$\Pi_2^L = H_2^*$ and $\Pi_1^F = L_1(p_2^H)$. If $\underline{p}_2 > \underline{p}_1$ the equilibrium is unique. If $\underline{p}_2 = \underline{p}_1$ there is an additional equilibrium in which $p_2 = \underline{p}_2$ and $p_1 = p_1^H$ (and thus a continuum of equilibria in which firm 2 mixes between $\underline{p}_2$ and p_2^H). In this second pure strategy equilibrium $\Pi_2^L = L_2(\underline{p}_2) = H_2^*$ and $\Pi_1^F = H_1(p_1^H) = H_1^*$.

(d) If $\underline{p}_2 < \underline{p}_1$ then $p_2 = \underline{p}_1$ and $p_1 = p_1^H$. Profits are $\Pi_2^L = L_2(\underline{p}_1)$ and $\Pi_1^F = H_1^*$.

Proof: (a) and (b) are straightforward. The proof of (c) is identical to the proof of Theorem 2(c) of Deneckere & Kovenock (1988, 1992). The proof of (d) is identical to the proof of Theorem 3(c) in the same papers. ■

Note that Figs 2 and 3 can also be used to gain insight into the nature of the regions where Proposition 2(a)–(d) hold. For the given cost pairs illustrated, both (a) and (b) of Proposition 2 coincide with the same lettered regions of Proposition 1. Region (c) of Proposition 1 is split into two regions ((c) and (d)) in Proposition 2, depending on the relative values of $\underline{p}_1$ and $\underline{p}_2$.

When the low cost firm is exogenously specified as a leader we obtain:

Proposition 3. *For any quadruple (k_1, k_2, c_1, c_2) with $c_2 > c_1$, the price setting game with firm 1 as a price leader has the following equilibria:*

(a) If $k_i \le r_i(k_j \mid c_i)$, $i = 1,2$, $j \ne i$, then both firms set $p = P(k_1 + k_2)$ and profits are given by $\Pi_i = k_i(P(k_1 + k_2) - c_i)$.

(b) If $c_2 \ge P(k_1)$ and $c_2 \ge \underline{p}_1$, firm 1 sets $p_1 = \min(p_1^L, c_2)$ and firm 2 sets any $p_2 \ge c_2$ (or any mixed strategy placing all mass at or above c_2), with no mass placed at c_2 if $p_1 = c_2$. Equilibrium profits are $\Pi_1^L = [\min(p_1^L, c_2) - c_1]\, d(\min(p_1^L, c_2))$ and $\Pi_2^F = H_2^* = 0$. If $\underline{p}_1 = c_2$ there is also an equilibrium in which firm 1 sets p_1^H (and therefore a continuum of equilibria in which firm 1 randomizes between $\underline{p}_1$ and p_1^H) and firm 2 matches. In this equilibrium $\Pi_1^L = H_1^*$ and $\Pi_2^F = L_2(p_1^H)$.

(c) If the conditions in (a) and (b) do not hold and $\underline{p}_2 \ge \underline{p}_1$ then there exists an equilibrium in which $p_1 = \underline{p}_2$, $p_2 = p_2^H$, $\Pi_1^L = L_1(\underline{p}_2)$ and $\Pi_2^F = H_2^*$. If $\underline{p}_2 > \underline{p}_1$ the equilibrium is unique. If $\underline{p}_2 = \underline{p}_1$ then there is an additional equilibrium in which $p_1 = p_1^H$ and $p_2 = p_1$ (and therefore a continuum of equilibria in which firm 1 mixes between $\underline{p}_1$ and p_1^H). In this second equilibrium $\Pi_1^L = H_1^*$ and $\Pi_2^F = L_2(p_1^H)$.

(d) If $\underline{p}_1 > \underline{p}_2$ then $p_1 = p_1^H$ and $p_2 = p_1$. Profits in this equilibrium are $\Pi_1^L = H_1^*$ and $\Pi_2^F = L_2(p_1^H)$.

The proof of this proposition, which we outlined above, is again a straightforward extension of Theorems 2 and 3 of Deneckere & Kovenock (1988, 1992). As in the case of the previous proposition, Figs 2 and 3 can again be used to gain insight into the regions where Proposition 3(a)–(d) hold. Regions (a) and (b) again coincide with the same lettered regions of

Proposition 1, while region (c) of Proposition 1 is split into two regions ((c) and (d)) in Proposition 3.

IV. ENDOGENOUS LEADERSHIP

The results of Section III allow us to compare the payoffs of the two firms in the simultaneous move game and in each of the exogenously determined leader-follower games. Let Π_i^S be the payoff to firm i in the simultaneous move game, Π_i^L the payoff to i as a price leader, and Π_i^F the payoff to i as a price follower. When $\lambda \equiv (k_1, k_2, c_1, c_2)$ is in the range where the equilibrium in the simultaneous move game is unique and in non-degenerate mixed strategies (region (c) of Proposition 1 and Figs 2 and 3) and when $\underline{p}_i < \underline{p}_j$,

$$\Pi_j^S = \Pi_j^L = \Pi_j^F$$

and

$$\Pi_i^S = \Pi_i^L < \Pi_i^F.$$

The strict inequality follows since $L_i(p)$ is strictly increasing on $[c_i, p_i^L)$ and for λ satisfying these conditions $p_i^L > p_j^H > \underline{p}_j$. Thus, for λ in this range firm j is indifferent as to which of the three games it plays, while firm i is indifferent between being a leader and moving simultaneously, but strictly prefers to follow.[9] When λ is such that both firms sell all of their capacity in the simultaneous move equilibrium (region (a) of Proposition 1 and Figs 2 and 3) both firms are indifferent between all three games. Finally, when λ is such that (b) of Proposition 1 pertains, if $c_i < c_j$ the high cost firm is driven out of the market in all three games (except when firm i leads and $\underline{p}_i = c_j$, in which case there are two pure strategy equilibria, one of which gives firm j a positive return). Thus, the high cost firm is indifferent as to the game it plays. The low cost firm always does at least as well as a follower (in which case there are multiple equilibria) than as a leader or in the simultaneous move game. However, if the high cost firm plays an equilibrium strategy as leader of pricing at its unit cost, then the low cost firm is indifferent between all three games as well. If (b) pertains and $c_i = c_j$ we are in the standard Bertrand region and both firms earn zero profit in the simultaneous move game and as leader. Again, the profit as follower depends on the equilibrium played. When the leader plays the equilibrium strategy of pricing at unit cost the follower's profit is zero so that, if this equilibrium strategy were always selected, the firms would be indifferent between all three games.

From the above discussion it seems reasonable to conclude that, in a model where leadership is endogenized, for λ such that (a) or (b) of Proposition 1 pertains, no consistent pattern of leadership would arise. Since all three exogenously specified timing games may have profits and prices which are very much alike, the specifics of the game form and equilibrium selection would largely dictate the actual sequencing of moves. We will therefore assume that profits (and prices) are those of a simultaneous move equilibrium in these cases. On the other hand, for λ satisfying the conditions of Proposition 1 (c), if $\underline{p}_i > \underline{p}_j$ we can expect to see leadership arise; firm j always has incentive to outwait firm i and force firm i to lead. Firm i has no incentive to delay setting its price. Hence we can expect firm i to become a leader.[10] Deneckere & Kovenock (1988, 1992) provide three models of the timing of price announcement in which endogenous leadership is obtained in this region. We summarize here the simplest of these models.

Take the time period under consideration to be the unit interval [0, 1], and divide the interval up into T ($T \geq 4$, even) periods each of length $\tau = \frac{1}{T}$, labelled $t = 0, \ldots, T-1$. Suppose that firm 1 may announce price at the beginning of even-indexed intervals and that firm 2 may do so at the beginning of odd-indexed intervals. Once a price is announced it must remain in effect for the duration of the unit interval. Assume that until both firms announce prices, both firms earn zero profits, but that demand remains $d(p)$ regardless of the point at which prices are announced.[11] Thus the profit expressions in (2.1) remain the profits obtained once prices are announced. Assume that firms discount profits at the rate $r > 0$ and let $\delta = e^{-r\tau}$. It is easily demonstrated that, within the relevant region, if $\underline{p}_i > \underline{p}_j$ and firm i is faced with a situation in which neither firm has set price, it will always set its leader price at the first opportunity. Thus, if the discount factor δ is not too small, firm j will always wait until firm i sets price, and follow with its best response. Regardless of which firm has the opportunity to announce price first, as the period between moves become shorter, a price leader equilibrium is obtained with firm i as a leader.

While this game has many simplifying assumptions, and does not address the question of when, if ever, simultaneous move games will be played, [12] it does illustrate how the consideration of endogenous timing may be used to determine when a robust pattern of leadership will arise and identify the price leader. We now apply the results of this timing game to analyze the effects of quotas and tariffs in a model in which a domestic market is supplied by a duopoly consisting of a domestic and a foreign firm. We will denote by $G(k_1, k_2, c_1, c_2)$ the endogenous timing game where the appropriate price leader equilibrium is played in region (c) of Proposition 1 and simultaneous move prices are set in regions (a) and (b).

V. THE EFFECTS OF QUOTAS (VERs)

The model developed in the previous sections is a natural framework for examining the effects of quotas and tariffs in a duopolistic setting. Suppose firm 1 is a foreign firm and firm 2 a domestic firm, each producing for the domestic market only, where demand satisfies A1. We assume that the firms are capacity-constrained price setters with given capacities and unit costs of production up to capacity. In the absence of intervention via a quota or tariff firms play the endogenous timing game $G(k_1, k_2, c_1, c_2)$.

The imposition of a quota at a level strictly less than the foreign firm's capacity restricts that capacity to the level of the quota. We shall call such a quota a 'binding quota'.[13] If such a binding quota, $k_1^q < k_1$, is imposed the firms play the game $G(k_1^q, k_2, c_1, c_2)$. The levying of a unit tariff at a fixed level, $t > 0$, is assumed to raise the unit cost of the foreign firm in producing the good to $c_1^t \equiv c_1 + t$. Thus, when a tariff is levied the firms play the game $G(k_1, k_2, c_1^t, c_2)$.[14]

To facilitate the analysis we will assume initially that in the absence of policy intervention the unit costs of the two firms are the same, $c_1 = c_2 = c$, although capacities may differ. As demonstrated in Deneckere & Kovenock (1988, 1992), in this case the firm with the larger capacity will become a price leader when λ lies in region (c) of Proposition 1.

Imposing a binding quota on the foreign firm shifts the capacity pair (k_1, k_2) parallel to the k_1-axis toward the k_2-axis in the (k_1, k_2) plane (see Fig. 4). Writing $\underline{p}_i$ as a function of the parameters (k_1^q, k_2), within region (c) $\underline{p}_1(k_1^q, k_2) \leq \underline{p}_2(k_1^q, k_2)$ if and only if $k_1^q \leq k_2$. Since unit costs up to capacity are not altered by the quota, price leadership is determined by the relative capacities following the quota. The effects of a quota on the endogenous timing game are summarized in the following proposition:

Proposition 4. *In $G(k_1, k_2, c, c)$ a binding quota set at k_1^q has the following effects on the identity of the endogenously determined price leader:*

(a) *If $k_i \geq d(c)$, $i = 1, 2$, (1) a quota $d(c) \leq k_1^q < k_1$ leaves no consistent pattern of leadership. (2) A quota such that $k_1^q < d(c)$ makes the domestic firm a price leader and the foreign firm a follower.*

(b) *If $k_i \leq r_i(k_j \mid c)$ $i = 1, 2$, $j \neq i$, a binding quota leaves no consistent pattern of leadership.*

(c) *If $k_2 > k_1$, $k_2 > r_2(k_1 \mid c)$, and $k_1 < d(c)$, (1) a binding quota such that $k_2 > r_2(k_1^q \mid c)$ does not change the identity of the leader, which remains firm 2. (2) A binding quota such that $k_2 \leq r_2(k_1^q \mid c)$ leads to no consistent pattern of leadership.*

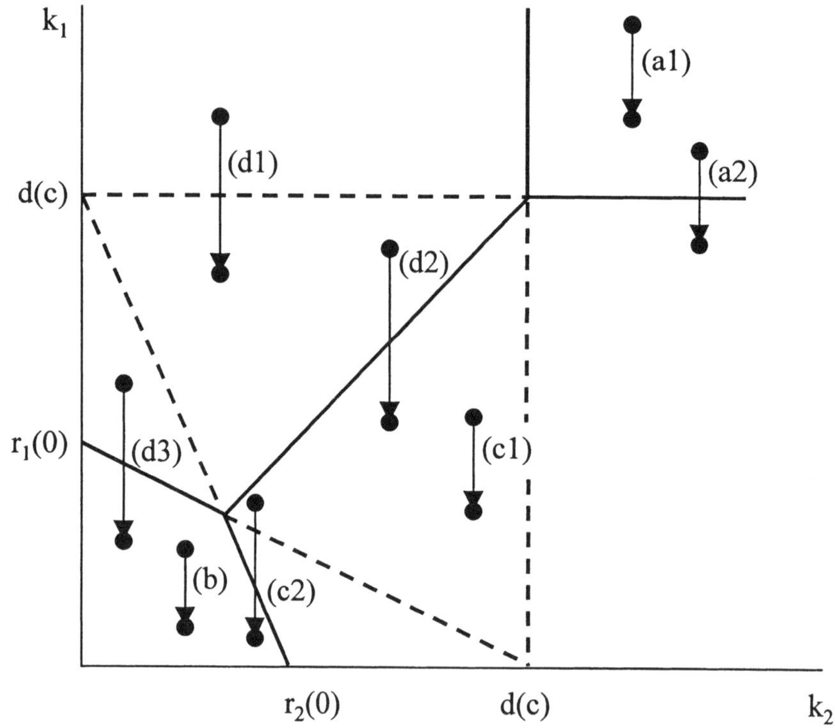

Fig. 4. The effects of quotas corresponding to the subcases of Proposition 4.

(d) *If $k_1 > k_2$, $k_1 > r_1(k_2 | c)$, and $k_2 < d(c)$, (1) a binding quota such that $k_1 > k_1^q > k_2$ and $k_1^q > r_1(k_2 | c)$ does not change the identity of the leader, which remains firm 1. (2) A quota such that $k_1^q < k_2 < k_1$ and $k_2 > r_2(k_1^q | c)$ induces a change in the identity of the leader; the domestic firm becomes a leader and the foreign firm a follower. (3) A quota such that $k_1^q \leq r_1(k_2 | c)$ and $k_2 \leq r_2(k_1^q | c)$ leads to no consistent pattern of leadership.*

Quotas corresponding to the various subcases of (a) through (d) are illustrated in Fig. 4. One major implication of Proposition 4 is that either firm may be an endogenous leader following the imposition of a quota, depending on the size of k_1^q and k_2. This stands in contrast to the model of Ito & Ono (1982) who, in their Proposition 3, conclude that the domestic firm will always become a price leader. It should be emphasized that Ito and Ono identify price leadership with behavioral assumptions which need not hold in the equilibrium of a

Stackelberg price setting game. As is evidenced by the above proposition these differences can be very important. The results in Proposition 4 also contrast with Harris' (1985) *assumption* that the domestic firm becomes a price leader when a VER is imposed. Harris justifies this assumption by arguing that the foreign firm's price followership must arise in order to insure that the VER is met. While endogenous timing does not always yield the outcome assumed by Harris, it is worth noting that the results of Proposition 4 provide support for Harris' assumption in the context in which it is made. Harris assumes that initially the firms have no capacity constraints (albeit in a model with imperfect substitutes).[15] If we take the output of each firm in the Bertrand equilibrium to be $(d(c)/2)$ in this case, then Harris' assumption of a VER at the 'free trade level' does indeed yield the domestic firm as an endogenous price leader. Thus, our model might be viewed as providing a game theoretic justification for the Harris leadership assumption, while extending the analysis to the complete capacity space.

We now examine the effects of a quota on the prices and profits of the firms. Harris (1985) and Krishna (1989) have drawn attention to these effects in analyzing the impacts of VERs. Our analysis indicates how their conclusions might be altered when endogenous conduct, in particular the timing of price announcements, is taken into account. In the continuation we will distinguish between a quota which alters the existence or identity of a price leader and one which does not. The former will be called a 'discrete quota' and the latter a 'marginal quota'. For example, if initially $k_1 > k_2$, $k_1 > r_1(k_2 | c)$, and $k_2 < d(c)$, a quota k_1^q such that $k_1^q < k_2$ will be discrete since it will alter the endogenous timing equilibrium from one in which firm 1 is a leader to one in which either firm 2 is a leader or no leadership arises. On the other hand, if $k_1^q > k_2$ and $k_1^q > r_1(k_2 | c)$, the quota is marginal; firm 1 remains a price leader.[16] The effects of a marginal quota are examined in the next proposition:

Proposition 5. *In $G(k_1, k_2, c, c)$ a marginal quota has the following effects on the market price and profits of the firms:*

(a) *If $k_i \geq d(c)$, $i = 1, 2$ or if $k_1 > k_2$ and $k_1 > r_1(k_2 | c)$ a marginal quota does not affect the price or profit of either firm.*
(b) *If $k_i \leq r_i(k_j | c)$ $i = 1, 2, j \neq i$, a marginal quota increases the price. The foreign firm makes lower profits and the domestic firm makes higher profits.*
(c) *If $k_2 > k_1$, $k_2 > r_2(k_1 | c)$, and $k_1 < d(c)$, a marginal quota increases the price. The domestic firm is always better off, but the foreign firm may be better or worse off. The effect of the quota on the foreign firm's profits depends on*

its capacity, the unit cost of production up to capacity, and the functional form of demand.

Proof. Let Π_i and Π_i^q denote the pre- and post-quota profits of firm i, $i=1, 2$.

(a) If $k_i \geq d(c)$, $i=1, 2$ then in the initial equilibrium $p=c$ and $\Pi_i=0$. If $k_1>k_2$, $k_1>r_1(k_2 | c)$ and $k_2<d(c)$, the initial equilibrium price is $p=p_1^H(k_2, c)=\text{argmax}_p(d(p)-k_2)(p-c)$, and profits are $\Pi_1=H_1^*(k_2, c)=H_1(p_1^H(k_2, c))$ and $\Pi_2=k_2(p_1^H(k_2, c)-c)$. Since each of these expressions is independent of k_1 within the relevant regions, a marginal quota has no effect.

(b) A binding quota increases price from $P(k_1+k_2)$ to $P(k_1^q+k_2)$. Hence $\Pi_2=k_2(P(k_1+k_2)-c)<k_2(P(k_1^q+k_2)-c)=\Pi_2^q$. Since k_1 lies below the Cournot best response to k_2, $\Pi_1=k_1(P(k_1+k_2)-c)>k_1^q(P(k_1^q+k_2)-c)=\Pi_1^q$.

(c) In this region of capacities firm 2 is an endogenous price leader. Before the quota is imposed the equilibrium price is $p=p_2^H(k_1, c)=\text{arg-max}_p(d(p)-k_1)(p-c)$, while equilibrium profits are $\Pi_1=k_1(p_2^H(k_1, c)-c)$ and $\Pi_2=H_2^*(k_1, c)=H_2(p_2^H(k_1, c))$. Over the relevant range of capacities

$$\frac{dp_2^H}{dk_1}=\left(\frac{d^2H_2(p_2^H)}{dp^2}\right)^{-1}<0$$

from the concavity of $H_2(p)$ (when it is non-zero). Since p_2^H, the maximizer of $H_2(p)$, remains the equilibrium price for parameter values (k_1, c) and (k_1^q, c), price increases with the quota. Differentiating firm 2's profit with respect to k_1 over the relevant range yields

$$\frac{d\Pi_2}{dk_1}=-(p_2^H(k_1, c)-c)<0$$

Hence, firm 2's profit will always increase with the imposition of a binding quota. Finally,

$$\frac{d\Pi_1}{dk_1}=(p_2^H(k_1, c)-c)+k_1\frac{dp_2^H}{dk_1}$$

The first term is positive over the relevant region while the second term is negative. As k_1 goes to zero (remaining in the relevant region) the second term approaches zero while the first is bounded away from zero. Therefore, for initial values of k_1 small enough, the profit of firm 1 decreases with the imposition of a binding quota. As k_1 approaches $d(c)$ (within the relevant region) the first term goes to zero, while the second is bounded away from zero as long as $d''(p)$ and $d'(p)$ are bounded from below at $p=c$.[17] Hence, for initial values of k_1 large enough, the profit of firm 1 increases with the imposition of a binding quota close to the initial capacity. ■

A simple example may help to illustrate the ambiguity of the effect of a marginal quota on the foreign firm's profit when the domestic firm is an endogenous price leader (part c). Suppose $d(p) = \max(0, 1 - p)$. Then $(d\Pi_1/dk_1) = (1 - c - 2k_1)/2$, so that $(d\Pi_1/dk_1) \gtreqless 0$ as $k_1 \lesseqgtr (1 - c)/2$. Hence a quota which reduces the foreign firm's capacity by a 'small' amount will have a negative effect on the firm's profit if k_1 is less than or equal to half of the demand at a price equal to the unit cost of production ($(1 - c)/2$), while it will have a positive effect if k_1 is above this level.

The traditional view of the effects of a (marginal) quota is that it harms the foreign firm and benefits the domestic firm. In their pioneering work, Harris (1985) and Krishna (1989) have emphasized that a quota can act as a device to facilitate collusion, benefitting both firms. Both the Harris and Krishna models evaluate the effect of a VER (or quota) set at or near the free trade level from an initial position with no capacity constraints. Our analysis of a marginal quota indicates that the conclusions of Harris and Krishna hold only for certain ranges of the parameters. Consideration of a discrete quota provides additional insight into the potential effects of such restrictions. In particular, since a discrete quota by definition affects the position of leadership in the market, the domestic firm may be made worse off and the foreign firm better off.

Proposition 6. *In $G(k_1, k_2, c, c)$ there exists a quota that makes the domestic firm worse off and the foreign firm better off if and only if $d(c) > k_2$, $k_1 \geq k_2$, and $k_2 > k_2^c$.*[18]

Proof. We demonstrate the 'if' part first. Under the stated conditions on capacities we know that firm 1 is initially a price leader and sets price $p_1^H(k_2, c) = \text{argmax}_p (p - c)(d(p) - k_2) = P(k_2 + r_1(k_2|c)) > P(k_1 + k_2)$. Thus, $\Pi_2 = (p_1^H(k_2, c) - c)k_2$ and $\Pi_1 = (p_1^H(k_2, c) - c)(d(p_1^H(k_2, c)) - k_2)$. By levying a quota k_1^q such that $k_1^q < k_2$ but k_1^q is arbitrarily close to k_2 we can create a capacity pair for which firm 2 is a price leader, setting a price $p_2^H(k_1^q, c) = P(r_2(k_1^q|c) + k_1^q) > P(k_1^q + k_2)$. (Under the stated assumptions a k_1^q satisfying these conditions will exist). At such a capacity pair (k_1^q, k_2), $\Pi_1^q = (p_2^H(k_1^q, c) - c)k_1^q$ and $\Pi_2^q = (p_2^H(k_1^q, c) - c)(d(p_2^H(k_1^q, c)) - k_1^q)$. Since k_1^q is arbitrarily close to k_2, $p_2^H(k_1^q, c)$ is arbitrarily close to $p_1^H(k_2, c)$. Thus, with $k_1^q > d(p_2^H(k_1^q, c)) - k_2$, $\Pi_1^q > \Pi_1$. Similarly, with $k_2 > d(p_2^H(k_1^q, c)) - k_1^q$, $\Pi_2 > \Pi_2^q$.

To prove the 'only if' part of the Proposition, note that Proposition 5 implies that a marginal quota can never achieve the stated effects on profits, so the quota must be discrete. The initial capacities cannot satisfy $k_i \geq d(c)$ $i = 1, 2$, because the equilibrium coincides with Bertrand and the domestic firm cannot possibly be made worse off than in this equilibrium. Furthermore, the initial capacities cannot both lie under the respective Cournot best response functions for, if so, any quota is marginal, and the domestic firm benefits. Finally, the initial capacities cannot satisfy $k_2 > k_1$, $k_2 > r_2(k_1 | c)$, and $k_1 < d(c)$ since a quota in this case is either marginal, or pushes capacities into the region under both Cournot best response functions, in which case it is again easily seen that the domestic firm benefits. This leaves the region of capacities where $k_1 \geq k_2$, $k_1 > r_1(k_2 | c)$ and $k_2 < d(c)$ as a potential location for initial capacities. But observe that any quota pushing (k_1^q, k_2) into the region under both Cournot best response functions cannot possibly help the foreign firm, since it would obtain $k_1^q(P(k_1^q + k_2) - c)$ rather than $r_1(k_2 | c)(P(r_1(k_2 | c) + k_2) - c)$. Hence the initial capacities must be such that there exists a quota which generates an equilibrium in which firm 2 is a leader. This is only possible if $k_2 > k_2^c$. ■

Proposition 6 is surprising and arises directly from our endogenization of price leadership. Imposing a quota shifts the equilibrium regime, which generates a discontinuous change in the profits of the firms.[19] However, this does not imply that anything can happen. Despite the variety of effects a quota may have on the distribution of profit across firms, a quota does lead to a more collusive outcome in the following sense: market price always increases or remains the same.

Proposition 7. *In $G(k_1, k_2, c, c)$ a quota (weakly) increases price.*

Proof. The claim for a marginal quota was demonstrated in Proposition 5. We complete the proof by demonstrating that the claim holds for all possible discrete quotas.

(a) Suppose initially $k_i \geq d(c)$, $i = 1, 2$. Then a discrete quota must make firm 2 a price leader, setting $p_2 = p_2^H(k_1^q, k_2, c, c) > c$. Since the price in the initial equilibrium is c, the quota increases price.

(b) Suppose $k_1 > k_2$, $k_1 > r_1(k_2 | c)$ and $k_2 < d(c)$. A discrete quota can lead to one of two situations: it can either make firm 2 a price leader (by generating a capacity pair (k_1^q, k_2) such that $k_1^q < k_2$ and $k_2 > r_2(k_1^q | c)$), or push capacities below the lower envelope of the Cournot best response functions ($k_1^q \leq r_1(k_2 | c)$ and $k_2 \leq r_2(k_1^q | c)$), where no leadership arises. In the latter case the equilibrium price is $p_1^H(k_1, k_2, c, c)$ before the quota and $P(k_1^q + k_2)$ after the quota. Since $k_1^q \leq r_1(k_2 | c)$, $p_1^H(k_1, k_2, c, c) = P(r_1(k_2 | c) + k_2) \leq P(k_1^q + k_2)$, so the quota cannot decrease price. In the former case, the quota shifts price from $p_1^H(k_1, k_2, c, c)$ to $p_2^H(k_1^q, k_2, c, c)$. Since $p_1^H(k_1, k_2, c, c) = P(k_2 + r_1(k_2 | c))$ and $p_2^H(k_1^q, k_2, c, c) = P(k_1^q + r_2(k_1^q | c))$, to prove the claim we need only show that $k_2 + r_1(k_2 | c) > k_1^q + r_2(k_1^q | c)$. To do so, note that $r_2(\cdot | c)$ and $r_1(\cdot | c)$ have the same functional form and that since the quota is assumed to make firm 2 a price leader, $k_1^q < k_2$. Hence, it is sufficient to show that $k + r_1(k \mid c)$ is increasing in k. To demonstrate this, note that from the concavity assumption in $A1$ $r_1(k \mid c) \equiv \text{argmax}_q\, q(P(q+k) - c)$ is the solution to the first-order condition $P(q+k) - c + qP'(q+k) = 0$. Totally differentiating this condition we obtain $(dq/dk) > -1$, so that $d[k + r_1(k \mid c)]/dk > 0$.

(c) The last case to be examined is when $k_2 > k_1$, $k_2 > r_2(k_1 \mid c)$, and $k_1 < d(c)$. In this case the only way that a quota can be discrete is if it generates a capacity pair under the lower envelope of the best response functions, where the market price is $P(k_1^q + k_2)$. Because firm 2 is initially a price leader, the price before the quota is $p_2^H(k_1, k_2, c, c) = P(k_1 + r_2(k_1 \mid c))$. From the assumption that the post-quota capacity pair is below both Cournot best response functions, $P(k_1^q + k_2) \geq P(k_1^q + r_2(k_1^q \mid c))$. Furthermore, since $k + r_2(k \mid c)$ is increasing in k (see part (b)) $P(k_1^q + r_2(k_1^q \mid c)) > P(k_1 + r_2(k_1 \mid c))$. Combining these two inequalities gives us the claim. ■

The above analysis assumes that firms have identical unit costs up to capacity. If we incorporate initial unit cost asymmetries into the model (see Fig. 3) an even richer set of outcomes may arise. For instance, if the foreign firm is the low cost firm, a quota may make the foreign firm a leader from an initial position in which the domestic firm was a leader. This arises because the boundary between the two regions of leadership (the set of capacities where $\underline{p}_1 = \underline{p}_2$ above the Cournot best response functions in capacity space) may bend back on itself, so that it is not a proper function of k_2 (see Fig. 3). Also, unlike the case of identical unit costs up to capacity, for certain asymmetric cost pairs (c_1, c_2) a quota can transform the equilibrium from an initial outcome in which firms play an unconstrained Bertrand equilibrium (with the high cost domestic firm driven out) to one in which both firms are capacity constrained in equilibrium or, alternatively, to either type of endogenous leader outcome. One implication of this result is that a quota can change the equilibrium from one in which a low cost foreign firm drives the domestic firm out of the market (and is thus a limit pricing monopolist) to one in which both firms are actively in the market and, yet, *both* firms may be better off. Finally, if the foreign firm is high cost a quota can alter the equilibrium from one in which it is driven out completely, to one in which it is able to enter the market (with both firms benefitting). In this case the quota makes the foreign firm less threatening as in the 'judo' result of Gelman & Salop (1983),[20] making it more profitable for the domestic firm to accommodate it.

A more complete analysis of the effects of a quota in a market with unit cost asymmetries is beyond the scope of this chapter. However, it should be evident from our analysis how to proceed on a case-by-case basis. We now move on to analyze the effects of tariffs.

VI. THE EFFECTS OF TARIFFS

A positive tariff t levied on imports raises the unit cost of the foreign firm in supplying the good to the domestic market to $c_1^t \equiv c_1 + t$. Thus, when a tariff is levied firms play the game $G(k_1, k_2, c_1^t, c_2)$. In the analysis of this section we again assume that firms start from a position of identical unit costs, so that $G(k_1, k_2, c, c)$ is the initial game played.

A tariff levied on the foreign firm will in some cases raise that firm's critical price $\underline{p}_1$ and in some cases leave $\underline{p}_1$ unaffected. To see this, note that if $k_2 > d(c)$ then $\underline{p}_1(k_1, k_2, c, c) = c$. Following a tariff $\underline{p}_1(k_1, k_2, c+t, c) = c+t > c$. Thus, in instances where the domestic firm has enough capacity to serve the whole market at the initial cost of the foreign firm, a tariff raises $\underline{p}_1$. If initially $k_1 \leq r_1(k_2 | c)$ then after a tariff is levied either $k_1 \leq r_1(k_2 | c+t)$ or $k_1 > r_1(k_2 | c+t)$ may hold; a tariff causes firm 1's Cournot best response function to shift down. If the first inequality holds $\underline{p}_1(k_1, k_2, c+t, c) = P(k_1 + k_2) = \underline{p}(k_1, k_2, c, c)$ and the tariff leaves $\underline{p}_1$ unchanged. If $k_1 > r_1(k_2 | c+t)$ then $\underline{p}_1(k_1, k_2, c+t, c) > P(k_1 + k_2)$, so that $\underline{p}_1$ increases due to the tariff. In the remaining regions of initial capacities, $\underline{p}_1(k_1, k_2, c, c)$ solves $L_1(p | k_1, c) = B_1(k_2, c)$ where $L_1(p | k_1, c_1)$ is the profit of firm 1 when it is low-priced with the dependence upon k_1 and c_1 stated explicitly and $B_i(k_j, c_i) \equiv \max_{p \in S_i}\{(p - c_i)(d(p) - k_j)\}$. Totally differentiating $\Psi(\underline{p}_1, c_1) \equiv L_1(\underline{p}_1 | k_1, c_1) - B_1(k_2, c_2) \equiv 0$, and making the appropriate adjustment to a directional derivative where $L_1(p)$ is not differentiable shows that $\frac{d\underline{p}_1}{dc_1} > 0$. Thus, $\underline{p}_1$ will increase when a tariff is levied unless the post-tariff value of $\underline{p}_1$ is $P(k_1 + k_2)$.

The domestic firm's critical price $\underline{p}_2$ is clearly unaffected by a tariff since this price is independent of the foreign firm's unit cost. The implications of these effects in determining the existence and identity of an endogenous leader are provided in the following proposition:

Proposition 8. *In $G(k_1, k_2, c, c)$ a positive tariff, $t > 0$, has the following effects on the identity of the endogenously determined price leader:*

(a) If $k_i \geq d(c)$, $i = 1, 2$ a tariff $t > 0$ leaves no consistent pattern of leadership (the asymmetric-cost Bertrand solution arises).

(b) If $k_i \leq r_i(k_j | c)$ $i = 1, 2$, $j \neq i$, a tariff $t > 0$ (1) leaves no consistent pattern of leadership if $k_1 \leq r_1(k_2 | c_1^t)$, (2) makes the foreign firm a price leader if $k_1 > r_1(k_2 | c_1^t)$ and $P(k_2) > c_1^t$, and (3) leads to no consistent pattern of leadership (with the foreign firm driven out of the market) if $P(k_2) \leq c_1^t$.

(c) If $k_2 > k_1$, $k_2 > r_2(k_1 | c)$ and $P(k_1) > c$, (1) a tariff $t > 0$ such that $P(k_2) \leq c_1^t$ and $\underline{p}_2(k_1, k_2, c_1^t, c) \leq c_1^t$ leads to no consistent pattern of leadership (the

asymmetric cost Bertrand solution results). (2) A tariff such that $P(k_2) > c_1^t$ and $\underline{p}_1(k_1, k_2, c_1^t, c) \geq \underline{p}_2(k_1, k_2, c_1^t, c)$ makes firm 1 a price leader. (3) A tariff for which capacities satisfy neither the conditions of (1) nor (2) leaves firm 2 in a position of leadership.

(d) If $k_1 \geq k_2$, $k_1 > r_1(k_2 \mid c)$ and $P(k_2) > c$, (1) a tariff such that $P(k_2) \leq c_1^t$ leads to no consistent pattern of leadership (Bertrand arises). (2) A tariff such that $P(k_2) > c_1^t$ leaves firm 1 in a position of leadership.

The potential effects of a tariff on the existence of leadership and the identity of a leader are illustrated in Fig. 5. A tariff shifts the common boundary of the regions where $\underline{p}_1 > \underline{p}_2$ and $\underline{p}_2 > \underline{p}_1$ (above $r_2(k_1 \mid c)$) to the left of the diagonal (with k_1 on the abscissa). After the tariff the boundary extends from the Cournot point corresponding to costs (c_1^t, c) to connect with the line $\phi(c_1^t, c) = \{k_1 : (c_1^t - c)d(c_1^t) = B_2(k_1, c)\}$ at the point where $k_2 = d(c_1^t)$. (For a proof

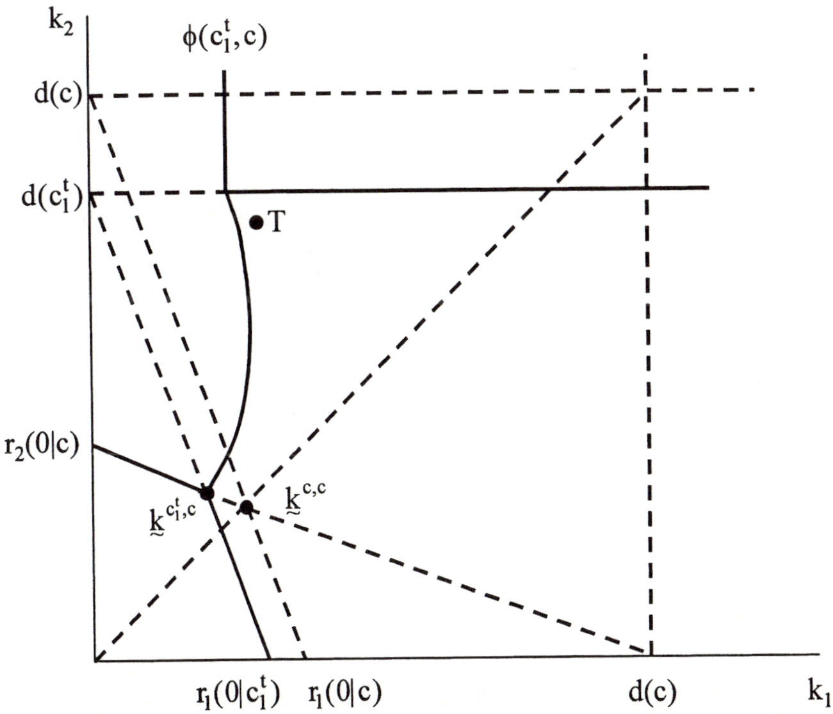

Fig. 5. The effects of a tariff on the regions of leadership.

of these assertions see Deneckere & Kovenock (1996), Theorem 3). Hence with the asymmetric cost structure generated by the tariff the large capacity firm is no longer necessarily the leader; it is possible for firm 1 to lead even when $k_1 < k_2$.

While a tariff may promote or inhibit leadership, the effects on profits are clear. A tariff always (weakly) hurts the foreign firm and (weakly) benefits the domestic firm.

Proposition 9. *In $G(k_1, k_2, c, c)$ a tariff always (weakly) decreases the foreign firm's profit and (weakly) increases the domestic firm's profit.*

Proof. Let Π_i and Π_i^t denote the pre- and post-tariff profits of firm i, $i = 1, 2$.

(a) When $k_i \geq d(c)$ $i = 1, 2$ then the initial equilibrium profits are $\Pi_i = 0$, $i = 1, 2$. After a tariff is levied, $c_1^t = c + t > c = c_2$, so $\Pi_1^t = 0$ and $\Pi_2^t = L_2(\min(p_2^L, c + t)) > \Pi_2 = 0$.

(b) When $k_i \leq r_i(k_j \mid c)$ $i = 1, 2$, $j \neq i$, in the original equilibrium $\Pi_i = k_i(P(k_1 + k_2) - c)$ $i = 1, 2$. If the condition of Proposition 8(b)(1) holds $\Pi_1^t = k_1(P(k_1 + k_2) - c_1^t) < \Pi_1$ and $\Pi_2^t = \Pi_2$. If Proposition 8(b)(2) holds $\underline{p}_1(k_1, k_2, c_1^t, c) > P(k_1 + k_2) = \underline{p}_2(k_1, k_2, c_1^t, c)$, so firm 1 becomes a leader and $\Pi_1^t < \Pi_1$ since $H_1(p)$ is decreasing in c_1 for every p such that $H_1(p) > 0$. Since $p_1^H(k_1, k_2, c_1^t, c) > \underline{p}_1(k_1, k_2, c_1^t, c) > P(k_1 + k_2)$, $\Pi_2^t = k_2(p_1^H(k_1, k_2, c_1^t, c) - c) > k_2(P(k_1 + k_2) - c) = \Pi_2$. Finally, if Proposition 8(b)(3) holds, $P(k_2) \leq c_1^t$, so $\Pi_2^t \geq k_2(P(k_2) - c) > k_2(P(k_1 + k_2) - c) = \Pi_2$. Clearly, $\Pi_1^t = 0 < \Pi_1$.

(c) When $k_1 \geq k_2$, $k_1 > r_1(k_2 \mid c)$ and $P(k_2) > c$, firm 1 is a leader in the initial equilibrium. A tariff such that Proposition 8(d)(1) holds implies $\Pi_1^t = 0$ and $\Pi_2^t = [\min\ (c_1^t, p_2^L) - c]d(\min(c_1^t, p_2^L))$. Clearly, $\Pi_1^t < \Pi_1$. Since $p_1^H(k_1, k_2, c, c) < P(k_2) \leq c_1^t$, $\Pi_2^t \geq (P(k_2) - c)k_2 > (p_1^H(k_1, k_2, c, c) - c)k_2 = \Pi_2$. If Proposition 8(d)(2) holds firm 1 remains a leader after the tariff is imposed. It is easily verified that $p_1^H = \text{argmax}_{p \in S_1}(p - c_1)(d(p) - k_2)$ increases due to the tariff. It follows that $\Pi_2^t = k_2(p_1^H(k_1, k_2, c_1^t, c) - c) > k_2(p_1^H(k_1, k_2, c, c) - c) = \Pi_2$. The profit of firm 1 decreases since $\Pi_1 = (p_1^H(k_1, k_2, c_1, c_2) - c_1)(d(p_1^H) - k_2)$ and $(d\Pi_1/dc_1) < 0$ (from the envelope theorem).

(d) When $k_2 > k_1$, $k_2 > r_2(k_1 \mid c)$ and $P(k_1) > c$ firm 2 is a leader in the absence of a tariff. A tariff raises $\underline{p}_1$ but leaves $\underline{p}_2$ unchanged. If Proposition 8(c)(1) holds $\Pi_1 > \Pi_1^t = 0$ and, since by assumption $P(k_2) \leq c_1^t$ and $\underline{p}_2(k_1, k_2, c_1^t, c) \leq c_1^t$, $\Pi_2^t = [\min(c_1^t, p_2^L) - c]d(\min(c_1^t, p_2^L)) \geq L_2(\underline{p}_2(k_1, k_2, c, c)) = \Pi_2$, with strict inequality if $c_1^t > \underline{p}_2$. If Proposition 8(c)(2) holds the tariff makes firm 1 a leader. Clearly firm 2 gains since $\Pi_2^t = k_2(p_1^H(k_1, k_2, c_1^t, c) - c) > k_2(\underline{p}_1(k_1, k_2, c_1^t, c) - c) \geq L_2(\underline{p}_2) = \Pi_2$. Firm 1's profit decreases. To see this, let $t^1 \leq t$ be a tariff such that $\underline{p}_1(k_1, k_2, c + t^1, c) = \underline{p}_2(k_1, k_2, c, c)$. Then $\Pi_1^{t^1} \leq (\underline{p}_1(k_1, k_2, c + t^1, c) - c - t^1)k_1 < k_1(p_2^H(k_1, k_2, c, c) - c) = \Pi_1$. Once this is established, the calculations of part (c) above show that the profit of firm 1 will decrease even further as the tariff is increased to t. Finally, if Proposition 8(c)(3) holds the tariff does not alter the equilibrium price. Thus, $\Pi_2^t = \Pi_2$ and $\Pi_1^t < \Pi_1$. ■

The effects of a tariff on profits outlined in Proposition 9 are consistent with the traditional view of how tariffs affect the market equilibrium. However, the effect of a tariff on price may be less consistent with traditional analysis. While for most tariffs levied the market price will remain the same or increase, a tariff which switches leadership from the domestic firm to the foreign firm may reduce price. Such a situation is indicated by the point T in Fig. 5. In $G(k_1, k_2, c, c)$ the equilibrium has firm 2 leading and setting price $p_2^H(k_1, k_2, c, c) > \underline{p}_2(k_1, k_2, c, c)$. The strict inequality follows from the fact that $k_2 > r_2(k_1 | c)$ and $P(k_1) > c_2$, and indeed we may expect the difference between p_2^H and $\underline{p}_2$ to be large since $P(k_1)$ is much larger than c_2 and k_2 is substantially larger than $r_2(k_1 | c)$. After the tariff is levied firm 1 is a price leader pricing at $p_1^H(k_1, k_2, c_1^t, c)$, which is close to $\underline{p}_1(k_1, k_2, c_1^t, c)$ since k_2 is close to $d(c_1^t)$. However, $\underline{p}_1(k_1, k_2, c_1^t, c)$ is in turn close to $\underline{p}_2(k_1, k_2, c_1^t, c) = \underline{p}_2(k_1, k_2, c, c)$, since (k_1, k_2) lies close to the boundary where $\underline{p}_1$ and $\underline{p}_2$ are equal. Hence $p_1^H(k_1, k_2, c_1^t, c) < p_2^H(k_1, k_2, c, c)$ and price decreases with the tariff. We summarize this analysis in:

Proposition 10. *A tariff generally increases market price or leaves the price the same. A tariff that switches leadership from the domestic firm to the foreign firm may lower price.*

VII. THE (NON) EQUIVALENCE OF QUOTAS AND TARIFFS

A quota and a tariff are usually defined to be equivalent if (1) they allow equal amounts of imports and (2) the market price is the same. An interesting question which arises in the analysis of quotas and tariffs is whether for every binding quota there is an equivalent tariff (and vice versa for positive tariffs). Deneckere & Kovenock (1989) have shown for the simultaneous move price setting game that, starting from an arbitrary initial position (k_1, k_2, c_1, c_2), if a binding quota (positive tariff) is levied such that in the resulting equilibrium neither firm is driven from the market, then there exists no positive tariff (binding quota) that generates the same equilibrium price distributions. In contrast to this result, the following proposition demonstrates that in the endogenous timing game G there are some initial parameter values (k_1, k_2, c, c) and some binding quotas for which equivalent tariffs exist:

Proposition 11. *In $G(k_1, k_2, c, c)$ any binding quota which generates an equilibrium in the pure strategy range under the Cournot best response functions ($k_1^q \le r_1(k_2 | c)$ and $k_2 \le r_2(k_1^q | c)$) has an equivalent tariff. Any marginal quota which does not alter the amount the foreign firm sells in equilibrium (a* *marginal quota which keeps the equilibrium in regions B or C of Fig. 6) is equivalent to a zero tariff. All other binding quotas have no equivalent tariff.*

Proof. We first show that a binding quota such that $k_1^q \leq r_1(k_2 | c)$ and $k_2 \leq r_2(k_1^q | c)$ has an equivalent tariff. Note that in the post-quota equilibrium imports are k_1^q and the price is $P(k_1^q + k_2)$. Starting from the initial position (k_1, k_2, c, c), given these assumptions, there is always a tariff large enough (or zero tariff iff $k_1^q = r_1(k_2 | c)$) so that $k_1^q = r_1(k_2 | c + t)$. Since in this case $k_2 < d(c + t)$, when such a tariff is levied (k_1, k_2, c_1^t, c) lies in the region where firm 1 is a price leader (see Deneckere & Kovenock (1996) Theorem 3). Hence, the equilibrium price after the tariff is levied is $p_1^H(k_1, k_2, c_1^t, c) = P(k_2 + r_1(k_2 | c + t)) = P(k_2 + k_1^q)$. Furthermore, the foreign firm's sales are k_1^q. This proves the first assertion. The second assertion is obvious; from Proposition 5 part (a) we know that the only regions where a marginal quota does not affect the amount of

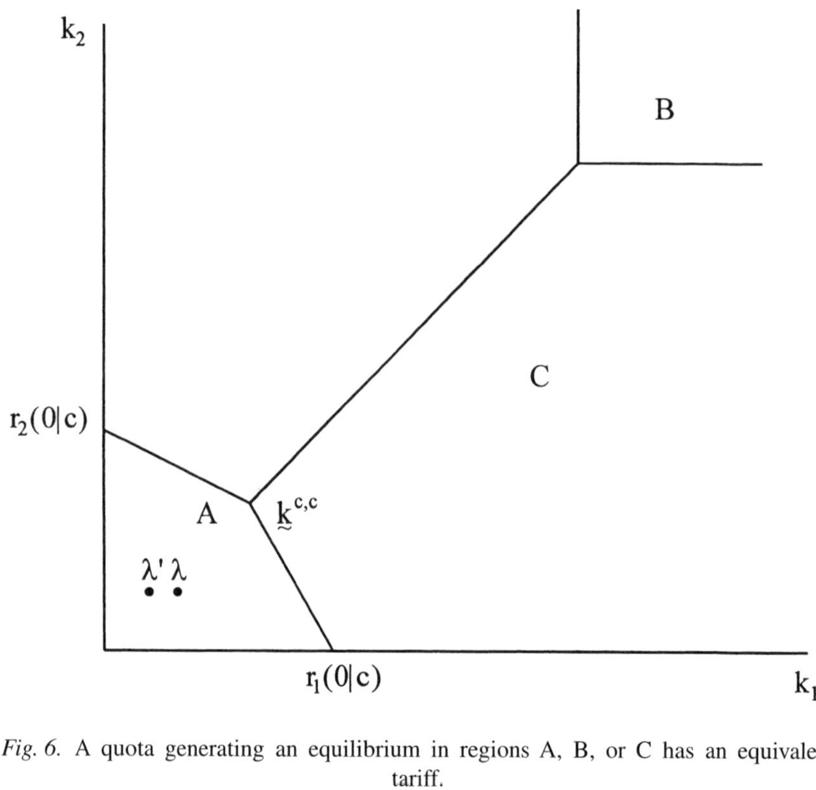

Fig. 6. A quota generating an equilibrium in regions A, B, or C has an equivalent tariff.

imports are those corresponding to regions B and C of Fig. 6. Finally, to show that all other quotas have no equivalent tariffs, note that a quota pushing the equilibrium into the range where firm 2 leads generates a price $p_2^H(k_1^q, k_2, c, c) > P(k_1^q + k_2)$, while firm 1 sells k_1^q. For a tariff to achieve the same result firm 2 cannot be a leader after the tariff for then firm 1 would sell $k_1 \neq k_1^q$. Hence, the post-tariff equilibrium must exhibit no leadership, or firm 1 must be a leader. The former case obviously cannot possibly generate sales k_1^q for firm 1. The latter case may generate imports of k_1^q if $k_1^q = r_1(k_2 | c + t)$, but this implies that $p_1^H = P(k_1^q + k_2) < p_2^H(k_1^q, k_2, c, c)$. ■

The results of Proposition 11 should be contrasted with some of the previous literature on the equivalence of quotas and tariffs. In Ito & Ono's (1982) model the incorporation of endogenous leadership is claimed to always lead to a higher price under a quota than under an import-equivalent tariff. Again, it should be emphasized that Ito and Ono make behavioral assumptions which need not hold in a Stackelberg price setting game (the follower always sells its supply at the price quoted by the leader). Harris (1985) and Krishna (1989) both emphasize the non-equivalence of quotas and tariffs (at least for substitute goods) for the initial conditions and quotas upon which they focus.[21] Our results indicate that the (non)equivalence of quotas and tariffs depends not only on the equilibrium concept used, but also on the cost and capacity assumptions made.

One interesting implication of the analysis of Proposition 11 is that a quota that reduces the foreign capacity only slightly may be equivalent only to a very severe tariff. This is the case for instance when the point $\lambda(k_1, k_2)$ in Fig. 6 is transformed by a quota to $\lambda' = (k_1^q, k_2)$. As is evident from the construction of the proof of the proposition the tariff needed to obtain equivalence is one for which $k_1^q = r_1(k_2 | c + t)$. Given that k_1^q lies well beneath the best response function $r_1(k_2 | c)$ this would require a very severe tariff. Thus, instruments which lead to identical domestic profit and consumer surplus may have disparate effects upon foreign profit.

The above analysis, of course, has ignored issues of government revenues from tariffs and quotas or how these instruments might be combined. Nonetheless it should be evident to the reader that these issues can be easily handled within the framework proposed.

VIII. AN EXPLICIT WELFARE COMPARISON

We conclude by providing an example of how our model facilitates welfare calculations. Our focus is on an economically relevant environment that is often conducive to protectionist measures, a situation in which there is excess domestic capacity.[22]

In our homogeneous good model, a quota[23] always reduces domestic welfare when it is imposed from an initial position in which the classical Bertrand equilibrium arises (both firms have capacities at or above $d(c)$). The explanation is straightforward. In the initial situation domestic consumer surplus is that corresponding to the area under the demand curve above a price equal to c. Both domestic and foreign profits are zero. Imposing a quota which reduces the foreign capacity below $d(c)$ leads to a situation in which the domestic firm is a price leader. The market price, p_2^H, lies above c and therefore the region under the demand curve between $p=c$ and $p=p_2^H$ is lost domestic consumer surplus. Domestic profit certainly cannot increase by more than this since this change places an upper bound on the change in total profit. Since the foreign profit was initially 0 and increases to $(p_2^H - c)k_1^q$ after the quota, the decrease in domestic consumer surplus must be strictly greater than the gain in domestic profit.

Hence, in the homogeneous good analogue of the situation examined by Harris and Krishna, a quota that reduces foreign capacity below $d(c)$ always reduces domestic welfare. The welfare implications of a quota are quite different in the other instance in which domestic excess capacity exists. When the domestic firm is initially an endogenous price leader (which implies that it alone has excess capacity), a quota that reduces foreign capacity by a small amount has an ambiguous effect on domestic welfare. If the foreign capacity is large, such a quota reduces domestic welfare; if the foreign capacity is small it increases domestic welfare.

Before demonstrating this, we should note that our analysis of the effect of a quota in this case is related to the work of Eldor & Levin (1990) on trade liberalization. In that paper, the authors assume that the domestic firm is a monopolist facing a competitive foreign fringe.[24] This contrasts with our model, which obtains similar behavior endogenously (for certain parameter values) in a price setting duopoly. When the domestic firm is an endogenous leader it maximizes its profit on the residual demand curve left after the foreign duopolist has sold all of its capacity.

Suppose that the domestic firm is initially a price leader. The initial domestic welfare is then

$$W = CS(p_2^H) + [d(p_2^H) - k_1](p_2^H - c),$$

where $CS(p_2^H)$ is the consumer surplus generated when all of the market demand is satisfied at a price p_2^H, and the second summand is domestic profit. Differentiating this expression with respect to k_1,

$$\frac{dW}{dk_1} = -d(p_2^H)\frac{dp_2^H}{dk_1} + \left[d'(p_2^H)\frac{dp_2^H}{dk_1} - 1 \right](p_2^H - c) + \left[d(p_2^H) - k_1 \right]\frac{dp_2^H}{dk_1}.$$

The first term is the effect of a change in foreign capacity on consumer surplus. The second and third terms give the effect on domestic profit. Cancelling and rearranging terms, we find

$$\frac{dW}{dk_1} = \left[d'(p_2^H)\frac{dp_2^H}{dk_1} - 1 \right](p_2^H - c) - k_1\frac{dp_2^H}{dk_1}. \tag{8.1}$$

To sign this expression, note that total differentiation of the first order condition for the domestic firm's profit maximization problem implies that

$$\frac{dp_2^H}{dk_1} = \left[d''(p_2^H)(p_2^H - c) + 2d'(p_2^H) \right]^{-1} < 0.$$

Inserting this into the first term of the right hand side of (8.1) implies

$$\frac{dW}{dk_1} < -\frac{1}{2}(p_2^H - c) - k_1\frac{dp_2^H}{dk_1}.$$

For k_1 sufficiently small this expression will be negative. Hence, for a low initial level of foreign capacity, a decrease in that capacity due to a quota will increase domestic welfare.[25]

To illustrate how large the range of capacities over which a quota enhances domestic welfare may be, suppose demand is $d(p) = \max(0, 1 - p)$. Prior to intervention the domestic firm sets the price $p_2^H = (1 - k_1 + c)/2$ and domestic welfare is the sum of consumer surplus, $\frac{1}{8}(1 + k_1 - c)^2$ and domestic profit $\frac{1}{4}(1 - k_1 - c)^2$. If the quota reduces foreign capacity by a sufficiently small amount, the domestic firm will remain a price leader, charging $p_2^H = (1 - k_1^q + c)/2$. Differentiating domestic welfare, W, with respect to k_1 we find

$$\text{sgn}\left(\frac{dW}{dk_1}\right) = \text{sgn}\left(k_1 - \frac{(1-c)}{3}\right).$$

Hence, for an initial foreign capacity below $\frac{1}{3}(1 - c)$, (which in this case happens to be the Cournot level of output), a quota that reduces foreign capacity by a small amount increases domestic welfare.

How important is this example of the effect of a quota in a market with excess domestic capacity? Of course, which of these two scenarios best describes the economic environments in which quotas are imposed is an empirical question that is not addressed in this chapter. However, if one takes the view that a quota is more likely to be imposed in a situation in which it increases domestic welfare,[26] the above comparisons support an interpretation that contrasts with the view of a quota as a facilitating practice. In the range of capacities where domestic welfare increases with a slight decrease in foreign capacity, foreign profit decreases (see Section V).

The difference in the interpretation of the nature of competition in the classical Bertrand equilibrium and the price leader equilibrium may also lend a clue as to which initial equilibrium is more relevant to the analysis of a quota's effects. In the classical Bertrand equilibrium (in which both firms have excess capacity) price competition is fierce and price is driven to marginal cost. In contrast, when the domestic firm is a price leader, the foreign firm produces to capacity and benefits from the price umbrella provided by the domestic firm. The domestic firm finds it optimal to act as a monopolist on the residual demand curve left after the sale of foreign capacity, and the price-cost margin is high. Unlike the classical Bertrand equilibrium, further increases in foreign capacity erode domestic profits.

In ending, we examine the domestic welfare implications of a tariff. Starting from the classical Bertrand equilibrium, a tariff leads to an increase in price. The domestic firm becomes a monopolist, limit pricing at $p_2 = c + t$. The standard distortion due to output contraction arises and domestic welfare falls.

In contrast, if the domestic firm is initially an endogenous price leader, a tariff which leaves the domestic firm in a leadership position increases domestic welfare. The domestic firm's monopoly price on the residual demand curve remains the same ($p_2^H = \frac{1}{2}(1 - k_1 + c)$ in the linear example above) and foreign output remains equal to foreign capacity. In this case, the only effect of the tariff is to extract t dollars in tariff revenue on each unit of foreign capacity. Of course, a large tariff may change the nature of leadership. In this case welfare may go up or down. Nonetheless, just as in the case of a quota, price leadership by the home firm admits welfare improving intervention, whereas the classical Bertrand equilibrium does not.

NOTES

1. See Scherer (1980) for a detailed account of important markets exhibiting price leadership.

2. Although we do discuss how our conclusions are altered if different initial costs are allowed.

3. Krishna allows for more general demand and also looks at complements. However, her characterization of the simultaneous move equilibrium for the case of substitutes does not hold for general capacity pairs. For a large part of the capacity space this characterization is still an open question. Krishna's results also do not hold for imperfect but close substitutes (this violates her concavity assumption).

4. For a discussion of the relative merits of efficient rationing and the most popular alternative, proportional rationing, see Davidson & Deneckere (1986).

5. This assumption is made for the purely technical reason of allowing standard equilibrium existence theorems to be used. Any alternative rule can be used with no substantive effect on the outcome of the game (see Deneckere & Kovenock (1989)). To break ties in the simultaneous move game when $c_1 = c_2$ we assume that firm 1 sells its capacity first.

6. In equilibrium, such a price could only be set if the firm did not sell anything at that price. Hence, our assumption removes imperfect equilibria.

7. When $\underline{p}_i = \underline{p}_j$ multiple equilibria arise in the leader-follower games.

8. There are also equilibrium strategies in which firm 2 plays a non-degenerate mixed strategy in this case.

9. When λ is in the range corresponding to region (c) of Proposition 1 but $\underline{p}_1 = \underline{p}_2$ there are multiple equilibria in each leader follower game. In this case we will assume that when firm i leads it sets p_i^H, so that both firms are indifferent between the simultaneous move game and leading, but strictly prefer to follow.

10. If $\underline{p}_i = \underline{p}_j$ in (c) we may arbitrarily assign leadership to firm 1.

11. A flow demand can easily be substituted with the same qualitative results.

12. See Deneckere & Kovenock (1988) for a game form in which both simultaneous and sequential price setting are permitted and price leadership arises for the same capacities and costs indicated here.

13. This terminology is widely used in the literature to refer to a quota which strictly reduces the quantity sold in the market by the foreign firm at given prices. In this model, since firms may not sell all of their capacity in equilibrium, this need not be the case.

14. It should be evident that this approach allows us to easily investigate the effects of intervention using both policy tools. Also, one could easily allow c_2 to be altered through subsidies.

15. See Furth & Kovenock (1993) for an analysis of the incentives to lead and follow in a capacity constrained price setting game with imperfect substitutes. Deneckere, Kovenock & Lee (1992) examine these incentives in a game with loyal consumer segments.

16. It is our intent that a quota which alters the situation from one in which capacities are large enough to generate the standard Bertrand equilibrium to one in which capacities lie below the relevant Cournot best response functions also be termed a discrete quota, even though the situation remains one of no leadership. Since when unit costs up to capacity are identical, the latter region of capacities cannot be obtained by imposing a quota from an initial position in the former region, we ignore this distinction here. Such a distinction should, however, be made when dealing with different unit costs of production up to capacity.

17. (A1) guarantees that this will hold except possibly when $c = 0$.

18. k_2^c denotes firm 2's output in the (unique) Cournot equilibrium when demand is $d(p)$ and the (constant) unit cost of production for both firms is c.

19. See Horstmann & Markusen (1992) for similar discontinuous regime changes in a different context.

20. Gelman and Salop assume that it is the high-cost firm which is a price leader in the 'judo' equilibrium. Deneckere & Kovenock (1992) show that at the high cost firm's optimal capacity in the Gelman-Salop model the low-cost firm would become an endogenous leader.

21. Krishna's definition of an import equivalent tariff does not seem to match our definition here, since imports under the quota in her model are a non-degenerate random variable and under the asserted import equivalent tariff are not. Given our definition there clearly does not exist a tariff which generates the same market equilibrium as the quota she examines.

22. For a description of several two-stage games of simultaneous capacity choice followed by endogenous price leadership that generate excess domestic capacity as an equilibrium outcome see Deneckere & Kovenock (1992). Price setting models seem to be more useful than quantity setting models in analyzing situations of excess capacity. This is due mainly to the difficulty in obtaining excess capacity as (subgame perfect) equilibrium behavior in the latter class of models.

23. Throughout this section we will assume that all quota rents are appropriated by the foreign firm. Hence, a quota is equivalent to a voluntary export restraint.

24. Eldor and Levin also look at a domestic Cournot oligopoly, but do not relate their results to the literature on quotas with price setting firms.

25. We assume in this discussion that such a decrease leaves capacities in a range where the domestic firm is a price leader. If a quota reduces foreign capacity to a point where it is below $r_1(k_2|c)$, the domestic firm is no longer an endogenous leader. Such a quota leads to lower domestic welfare than a quota satisfying $k_1^q = r_1(k_2|c)$. This can be seen by noting that a small decrease in foreign capacity beyond this point reduces domestic consumer surplus by $(k_1^q + k_2)$ times the resulting price change, but increases domestic profit by k_2 times the price change.

26. We are well aware of the strong political economic arguments that might not lend support to this view.

ACKNOWLEDGMENTS

We are grateful to Marie Thursby and Casper de Vries for helpful discussions and comments. We have also benefitted from discussions with seminar participants at the International Conference on Mergers, Oligopoly, and Trade in Aix-en-Provence (June, 1990) and the Midwest International Economics Meetings in Champaign (November, 1990). Deneckere acknowledges financial support through National Science Foundation Grant SES–8619012 and the Kellogg Graduate School of Management's Beatrice/Esmark Research Chair. Kovenock has benefitted from financial support through Erasmus Universiteit Rotterdam, an Ameritech Foundation Summer Faculty Research Grant, and the Jay N. Ross Young Faculty Scholar Award. Sohn gratefully acknowledges the financial support of a David Ross Dissertation Grant and an Ameritech Foundation Dissertation Fellowship. The opinions expressed in this chapter are those of the authors who bear full responsibility for any errors or omissions.

REFERENCES

Bhagwati, J. (1965). On the Equivalence of Tariffs and Quotas. In: R. E. Baldwin et al. (Eds), *Trade, Growth, and the Balance of Payments: Essays in Honor of Gottfried Haberer* (pp. 53–67). Chicago: Rand McNally.

Davidson, C., & Deneckere, R. (1986). Long-Run Competition in Capacity, Short-Run Competition in Price, and the Cournot Model. *The Rand Journal of Economics, 17*(3), 404–415.

Dean, J., & Gangopadhyay, S. (1992). Strategic Trade Practices in the Presence of a VER. *International Economic Review, 33*(3), 645–659.

Deneckere, R., & Kovenock, D. (1988). *Price Leadership.* Northwestern University, Center of Mathematical Studies in Economics and Management Science, Discussion Paper No. 773.

Deneckere, R., & Kovenock, D. (1989). *Capacity-Constrained Price Competition When Unit Costs Differ.* Northwestern University, Center for Mathematical Studies in Economics and Management Science, Discussion Paper No. 861.

Deneckere, R., & Kovenock, D. (1992). Price Leadership. *Review of Economic Studies, 59,* 143–162.

Deneckere, R., & Kovenock, D. (1996). Bertrand-Edgeworth Duopoly with Unit Cost Asymmetry. *Economic Theory, 8*(1), 1–25.

Deneckere, R., Kovenock, D., & Lee, R. (1992). A Model of Price Leadership Based on Consumer Loyalty. *Journal of Industrial Economics, 40,* 147–156.

Dixon, H. (1987a). The General Theory of Household and Market Contingent Demand. *The Manchester School, 55*(3), 287–304.

Dixon, H. (1987b). Approximate Bertrand Equilibria in a Replicated Industry. *Review of Economic Studies, 54*(1), 47–62.

Eldor, R., & Levin, D. (1990). Trade Liberalization and Domestic Monopoly: A Welfare Analysis. *International Economic Review, 31*(4), 773–782.

Furth, D., & Kovenock, D. (1993). Price Leadership in a Duopoly with Capacity Constraints and Product Differentiation. *Journal of Economics, 57,* 1–35.

Gelman, J., & Salop, S. (1983). Judo Economics: Capacity Limitation and Coupon Competition. *Bell Journal of Economics, 14,* 315–325.

Harris, R. (1985). Why Voluntary Export Restraints are 'Voluntary'. *Canadian Journal of Economics, 18*(4), 799–809.

Horstmann, I., & Markusen, J. (1992). Endogenous Market Structures in International Trade (Natura-Facit-Saltum). *Journal of International Economics, 32*(1–2), 109–129.

Ito, M., & Ono, Y. (1982). Tariffs, Quotas, and Market Structure. *Quarterly Journal of Economics, 97,* 295–305.

Kreps, D., & Scheinkman, J. (1983). Quantity Pre-commitment and Bertrand Competition Yields Cournot Outcomes. *Bell Journal of Economics, 14*(2), 326–337.

Krishna, K. (1989). Trade Restrictions as Facilitating Practices. *Journal of International Economics, 26,* 251–270.

Levitan, R., & Shubik, M. (1972). Price Duopoly and Capacity Constraints. *International Economic Review, 13,* 111–123.

Osborne, M., & Pitchik, C. (1986). Price Competition in a Capacity-Constrained Duopoly. *Journal of Economic Theory*, *38*, 238–260.

Scherer, F. (1980). *Industrial Market Structure and Economic Performance* (2nd ed.). Chicago, IL: Rand McNally.

STRATEGIC ADVERTISING AND PRICING IN E-COMMERCE

Dale O. Stahl

ABSTRACT

We develop a model of pricing and advertising for e-commerce. N identical sellers, selling an homogeneous product, each choose a price and a web-based advertising policy. Buyers have usual demand functions and search optimally on the Internet. A Symmetric Nash Equilibrium (SNE) is derived and characterized. If the marginal cost of advertising exceeds a critical level, then the monopoly price with zero advertising is the unique SNE; otherwise, there is a mixed-strategy SNE which can be expressed as a price distribution and an advertising policy conditional on price. Remarkably, as the number of sellers increases without limit, the SNE converges to the monopoly price and advertising per seller vanishes. On the other hand, the mean price at which sales occur is less than the monopoly price. Still for some advertising technologies, total social welfare is declining in the number of sellers. This results cast doubt on the consumer benefits of e-commerce.

I. INTRODUCTION

Electronic commerce (e-commerce) has been experiencing tremendous growth over the last year, and it is not likely to slow given its exemption from sales taxes. Pundits promise lower prices for consumers due to lower search costs, at the same time that critics worry about privacy and the ability of savy

Industrial Organization, Volume 9, pages 69–100.

ISBN: 0-7623-0687-4

e-businesses to price discriminate on the basis of surreptitiously obtained private information. Further, since the cost of setting up a web site is considerably less than the cost of setting up a network of retail stores, more firms should enter creating downward competitive pressure on prices.

One of the salient changes in the e-commerce market is advertising. Not only is setting up a web site a form of advertising itself, but the content can be tailored to the preferences of the viewer (even without his/her knowledge of this bias). Non-solicited advertising can be targeted to narrowly defined consumer groups, more so than zip-code filtered mailings and phone solicitations. Since push advertising on the web rarely contains price information, the interested consumer must still visit a firm's web site to obtain specific price information and execute a purchase. While there are a growing number of shopping aides that claim to find the best bargains, the coverage of firms is incomplete and the reputation is unknown.

Another distinguishing feature of e-commerce is the elimination of the need for a physical trip to a firm's store to complete a purchase transaction. In the old literature on search, these physical trip costs were the major (if not sole) component of search costs. For e-commerce, the relevant search cost is the time spent browsing to obtain price information. Given the current and expected levels of congestion on the Internet, these search costs are non-negligible. Even with the current browsing technology, the subsequent cost of executing a purchase is very small in comparison to the initial search costs.

These stylized facts form the basis of the search model studied in this chapter. Given a reasonable model of e-commerce search and advertising, the central question that we investigate is how the competitiveness of the resulting market depends on the key parameters: search costs, advertising costs, and the number of firms. First we review the existing literature.

Economists have long been interested in understanding how markets work with imperfect information. Rothschild (1975) provided a critical survey, pointing out that most of the literature is concerned with only one side of the market – such as buyer search from a known distribution of seller prices. A complete analysis would include optimal buyer and seller behavior simultaneously determining a market equilibrium. A bothersome result is that if sellers do not advertise and search is costly, then the unique equilibrium is monopolistic (Diamond, 1971). Even with a mass of zero-search-cost buyers, the equilibrium distribution of prices converges to the monopoly price as the number of sellers increases (Stahl, 1989). But surely, it seems, sellers would have an incentive to advertise lower prices, thereby undermining the monopolistic equilibrium.

In a seminal article, Butters (1977) studied how sellers can influence buyer information by advertising. He demonstrated the existence of equilibrium price dispersion and actually presented an analytic solution and comparative statics results. He presented only a limit model with infinitely many buyers and sellers, so the influence of the number of buyers and sellers could not be explored. He also considered only the case of unit demands for buyers, and a specific advertising technology. On the other hand, Butters did consider both the case with no buyer search, and the case with a fixed buyer reservation price search rule.

The first case with no buyer search (buyers not receiving advertisements purchase nothing) was extended in Stahl (1994) to encompass any finite number of sellers and buyers, a general buyer demand function, and a general advertising technology. A unique Symmetric Nash Equilibrium (SNE) was derived in which the level of advertising was less than the socially optimal level except when buyers have unit demand functions (in which case, it is efficient). With no buyer search, sellers do not have the alternative of simply charging the monopoly price to the random share of buyers who stop at their store first, foregoing sales to other buyers. Thus, one force in the direction of monopoly pricing is absent. Hence, the SNE price distribution is non-degenerate with no tendency (as the number of sellers increases) to converge to monopoly pricing; on the other hand, it does not converge to marginal cost pricing either. Remarkably, for some advertising technologies, total welfare can decrease as the number of sellers increases.

This chapter extends the second case of Butters with buyer search to encompass any finite number of sellers and buyers, a general buyer demand function, optimal sequential buyer search, and a general advertising technology. We develop a complete game-theoretic solution (in mixed- strategies), and conduct comparative statics in terms of the advertising technology and the number of buyers and sellers. An explicit solution is presented for the special case of linear advertising costs.

Our search model, however, differs substantially from that of Butters'. He assumed that buyers who do not receive advertisements distribute themselves among sellers in the same proportion as the buyers who do receive advertisements. Butters defends this assumption on two grounds. First, he dismisses the model we will analyze as 'unpalatably complicated'. Second, he suggests that 'word-of-mouth' among buyers would result in the uninformed following the pattern of the informed. In contrast, the search model of this chapter adheres to the traditional optimal search literature despite its complications. Since the fundamental purpose of our inquiry is to study the effects of information transmission and acquisition technologies on the strategic outcomes, we shy from bold assumptions about leakage of information via word-of-mouth. Rather, we suggest that such word-of-mouth phenomena as well as all kinds of web sources are captured in our model with an appropriate interpretation of the parameter that represents advertising effectiveness.

Our model has a finite number of sellers and buyers. Sellers offer a homogeneous product, and have identical technologies. Each seller simultaneously chooses a price and an advertising level. Buyers have identical demand functions and positive search costs. Conditional on the advertisements received and the SNE price distribution, a buyer decides whether simply to buy from the seller with the lowest advertised price, or to search further. The optimal buyer search rule is characterized by a reservation price.

A SNE can be characterized by a distribution of prices and an advertising policy conditional on the realization of a seller's own price. The optimal advertising effort is a decreasing function of the realized price. The upper bound of the SNE price distribution is the lessor of the monopoly price and the buyer's reservation price. The SNE price distribution is atomless except possibly at the upper bound, in which case the advertising effort must be precisely zero at the upper bound.

If the marginal advertising costs exceed a critical level, the unique SNE is the pure-strategy SNE at the monopoly price with no advertising. However, for lower marginal advertising costs, there is no pure-strategy SNE, but there is a mixed-strategy SNE; i.e. there is price dispersion.

As the number of sellers increases, the advertising effort per seller vanishes at all prices, and the SNE price distribution converges to the monopoly price. Nonetheless, the total industry level of advertising remains strictly positive, and the distribution of the prices at which sales actually occur converges to a non-degenerate distribution bounded above marginal costs with an atom at the monopoly price. As in the model without buyer search (Stahl, 1994), for some advertising technologies, total welfare can be decreasing in the number of sellers. In the special case of unit buyer demand, there is over-advertising. More generally there can be over – or under-advertising at the SNE.

As the cost of advertising shrinks, the SNE converges towards marginal-cost pricing, but as the search cost shrinks, the SNE stays bounded above marginal-cost pricing. Thus, the two channels of information transmission have different impacts in that as the comparative advantage shifts towards buyer search, sellers curtail advertising; consequently buyers face higher prices than when search costs are higher but advertising costs are very low.

The chapter is organized as follows. Section II presents the basic model and derives optimal buyer and seller behavior. Section III derives the SNE, and

Section IV presents explicit solutions for linear and quadratic advertising costs. Section V presents results on the asymptotic behavior of the SNE. Section VI considers the welfare implications. Finally, Section VII summarizes the results and compares them with related work. All proofs are relegated to an Appendix.

II. THE MODEL

There are N identical sellers all offering the same homogeneous good. With identical linear production technologies, there is no loss of generality in taking marginal costs to be zero. Let p_j denote the price set by seller j. Alternatively, interpret p_j as net of marginal costs.

Advertising efforts generate a distribution of informed and uninformed buyers (before any buyer search). Let α_j denote the fraction of buyers who become informed about j's price. Sellers are assumed to have identical advertising technologies, which can be represented by a cost function $\gamma(\alpha, M)$ that gives the cost per buyer of informing a fraction α of M buyers. For example, given the classic urn technology assumed by Butters, we would have $\gamma(\alpha, M) = b\ln(1-\alpha)/\ln(1-1/M)$, where b is the cost per advertisement (hereafter abbreviated 'ad').[1] Unless explicitly considering the effects of M, we will suppress this argument and write $\gamma(\alpha)$. We assume that $\gamma(\alpha)$ is strictly convex and twice continuously differentiable with $\gamma'(0) > 0 = \gamma(0)$.

There are M identical buyers with demand functions D(p). Alternatively, interpret D(p) as the average demand function with idiosyncratic characteristics distributed independently of all other variables. Define the revenue function $R(p) \equiv pD(p)$. We assume that R(p) is continuously differentiable with a unique maximum at $\hat{p}$, and is strictly increasing for all $p < \hat{p}$. Further, we assume that $0 \leq D(p) < \infty$ for all $p \in (0, \hat{p})$.

II.1. Buyer Search

Buyers receive information about seller prices through seller ads and individual search efforts. Each search costs a fixed amount $c > 0$.[2] Let $F_0(p_j)$ denote the posterior probability distribution of the price charged by seller j conditional on not having received an ad from seller j. If p^* is the lowest price observed so far, then the net benefit of an additional search is

$$H(p^*, c, F_0) = \int_0^{p^*} D(p)F_0(p)\,dp - c. \tag{1}$$

We assume perfect recall. It is well-known (e.g. Kohn & Shavel, 1974) that the optimal search rule is characterized by a stationary reservation price, r, implicitly defined by

$$H(r, c, F_0) = 0, \tag{2}$$

if a root exists; otherwise, we define the reservation price to be $+\infty$. Note that $c > 0$ implies $r > 0$. For later use, note also that if F_0 is concentrated at one price, say $\tilde{p}$, then $c > 0$ implies $r > \tilde{p}$. Similarly, if $F_{0,n}$ is a sequence converging weakly to a distribution concentrated at $\tilde{p}$, then there is an n' such that for all $n > n'$, $r > \tilde{p}$.

Buyers search if and only if the lowest price they are aware of (either through ads or prior search) is more than r. After search is terminated, a buyer purchases an amount $D(p_j)$ from the lowest priced seller (say j) she has found. A buyer who has received no ads and sampled no sellers is assumed to select a seller at random (say k) and purchase an amount $D(p_k)$, provided $p_k \leq r$.[3]

II.2. Seller Advertising and Pricing Decisions

A SNE is a triplet $\{F(\cdot), \alpha(\cdot), r^*\}$ such that (i) when any subset of $N-1$ sellers choose prices from the probability distribution $F(\cdot)$ and advertising effort $\alpha(p)$, then the remaining seller has no incentive to deviate, and (ii) r^* is the optimal buyer reservation price given $\{F(\cdot), \alpha(\cdot)\}$.

In this section, we take the buyer reservation price r as given, and consider seller pricing decisions. To be explicit, we denote the price distribution by $F(p; r)$. Let P denote the maximum price in the support of $F(p)$, and let B denote the minimum price in the support.[4] Ultimately, we will derive a *consistent reservation price* r^* that is optimal for $F(p; r^*)$.

Take a random buyer, and let $\varphi(p_j)$ denote the probability that p_j is the lowest price known by this buyer conditional on knowing that j is charging p_j. Let β denote the probability that a random buyer is uninformed about the other $N-1$ sellers. Consider seller j's decision about advertising effort α_j at price $p_j \leq r$. The expected profits per buyer (i.e. total expected profits divided by M) can be expressed as

$$E\pi(p_j, \alpha_j) = R(p_j)\{\alpha_j\varphi(p_j) + (1-\alpha_j)\beta/N\} - \gamma(\alpha_j). \tag{3}$$

$\partial E\pi(p_j, \alpha_j)/\partial\alpha_j = R(p_j)\{\varphi(p_j) - \beta/N\} - \partial\gamma/\partial\alpha_j$, and $\partial^2 E\pi(p_j,\alpha_j)/(\partial\alpha_j)^2 = -\partial^2\gamma/(\partial\alpha_j)^2 < 0$. Thus, $E\pi(p_j,\alpha_j$is strictly concave in α_j, which implies that there exists a unique function $\alpha(p_j$that maximizes Eq. (3). The parameters (such as r) that affect $\alpha(\ )$ will be made explicit later.

Therefore, the strategy space for sellers can be reduced to a price distribution $F(p; r)$ and a function $\alpha(p)$, with the interpretation that sellers randomly choose a price p_j and then choose $\alpha(p_j)$.

Let $A(p_j; r)$ denote the probability that a random buyer is informed that seller j has a price $p \leq p_j$. Given $F(p; r)$ and $\alpha(p)$,

$$A(p_j; r) = \int_B^{p_j} \alpha(p)\, dF(p; r). \tag{4}$$

It is notationally convenient to define $\bar{\alpha} \equiv A(P; r)$, the expected value of α. Note that $\bar{\alpha}$ is also the probability that a random buyer is informed about (say) seller j. Hence, in Eq. (3), $\beta = (1 - \bar{\alpha})^{N-1}$: the probability that a random buyer is uninformed about the $N - 1$ sellers playing $F(p; r)$ and $\alpha(p; r)$. Further, if $F(p; r)$ contains no atoms at or below p_j, then in Eq. (3), $\varphi(p_j) = [1 - A(p_j; r)]^{N-1}$.

We will now state a result that is very useful because it leads to an expression for payoffs that is continuous function.

Lemma 1. If a SNE $F(p; r)$ contains an atom at $\tilde{p} > 0$, then $\alpha(\tilde{p}) = 0$.

In other words, if $F(p; r)$ contains an atom at any price, then that price will not be advertised. Intuitively, if F contains an atom at $\tilde{p} > 0$, then a seller can increase profits by undercutting $\tilde{p}$ slightly and advertising the lower price.

Next, observe that there can be no atom at $\tilde{p} = 0$, because then expected profits would be precisely zero, but since $r > 0$, a seller would make a positive profit at positive prices up to r. Coupled with this fact, an immediate corollary of Lemma 1 is that $A(\cdot\,; r)$ is continuous over the relevant domain. Hence, for $p_j \leq r$, we can rewrite Eq. (3) as

$$E\pi(p_j, \alpha_j = R(p_j)\{\alpha_j[1 - A(p_j; r)]^{N-1} + (1 - \alpha_j)(1 - \bar{\alpha})^{N-1}/N\} - \gamma(\alpha_j). \tag{5}$$

Two very useful and important result follows. First, SNE expected profits, $E\pi^*$, are strictly positive, because a store charging $r > 0$ will always capture a share of the completely uninformed buyers, and the latter is always positive because setting $\alpha = 1$ would generate negative profits. Second, the maximum price in the support of the SNE price distribution is the lessor of the monopoly price $\hat{p}$ and the buyers' reservation price r. Intuitively, prices higher than r are suboptimal because buyers will always search further. Prices above the monopoly price can never be optimal because revenues are declining. If every seller prices below $\min(r, \hat{p}\}$, then since expected profits, Eq. (5), are continuous, a seller can raise prices slightly without losing any customers, thereby increasing sales.

Lemma 2. (a) $E\pi^* > 0$, and (b) $P = \min\{r, \hat{p}\}$.

One of the implications of Lemma 2 is that buyers will not engage in any real search. If they receive any information, then with probability one the known prices are at or below their reservation price, so they will simply go to the lowest priced seller and purchase. If they are completely uninformed, then they will pick a seller at random and purchase since the price they find will be at or below their reservation price with probability one.[5] In Section III, we will completely characterize the SNE $F(p; r)$ and $\alpha(p)$.

III. DERIVATION OF SNE

We first state a number of properties of a SNE. Initially, we will take the buyer reservation price r as exogenously fixed

Lemma 3. The support of $F(p; r)$ is a connected interval $[B, P]$.

Lemma 4. There exists at most one price p' in the support of $F(p; r)$ such that $\alpha(p') = 0$.

Lemma 5. If there is an atom in $F(p; r)$, then it must be at P.

In other words, the price distribution must be atomless below P with connected support; there cannot be multiple price in the support of $F(p; r)$ that are not advertised; and if there is an atom at P, then that price will not be advertised. We next consider the existence of pure-strategy SNE.

III.1. Pure-Strategy SNE

Recall that if there is a pure-strategy at price P, then by Eqs (1) and (2), $r > P$. It then follows from Lemma 2 that the only possible consistent pure-strategy price is the monopoly price $\hat{p}$. By Lemma 1, it must be optimal not to advertise this price, otherwise undercutting would be profitable. Thus, we need to find conditions under which it is not optimal to advertise given every seller chooses the monopoly price. Intuitively, for sufficiently high advertising costs, it will not be optimal to advertise. Then, costly search implies monopoly prices (Diamond, 1971). More precisely:

Theorem 1. If $\gamma'(0) \geq [(N-1)/N]R(\hat{p})$, then $F\{\hat{p}\} = 1$ and $\alpha(\hat{p}) = 0$ is the unique consistent SNE; otherwise, there does not exist a pure-strategy SNE.

A pure-strategy SNE exists if and only if the marginal advertising cost is sufficiently high relative to monopoly revenues; otherwise, there is no pure-strategy SNE. In particular, note that if $\gamma'(0) \geq R(\hat{p})$, then $\hat{p}$ is a pure-strategy SNE for all $N \geq 2$. If $\gamma'(0) < R(\hat{p})/2$, then there does not exist a pure-strategy SNE for any $N \geq 2$. In the interim range, there exists an N' such that for all

$N \leq N'$, there is a unique (pure-strategy) SNE at $\hat{p}$, and for all $N > N'$ there is no pure-strategy SNE. Thus, entry of sellers appears to reduce the likelihood of pure monopoly pricing. [We will re-examine the effects of seller entry in Sections V and VI.]

III.2. Derivation of Mixed-Strategy SNE

In the remainder of the chapter, unless otherwise stated, we assume $\gamma'(0) < [(N-1)/N]R(\hat{p})$, so there does not exist a pure-strategy SNE. The first step is to derive an expression for $[1 - A(p; r)]^{N-1}$. It is notationally convenient to define $\alpha_P \equiv \alpha(P)$. A necessary condition of a NE is that $E\pi(p, \alpha(p))$ be equal to a constant for all $p \in [B, P]$. Using Eq. (5), $E\pi(P, \alpha_P) = R(P)\{\alpha_P + (1\text{-}\alpha_P)/N\}(1 - \bar{\alpha})^{N-1} - \gamma(\alpha_P)$, and hence,

$$[1 - A(p; r)]^{N-1} = \left[\left(\frac{R(P)}{R(p)}\right)\right]\left(\alpha_P + \frac{(1-\alpha_P)}{N}\right) - \frac{(1-\alpha(p))}{N}\Bigg]\left(\frac{(1-\alpha)^{N-1}}{\alpha(p)}\right)$$
$$+ \frac{\gamma[\alpha(p)] - \gamma(\alpha_P)}{R(p)\alpha(p)}. \qquad (6)$$

Given a solution for $\alpha(p)$, Eq. (6) can be used to calculate $A(p; r)$. But before that, we will use Eq. (6) to derive a parametric solution for $\alpha(p)$.

Since $E\pi(p, \alpha_j)$, Eq. (5), is strictly concave in α_j, the Kuhn-Tucker conditions are necessary and sufficient to characterize the optimal α_j:

$$\partial E\pi(p, \alpha_j/\partial\alpha_j = R(p)\{[1 - A(p; r)]^{N-1} - (1-\bar{\alpha})^{N-1}/N\} - \gamma'(\alpha_j) = 0, \qquad (7)$$

for interior optima; otherwise, $\partial E\pi(p,0)/\partial\alpha_j \leq 0$ implies $\alpha_j = 0$, and $\partial E\pi(p, 1)/\partial\alpha_j \geq 0$ implies $\alpha_j = 1$.

Substituting Eq. (6) into Eq. (7) yields

$$\alpha_j\gamma'(\alpha_j) - \gamma(\alpha_j) = [R(P)[\alpha_P + (1-\alpha_P)/N] - R(p)/N]\cdot(1-\bar{\alpha})^{N-1} - \gamma(\alpha_P), \qquad (8)$$

for interior optima. The 'corner' conditions translate to: (i) if the right-hand-side (r.h.s.) of Eq. (8) is no less than $\gamma'(1) - \gamma(1)$, then $\alpha_j = 1$ is the optimal solution; (ii) if the r.h.s. is non-positive, then $\alpha_j = 0$ is the optimal solution. The relevant parameters of Eq. (8) are P, $\bar{\alpha}$, and α_P. Using Lemma 2, $(r, \bar{\alpha}, \alpha_P)$ is a sufficient set of parameters so Eq. (8) and the corner conditions define a unique solution for α_j which we denote $a(p; r, \bar{\alpha}, \alpha_P)$. We will see momentarily that there is a one-to-one relationship between $\bar{\alpha}$ and α_P, so (r, α_P) will be a necessary and sufficient set of parameters.

Note that the left-hand side (l.h.s.) of Eq. (8) as a function of α_j begins at the origin and increases with slope $\alpha_j\gamma''$, while the r.h.s. is strictly decreasing in p up to $\hat{p}$. Hence, for interior solutions, $a(\cdot; r, \bar{\alpha}, \alpha_P)$ is a strictly decreasing function. This result is rather intuitive. Sellers want to advertise low prices more than high prices because the chance of being the winner (and hence the payoff to advertising) is greater when they set low prices.

We will now develop a characterization of the optimal $\alpha(p)$ that will permit us to eliminate $\bar{\alpha}$ from the parameter list.

Lemma 6. Eq. (7) must hold as a strict equality at $p=P$:

$$\gamma'(\alpha_P)=\left(\frac{N-1}{N}\right) R(P)\,(1-\bar{\alpha})^{N-1} \tag{9}$$

Using Eq. (9) to determine $\bar{\alpha}$ in terms of α_P, denoted $\bar{\alpha}(\alpha_P)$, we can now express the optimal advertising level as $\alpha(p; r, \alpha_P) \equiv a[p; r, \bar{\alpha}(\alpha_P), \alpha_P]$.

The requirement that Eq. (9) have a positive solution puts a lower bound on P, as follows.

Lemma 7. Given $\gamma'(0)<[(N-1)/N]R(\hat{p})$, a SNE must have $P>R^{-1}[\gamma'(0)N/(N-1)]$; hence, $r>\hat{r}\equiv R^{-1}[\gamma'(0)N/(N-1)]$.

It is useful to examine a graphical method of determining $\bar{\alpha}(\cdot)$ as presented in Fig. 1. The upward sloping function is $\gamma'(\alpha)$, and the downward sloping function is $[(N-1)/N]\cdot R(P)(1-\alpha)^{N-1}$, which are the left and right hand sides of Eq. (9) respectively. The functions have been drawn under the assumption that $\gamma'(0)<[(N-1)/N]R(P)$ so there is a unique interior intersection point. For any α_P, start on the horizontal axis at $\alpha=\alpha_P$ and move vertically to the $\gamma'(\,)$ function; then move horizontally to the other function. The final horizontal position is $\bar{\alpha}(\alpha_P)$. It is easy to see that $\bar{\alpha}(\cdot)$ is strictly decreasing with a maximum value of $a_H\equiv\bar{\alpha}(0)$. The unique fixed point of $\bar{\alpha}(\cdot)$ is denoted a_L. Observe that $0<a_L<a_H$.

It should be apparent that since $\alpha(\cdot; r, \alpha_P)$ is non-increasing, it must be that $\alpha_P\leq\bar{\alpha}(\alpha_P)$. Figure 1 illustrates how this requirement circumscribes the feasible range for α_P. For any $\alpha_P\in[0, a_L]$, as indicated by the dotted lines in Fig. 1, $\bar{\alpha}(\alpha_P)$ lies in the interval $[a_L, a_H]$, which is compatible with a SNE solution. On the other hand, for any $\alpha_P>a_L$, $\bar{\alpha}(\alpha_P)>\alpha_P$, and hence is not compatible.

Having a parametric solution $\alpha(p; r, \alpha_P)$, we can use Eq. (6) to define a parametric solution $A(p; r, \alpha_P)$. Next, to determine the lower bound B, note that we must have $A(B; r, \alpha_P)=0$, and then solve Eq. (6) for B in terms of $\alpha(\cdot; r, \alpha_P)$. Let $\alpha_B\equiv\alpha(B; r, \alpha_P)$, and define

$$L(B)\equiv R(B)-r(B)/\alpha_B=0, \tag{10}$$

where

$$r(B) \equiv \left[R(P)\left(\alpha_P + \frac{1-\alpha_P}{N} \right) - R(B)\,\frac{1-\alpha_B}{N} \right]\left(\frac{N\gamma'(\alpha_P)}{(N-1)R(P)} \right) + \gamma(\alpha_B) - \gamma(\alpha_P),$$

so $L(B)=0$ iff $A(B; r, \alpha_P)=0$. Hence, Eq. (10) defines has a unique solution $B(r, \alpha_P) \in (0, P)$.[6]

We will now derive a parametric solution for the price distribution which we will denote $F(p; r, \alpha_P)$. Differentiating Eq. (4) with respect to p_j, we see that $\partial A(p_j; r, \alpha_P)/\partial p_j = \alpha(p_j; r, \alpha_P)F'(p_j; r, \alpha_P)$, where F' denotes the probability density function. The easiest way to derive an expression for $\partial A/\partial p$ is to differentiate Eq. (5) with respect to p_j, using the Envelope Theorem to get

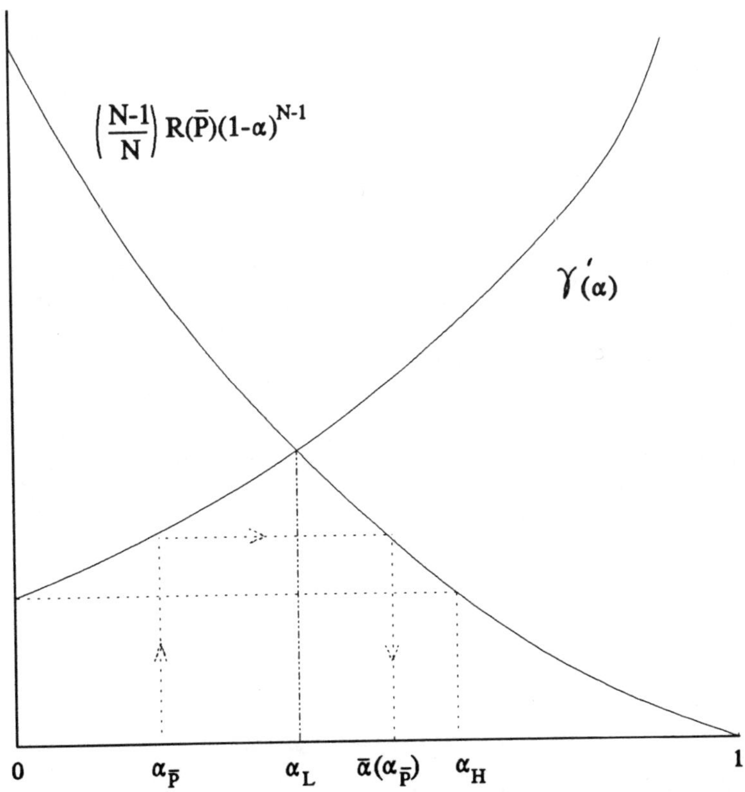

Fig. 1. Graph of Equation 9.

$$\frac{\partial A(p_j; r, \alpha_P)}{\partial p_j} = \frac{R'(p_j)}{R(p_j)} \cdot \left[\frac{\alpha[1 - A(p_j; r, \alpha_P)]^{N-1} + (1-\alpha)\gamma'(\alpha_P)/[(N-1)R(P)]}{(N-1)\alpha[1 - A(p_j; r, \alpha_P)]^{N-2}} \right], \tag{11}$$

where $\alpha = \alpha(p_j; r, \alpha_P)$. We can see that $\partial A(p; r, \alpha_P)/\partial p \geq 0$, and hence, $F'(p_j; r, \alpha_P) = [\partial A(p_j; r, \alpha_P)/\partial p_j]/\alpha(p_j; r, \alpha_P) \geq 0$. Integrating:

$$F(p_j; r, \alpha_P) \equiv \int_B^{p_j} \frac{\partial A(p; r, \alpha_P)/\partial p_j}{\alpha(p; r, \alpha_P)} \, dp \tag{12}$$

Note that by construction, $F(B; r, \alpha_P) = 0$ and the density is non-negative. The only remaining requirement is the endpoint condition that $F(p; r, \alpha_P) \leq 1$.

Theorem 2. Given $r \geq \hat{r}$, there exists an α_P such that $F(p; r, \alpha_P)$ and $\alpha(p; r, \alpha_P)$ defined by Eqs (6–12) are SNE strategies conditional on r.

Thus, we have a method of computing SNE strategies (conditional on buyer reservation price $r \geq \hat{r}$). Taking α_P as a parameter, we first compute a parametric advertising solution $\alpha(p; r, \alpha_P)$. Then, we use $\alpha(p; r, \alpha_P)$ to compute a parametric price distribution. Next, we adjust α_P until $F(P; r, \alpha_P) \leq 1$ and $\alpha_P[1 - F(P; r, \alpha_P)] = 0$.

We turn now to the question of uniqueness. First observe that there cannot be two SNE both with atoms at P, because an atom at P requires $\alpha_P = 0$, which in turn defines a unique solution to Eq. (8), and hence a unique price distribution as well. Whether or not there can be multiple SNE (not more than one with an atom at P) is an open question. To shed some light on this question, it is useful to know how $\alpha(p; r, \alpha_P)$ depends on α_P. Substituting Eq. (9) into Eq. (8) and differentiating with respect to α_P reveals that $\alpha(p; r, \alpha_P)$ is strictly increasing in α_P for all $\alpha_P \in (0, a_L)$. Thus, recalling Eq. (12), the direct effect of increasing α_P is to decrease $F(p; r, \alpha_P)$. Unfortunately, I have not been able to prove that the total effect is always negative. For uniqueness, it would suffice to show that $\partial F(p; r, \alpha_P)/\partial \alpha_P < 0$ when evaluated at any α_P such that $F(p; r, \alpha_P) = 1$. I suspect that for some revenue functions and advertising cost functions, there are multiple SNE. However, there will also be a wide class of revenue functions for which there will be a unique SNE.

III.3. Existence of a Consistent SNE

In Section III.2, we took the buyer reservation price r as exogenously fixed. Now we want to close the model by finding a consistent r* such that the seller price distribution $F(p; r^*)$ induces r^* as the buyer reservation price. Conditional on not receiving an ad from a seller, the posterior probability distribution on that seller's price is

$$F_0(p; r) \equiv \frac{\int_0^p [1 - \alpha(x; r)]dF(x; r)}{1 - \bar{\alpha}} = \frac{F(p; r) - A(p; r)}{1 - \bar{\alpha}} \quad (13)$$

Let $B_0 \equiv \inf\{p \mid \alpha(p) < 1\}$.[7] Then F_0 has support $[B_0, P]$ and is skewed towards P relative to F(p).

First, observe that for $r \geq \hat{p}$, $P = \hat{p}$, so $F(p; \cdot)$ is independent of $r \geq \hat{p}$; hence, $H[\hat{p}, c, F_0(\cdot; r)]$ is constant for $r \geq \hat{p}$. Consequently, if $H[\hat{p}, c, F_0(\cdot . \hat{p})] \leq 0$, then there exists a consistent buyer reservation price $r^* \geq \hat{p}$. This will occur, for example, if c is sufficiently high. On the other hand, if $H[\hat{p}, c, F_0(\cdot, \hat{p})] > 0$, then a consistent reservation price (if one exists) must lie in the open interval $(0, \hat{p})$.

Second, observe that the buyer reservation price enters the seller SNE strategy choices only via the term $R[\min\{r, \hat{p}\}]$. Recalling that R( ) is continuously differentiable, if there were a unique conditional SNE for every r, then the existence of a consistent r^* would be almost trivial. However, we must take account of the possibility that the SNE asserted by Theorem 2 may depend discontinuously on r.

Theorem 3. There exists a fully consistent SNE. That is, there exists a parameter α_P^* and a buyer reservation price r^* that is consistent with $F(p; r^*, \alpha_P^*)$ of Theorem 2.

The proof uses a fixed point argument to find a P^* and α_P^* simultaneously such that $H[P^*, c, F_0(\cdot; P^*, \alpha_P^*)] \leq 0$ and $F_0(P^*; P^*, \alpha_P^*) \leq 1$, with strict inequalities implying that $P^* = \hat{p}$ and $\alpha_P^* = 0$ respectively. Hence, there exists a consistent reservation price $r^* \geq P^*$.

IV. EXPLICIT SOLUTIONS FOR SPECIAL CASES

IV.1. Linear Advertising Costs

While strict convexity of advertising costs $\gamma(\alpha)$ played a crucial role in the above derivation of the mixed-strategy SNE, we can derive a unique SNE for the case when $\gamma(\alpha)$ is linear. Accordingly, in this subsection we assume that $\gamma(\alpha) = \gamma'\alpha$, for some $0 < \gamma' < [(N-1)/N]R(\hat{p})$. Equation (7) now produces a knife-edge: there is some price p^* such that if $p_j < p^*$, then $\alpha_j = 1$, and if $p_j > p^*$, then $\alpha_j = 0$. But since by Lemma 5, P is the only price such that $\alpha(P; r) = 0$, we must have $\alpha(p; r) = 1$ for all $p < P$. Then, from Eq. (13), $F_0(p) = 0$ for all $p < P$, so by Eq. (1) and Lemma 2, we must have $P = \hat{p}$, the monopoly price.

If there were no atom at $\hat{p}$, then $\bar{\alpha} = 1$, which would be inconsistent with Eq. (9). Hence, there must be an atom at $\hat{p}$, and Eq. (9) determines $\bar{\alpha} = 1 - \{N\gamma'/[(N-1)R(\hat{p})]\}^{1/(N-1)}$. Since $\alpha(p; r) = 1$ for all $p < \hat{p}$, $(1 - \bar{\alpha})$ is also the mass of the atom at $\hat{p}$. Then, expected profits are $E\pi(\hat{p},0) = R(\hat{p})(1-\bar{\alpha})^{N-1}/N$. Since $\alpha = 1$ for $p < \hat{p}$, $A(p; r) = F(p; r)$; hence, $E\pi(p,1) = R(p)[1 - F(p; r)]^{N-1} - \gamma'$. Setting $E\pi(p,1) = E\pi(\hat{p},0)$ and solving for $F(p; r)$ yields

$$F(p; r) = 1 - \left(\frac{N\gamma'}{(N-1)R(p)}\right)^{1/(N-1)} \tag{14}$$

Observe that $\lim_{p \to \hat{p}} F(p; r) = \bar{\alpha}$, as required.The lower bound B is determined by setting $F(B; r) = 0$, which yields $B = R^{-1}\,[N\gamma'/(N-1)]$.

Sellers choose the monopoly price, $\hat{p}$, with positive probability and do not advertise that price (because the net marginal benefits to advertising are zero). In the event that all sellers choose $\hat{p}$, all the buyers are uninformed and simply distribute themselves uninformly over the sellers. In all other events, some seller chooses a lower price and all buyers are fully informed about these lower prices, so only the lowest-priced seller has customers. A seller must trade-off the likelihood of being the lowest-priced seller and getting all the business with the likelihood of all sellers choosing $\hat{p}$ and sharing the business. The price distribution, Eq. (14), equates the expected profits of these alternatives.

Note that as γ' increases toward $[(N-1)/N]R(\hat{p})$, the lower bound B converges to $\hat{p}$, and the mass of the atom at $\hat{p}$ converges to 1. Thus, the price distribution converges pointwise to the degenerate distribution at the monopoly price, which is the unique solution for all $\gamma' \geq [(N-1)/N]R(\hat{p})$. Conversely, as $\gamma' \to 0$, the lower bound B converges to 0, the mass of the atom at $\hat{p}$ converges to 0, and $F(p)$ converges to 1 for all $p > B$. Thus, the price distribution converges weakly to the degenerate distribution at 0 [i.e. Bertrand pricing].

IV.2. Quadratic Advertising Costs

Suppose $\gamma(\alpha) = \gamma'\alpha + \gamma''\alpha^2/2$, for positive constants γ' and γ''. We can derive a closed form expression for the optimal advertising policy:

$$\alpha(p; \alpha_p) \equiv \left\{\left[R(P)\left(\alpha_P + \frac{1-\alpha_P}{N}\right) - \frac{R(p)}{N}\right]\left(\frac{2N(\gamma' + \gamma''\alpha_P)}{(N-1)R(P)\gamma''}\right) - \frac{2\gamma'\alpha_P}{\gamma''} - (\alpha_P)^2\right\}^{1/2}, \tag{15}$$

subject to $\alpha(p; \alpha_P) \in [0,1]$. We then seek a α_P such that $F(P; r, \alpha_P) \leq 1$ and $\alpha_P \cdot [1 - F(P; r, \alpha_P)] = 0$.

The behavior of the SNE $\alpha(p)$ is illustrated in Fig. 2. The example assumes search cost sufficiently high so $P = \hat{p}$, linear demand $D(p) = 1 - p$, $N = 10$, and $\gamma' = 0.01$. For sufficiently small γ'', the SNE has an atom at P and $\alpha_P = 0$. The SNE advertising level has $\alpha(p) = 1$ for low prices and sharply declines towards zero for prices near P. As $\gamma'' = 0$, the range of fully advertised prices ($\alpha = 1$) increases to $\lfloor B, P)$; i.e. the advertising solution converges to that for linear advertising costs. As γ'' increases, advertising is curtailed on more and more prices; i.e. $\alpha(p)$ rotates downward and counterclockwise. For sufficiently high γ'', $\alpha_P > 0$, and as γ'' increases further, $\alpha(p)$ declines towards zero.

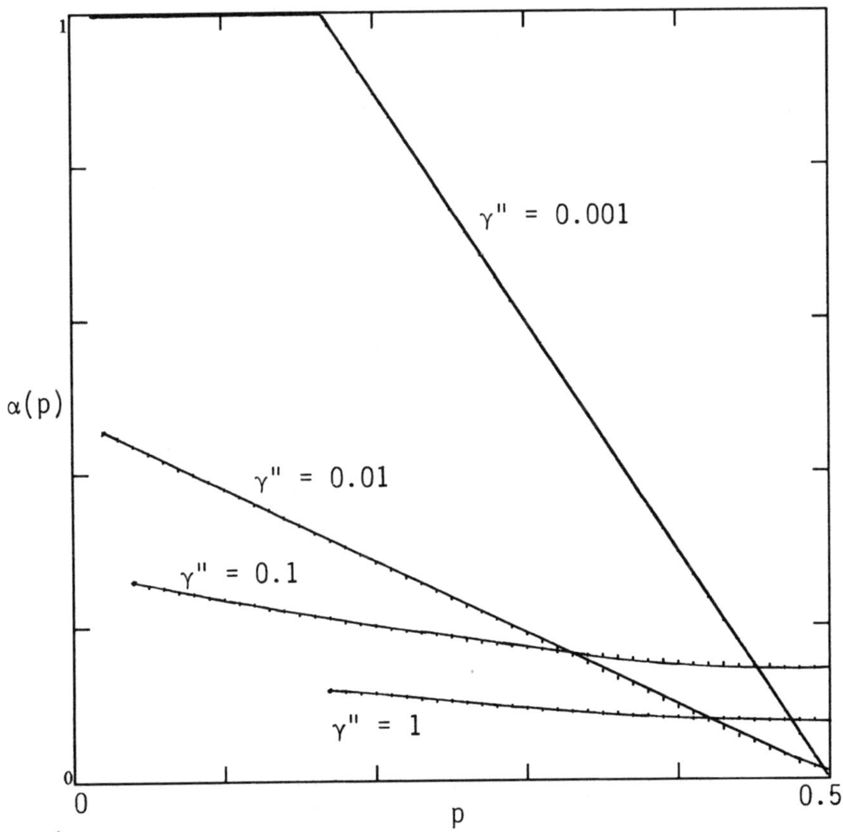

Fig. 2. Advertising Policies for Quadratic Costs.

Figure 3 displays several SNE price density functions for this example. As $\gamma'' \rightarrow 0$, the lowest price B declines towards 0.0112, and eventually an atom forms at $\hat{p}$ which increases in mass towards 0.2925. The limit as $\gamma'' \rightarrow 0$ is the SNE for the linear advertising case presented above. Notice that for $\gamma'' \leq 0.01$, the density is 'U-shaped'; thus, a seller is more likely to charge a high price or a low 'sale' price, when advertising costs do not increase too sharply.

V. ASYMPTOTIC BEHAVIOR

We are interested in how the qualitative behavior of a consistent SNE is affected by the number of buyers and sellers, advertising costs and search costs.

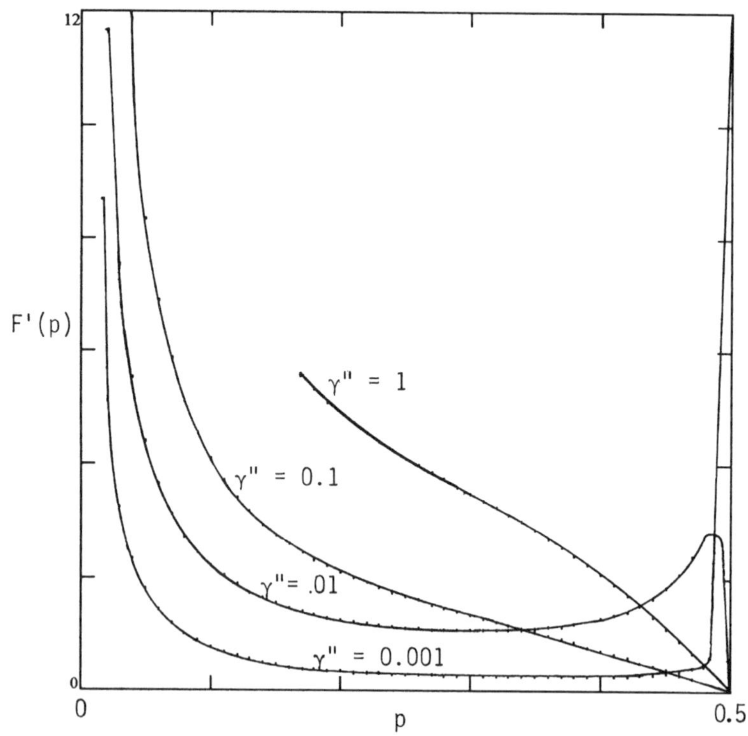

Fig. 3. Probability Density Functions of Prices for Quadratic Costs.

Assuming that $\gamma(\alpha, M)$ and $\partial\gamma(\alpha, M)/\partial\alpha$ increase monotonically to limits $\gamma(\alpha, \infty)$ and $\gamma'(\alpha, \infty)$, the limit set of SNE is completely characterized by the limit marginal advertising costs $\gamma'(\alpha, \infty)$.

From Fig. 1, it is apparent that both a_L and a_H decrease, which suggests that α_P and hence $\alpha(p; r)$ decrease. Unfortunately, due to the potential non-uniqueness of the SNE, we cannot derive any general comparative statics about the price distribution, other than that a well-defined limit distribution exists.

V.1. The Effect of the Number of Sellers

We will prove the remarkable result that the price distribution converges to the monopoly price as $N \rightarrow \infty$. It is easiest to see this result for the case of linear advertising costs. Following this, we will derive this result for the general case. Recalling from Theorem 1, that pure monopoly pricing is the unique SNE when $\gamma'(0) \geq R(\hat{p})$, we will focus exclusively on the remaining case when $\gamma'(0) < R(\hat{p})$.

From Eq. (14), given linear advertising costs $\gamma' > 0$,[8] as $N \rightarrow \infty$, $F(p; r) \rightarrow 0$ for all $p < P$, and the mass at P approaches unity. While B declines monotonically to $R^{-1}[\gamma'/R(P)]$, the SNE price distribution converges to the degenerate distribution with unit mass at P. But then recalling the remarks of Section II.1, there exists an N' such that for all $N > N'$, $r > P$, so by Lemma 2, $P = \hat{p}$. In other words, the upper bound of the price distribution hits the monopoly price for sufficiently many (finite) sellers, and becomes more concentrated at the monopoly price as the number of sellers increases without bound.[9]

To develop this result for the case of strictly convex advertising costs, we proceed with a number of lemmas. Note that while there may be multiple SNE for various values of N, the results apply to any SNE, so we will be showing not just that some SNE converges to monopoly pricing, but that *all* SNE converge to monopoly pricing.

Lemma 8.. As $N \rightarrow \infty$, $\bar{\alpha}$ and α_P converge to 0, $(1 - \bar{\alpha})^{N-1} \rightarrow \gamma'(0)/R(P)$, and $N\bar{\alpha} \rightarrow \ln[R(P)/\gamma'(0)]$.

From Lemma 8, we see that while the average effort of advertising per seller $\bar{\alpha}$ vanishes as $N \rightarrow \infty$, the aggregate average effort $N\bar{\alpha}$ does not. Moreover, the probability that a typical buyer is totally uninformed, $(1 - \bar{\alpha})^{N-1}$, remains bounded below 1, given $\gamma'(0)/R(P) < (N - 1)/N$.

Lemma 9. As $N \rightarrow \infty$, $\alpha(p) \rightarrow 0$, but $N\alpha(p) \rightarrow \infty$ for all $p < P$.

In other words, the advertising effort per seller shrinks to zero for all prices, but slowly enough so the aggregate advertising effort increases without bound for all $p<P$. Note that this result does not imply that expected advertising expenditures increase without bound, because expected advertising expenditures per seller are $E\gamma \equiv \int \gamma[\alpha(p)]dF(p)$, which clearly depend not only on $\alpha(p)$ but also on $F(p)$. Indeed, given Lemma 9, to keep $E\gamma$ bounded, it must be that $F(p)$ has less and less probability mass at $p<P$. This conclusion is formalized in the Theorem 5 below, whose proof uses a different insight. But first it is useful to present one last Lemma concerning the distribution of received prices.

Lemma 10. As $N \rightarrow \infty$, $[1 - A(p)]^{N-1}\, \gamma'(0)/R(p)$ and $A(p) \rightarrow 0$ for all p.

In other words, the probability that a typical buyer will be informed that seller j has a price $p_j \leq p$, $A(p)$, converges to zero as $N \rightarrow \infty$. On the other hand, because the aggregate advertising effort does not vanish, the probability that p_j is the lowest price of the other $N - 1$ sellers about whom a typical buyer is informed, $[1 - A(p_j)]^{N-1}$, is bounded above zero.

Finally, we come to our main results concerning the effect of seller entry on the SNE mixed-strategies.

Theorem 4. As $N \rightarrow \infty$, $F(p) \rightarrow 0$ for all $p<P$, and $B \rightarrow R^{-1}[\gamma'(0)]$.

Thus, we see that the SNE price distribution converges to the degenerate distribution with all mass at one price P. The next question concerns the effect of N on this price P and the mass at P.

Theorem 5. There is an N' such that for all $N>N'$, there is a unique SNE, $P=\hat{p}$, and $F\{P\}>0$.

Thus, Theorem 5 together with Theorem 4, establishes that the SNE converges to pure monopoly pricing as the number of sellers increases without bound. Indeed, once entry exceeds some finite level, there is an atom in the price distribution at the monopoly price, and the mass of that atom increases to 1 as $N \rightarrow \infty$.

The reason for this remarkable result is that sellers always have the option of being content with their captive share of buyers who are totally uninformed:asymptotically, a mass of $\gamma'(0)/R(\hat{p})$ buyers. On the other hand, competing for buyers requires expenditures for advertising and a cut in price, and these efforts are rewarded only if the seller has the lowest price in the information set of a typical buyer.[10]

This result seems counterintuitive if one focuses on the limit as $N \rightarrow \infty$: monopoly pricing with no advertising. Surely (you might say) advertising a price just below the monopoly price would be profitable. Indeed, the limit strategies are not equilibrium strategies for the model with $N = \infty$. It appears that (as in Stahl (1996)) there does not exist a SNE for the limit model with $N = \infty$. However, Theorems 4 and 5 are about the asymptotic behavior of the model with a large but finite number of sellers. To understand better how the limit model and the large-but-finite model differ, we can extract more insights from the foregoing Lemmas.

First, let's examine the advertising effort. From Lemma 8, we observe that the average effort of advertising per seller $\bar{\alpha}$ vanishes as $N \rightarrow \infty$. Expected advertising expenditures per seller are $E\gamma \equiv \int \gamma[\alpha(p)]dF(p)$. Using Lemma 9, $E\gamma$ vanishes as $N \rightarrow \infty$. Thus, average advertising effort and expenditures shrink towards zero with seller entry. On the other hand, by Lemma 8, the aggregate average effort $N\bar{\alpha}$ is bounded above zero. Similarly, total industry advertising expenditures $NE\gamma \approx N\bar{\alpha}\gamma'(0)$, which, by Lemma 8, converges to $\gamma'(0)\ln[R(\hat{p})/\gamma'(0)] > 0$. Because of this aggregate advertising effort, the probability that a buyer is completely uninformed, $(1 - \bar{\alpha})^N$, converges to $\gamma'(0)/R(\hat{p}) \in (0,1)$, despite the fact that $\bar{\alpha} \rightarrow 0$. Thus, there is always some positive mass of uninformed buyers, *and* a positive mass of (partially) informed buyers.

Given that there is always a positive mass of informed buyers, it does not follow that by slightly undercutting the monopoly price a seller can gain a discrete increase in sales. The seller must still be the lowest-priced seller among those about whom the buyer is informed. The probability that p_j is the lowest price among the sellers about whom a typical buyer is informed is $[1 - A(p_j]^{N-1}$, which, by Lemma 10 converges to $\gamma'(0)/R(p_j$. Thus, the marginal revenue gained is approximately $\gamma'(0)$, while the direct marginal advertising cost is $\gamma'(0)$, so the net direct marginal benefit of additional advertising $p_j < \hat{p}$ is zero. But then we must also take account of the foregone profits from the mass of uninformed buyers, so the total net marginal benefit from more advertising at low prices is non-positive. Thus, we see that even though $F(p)$ may be arbitrarily close to the degenerate distribution at $\hat{p}$, and even though the advertising effort per seller is arbitrarily small, it does not follow that selecting and advertising a price below $\hat{p}$ would be profitable. Hopefully this discussion makes Theorems 4 and 5 more understandable.

We can also deduce the effect of entry on SNE profits. Expected profits per seller are $E\pi^* \equiv E\pi(P,\alpha_P) = R(P)\{\alpha_P + (1 - \alpha_P)/N\}(1 - \bar{\alpha})^{N-1} - \gamma(\alpha_P) = [\alpha_P\gamma'(\alpha_P) - \gamma(\alpha_P)] + \gamma'(\alpha_P)/(N - 1)$. Then, by Theorem 5, there is an N' such that for all $N > N'$, $\alpha_P = 0$, so $E\pi^* = \gamma'(0)/(N - 1)$. Therefore, $E\pi^* \rightarrow 0$, as $N \rightarrow \infty$. On the other hand, total industry profits, $NE\pi^*$, converge to $\gamma'(0) > 0$.

V.2. Multiplicative Changes in Advertising Costs

To examine the effects of advertising costs, we consider multiplicative changes in the advertising cost function $\gamma(\alpha,M)$. Fix some $\gamma_0(\alpha)$ and let $\gamma(\alpha) \equiv \lambda\gamma_0(\alpha)$ for $\lambda > 0$.

Theorem 6. As $\lambda \to 0$, $\bar{\alpha} \to 1$, $B \to 0$, and $F(p; r^*) \to 1$ for all $p > 0$.

In other words, advertising saturates the market and the price distribution weakly converges to the degenerate distribution at $p = 0$ (the Bertrand outcome). Note that by Eq. (1), $r > 0$, so there is a positive lower bound on P. Since $\bar{\alpha} \to 1$, it follows that $\alpha_B \to 1$, but, when $\gamma_0(1) = \infty$, α_B approaches 1 at a rate slow enough so $\lambda\gamma_0(\alpha_B) \to 0$.

As marginal advertising costs increase, eventually $\gamma'(0) \geq ((N-1)/N]R(\hat{p})$, so by Theorem 1, the SNE converges to pure monopoly pricing. In general, the SNE lies between the Bertrand outcome and the monopoly outcome.

V.3. The Effect of Search Costs

It is interesting to ask how the SNE behaves as the buyer search costs become arbitrarily small, and to compare these results with the case of very low advertising costs. Recall that without advertising (Diamond, 1971), the unique equilibrium is monopoly-pricing, so there is a sharp discontinuity at $c = 0$. With advertising, we will see that there is still a discontinuity, albeit of smaller magnitude.

Theorem 7. As $c \to 0$, (a) B and P converge to $\hat{r} \equiv R^{-1}[\gamma'(0)(N-1)/N]$, and (b) $\alpha(p) \to 0$.

In other words, the SNE price distribution converges strongly to a single atom at $\hat{r} < \hat{p}$.[11] This follows directly from Eq. (1). Then, from Fig. 1 we can see that advertising vanishes.

This result stands in sharp contrast to Theorem 4: that marginal-cost pricing prevails as advertising costs diminish. We see that as the comparative advantage in information transmission switches from sellers to buyers, sellers drastically curtail their advertising efforts.

VI. WELFARE EFFECTS

A crucial factor in determining welfare effects is the cumulative probability distribution of the price at which buyers purchase. First, the probability distribution of the lowest known price is just $1 - (1 - A(p)]^N$. But there is a probability of $(1-\bar{\alpha})^N$ that a buyer will be uninformed, in which case the buyer faces $F_0(p)$. Letting G(p) denote the total cumulative distribution of the price at which buyers purchase,

$$G(p) = 1 - [1 - A(p)]^N + (1-\bar{\alpha})^N F_0(p) \tag{16}$$

At the upper bound P, $G(P) = 1 - (1-\bar{\alpha})^N[1 - F_0(P)]$, which is identically 1, since $F_0(P) = 1$. But note that if F has an atom at P, there is also an atom in G( ) at P of mass $(1-\bar{\alpha})^N F_0\{P\}$.

Of particular interest is the behavior of G(p) as $N \to \infty$. From Lemma 8 and Theorem 5, the last term of Eq. (16) converges to zero for all $p < P$. From Eq. (6) and Lemma 10, $(1 - A(p)]^N \to \gamma'(0)/R(p)$, which is strictly less than 1 for all $p \in (B, P]$. Hence, G(p) converges to a distribution with an atom of mass $\gamma'(0)/R(P)$ at the upper bound. Indeed, from Theorem 6, there exists an N′ such that for all $N > N'$, F( ) and hence G( ) have atoms at P.

The appropriate welfare measures are consumer surplus, producer surplus, and total surplus. Let $CS(P) \equiv \int_P^\infty D(p)dp$. Then, we can express expected consumer surplus as

$$ECS \equiv CS(P) + \int_B^P D(p)G(p)dp. \tag{17}$$

Expected producer surplus is just aggregate profits $NE\pi$. Expected total surplus can be expressed as

$$ETS \equiv CS(P) + R(P) - \int_B^P pD'(p)G(p)dp - N\int_B^P \gamma(\alpha(p)]dF(p). \tag{18}$$

We can see that anything that causes G(p) to increase (in the sense of stochastic dominance) will unambiguously increase ECS. However, except for a very special case, the effects on $NE\pi$ and ETS are far from obvious.

Consider the special case of unit demand for prices less than or equal to $\hat{p}$, and zero demand otherwise. Then, since $D' = 0$, G(p) is irrelevant to ETS. In fact, it is immediate from Eq. (18) that the social optimum is a zero level of advertising. The price merely allocates the maximum total surplus between buyers and sellers. Therefore, we can conclude that there is *over-advertising* in the SNE in the special case of unit demands. It is also clear that the demands could be perturbed slightly so $D'(p) < 0$, and still preserve the over-advertising conclusion.

To derive more definite welfare results, we turn to the case of linear advertising costs. The solution for the finite case with linear advertising costs was derived in section IV.1. Recall that there is an atom in the SNE price distribution F( ) at P of mass $f \equiv (1-\bar{\alpha})^{N-1} = \{N\gamma'/[(N-1)R(P)]\}^{1/(N-1)}$ at P.

First, consider entry of sellers. A key factor is how the mass at P in G( ) behaves. This mass is equal to $(1-\bar{\alpha})^N = \{N\gamma'/[(N-1)R(P)]\}^{N/(N-1)}$. We should like to know whether this atom is increasing or decreasing in mass as a function of N. It is straightforward to verify that $(1 - \bar{\alpha})^N$ is increasing in N for all $N \geq 2$ if $\gamma'/R(P) \leq 1/2e \simeq 0.184$. In other words, if the marginal advertising costs are small in comparison to monopoly revenues per buyer, then seller entry makes it more likely that buyers will be uninformed. Consequently, some buyers may be made worse off.

For example, if $D(p) = 1 - p$, and $\gamma' \leq 0.0125$, then ECS, $NE\pi$ and ETS are monotonically decreasing and total industry advertising expenditures, $NE\gamma$, are increasing in N for all $N \geq 2$. For $\gamma' = 0.125$, $NE\pi$, ECS and ETS are decreasing while $NE\gamma$ is increasing for all $N \geq 2$. For intermediate ranges of γ', $NE\pi$, ETS and $-NE\gamma$ are decreasing, while ECS first increases and then decreases in N. In other words, while buyers can sometimes benefit from entry, advertising expenditures typically increase more than enough to offset the buyers' benefits, resulting in a decline in social welfare.

Second, consider multiplicative changes in advertising costs. It is obvious that the mass is increasing in advertising costs, γ'. Indeed, since F( ) collapses to the degenerate distribution at $\hat{p}$ for sufficiently large γ', G( ) also collapses to the monopoly price. Thus, for all $p < P$, G(p) is decreasing, so ECS is also decreasing. Total profits, $NE\pi = R(P)f^{N-1} = N\gamma'/(N - 1)$, so total profits are increasing with γ'. Total advertising costs $N\int\gamma[\alpha(p)]dF = N\gamma'(1 - f)$. Using the expression for f and differentiating with respect to γ', we find that total advertising costs are increasing (decreasing) as $\gamma'/R(P) < (>) \ (1 - 1/N)^N < 1/e$.

To study the welfare effects of search costs, we need to subtract c from ECS. Then, we can write $d(ECS\text{-}c]/dc = \int_B^{\hat{p}} D(p)\,[\partial G(p)/\partial c]dp - 1$. Trivially, when advertising costs are linear, since the SNE is independent of c, buyers are unambiguously better off with lower search costs. The conclusion is not so immediate when advertising costs are strictly convex. From Theorem 4, we know that the SNE concentrates mass at $\hat{r}$ and $\alpha \rightarrow 0$. Hence for $c \approx 0$, from Eq. (16), $\partial G/\partial c \approx \partial F_0/\partial c$, and from Eq. (1), $\int_B^{\hat{p}} D(p)[\partial F_0/\partial c] - 1 < 0$. Thus, buyers benefit from reductions in search costs when search costs are already very low. On the other hand, ECS-c is not necessarily monotonic in c. Indeed, the effect of advertising cutbacks and a rising lower bound on prices (B) can result in buyers being worse off after an exogenous reduction in search costs.

VII. CONCLUSION

We have studied a model of seller pricing behavior in which buyers can become informed through seller advertising and direct search. Thus, the amount of information buyers have is an endogenous phenomena. Our representation of advertising is very general and can encompass word-of-mouth and other leakage effects beyond simple broadcasting of ads. In particular, this model is well-suited to represent e-commerce. We completely characterized the Symmetric Nash Equilibria.

If the marginal advertising cost is too high, then the monopoly price with zero advertising is the unique SNE; otherwise, there is a mixed-strategy SNE which can be expressed as a price distribution and an advertising policy conditional on price. The price distribution has a connected support, and is atomless below the highest price in the support. There can be an atom at the highest price in the support, in which case, it is optimal to refrain from advertising that price. The advertising effort is decreasing in price, so lower prices are advertised more intensively than higher prices.

As advertising costs become negligible, the SNE price distribution converges to the Bertrand price, and as advertising costs increase, the SNE price distribution converges to the monopoly price. On the other hand, as search costs become negligible, the SNE price distribution converges to a price higher than marginal production costs. Thus, the two information channels have different effects on the outcome.

The comparative statics on the effect of the number of sellers is most interesting. There is a finite number N' such that if there are more sellers than N', then there is an atom at the monopoly price. As the number of sellers increases without bound, advertising per seller vanishes and the SNE price distribution converges to the monopoly price.

To analyze the welfare effects, we derived the distribution of the price at which sales actually occur. This distribution is considerably skewed towards lower prices since it is a high order statistic. In particular, as the number of sellers increases without bound, this distribution converges to a non-degenerate distribution with a mean less than the monopoly price; on the other hand, the limit distribution has an atom at the monopoly price. We presented examples for a variety of demand functions and linear advertising costs in which welfare declined with entry of sellers.

The question of the efficiency of the SNE advertising effort is difficult to fully analyze. In the case of constant demand, there clearly is over-advertising. For other demand functions, there can be either over- or under-advertising.[12]

Comparing this chapter with Stahl (1989), both find that pricing becomes more monopolistic with entry, and both find that welfare can decrease with entry. Thus, *allowing sellers to advertise does not induce more competitive pricing*. As the number of sellers increases, a positive proportion of the buyers remain uninformed, so the asymptotics are similar.[13]

These results stand in sharp contrast to Cournot models (e.g. MasColell, 1982) and to Bertrand models with and without capacity constraints (e.g. Allen

& Hellwig, 1986) which find that prices become more competitive as the number of firms increases. The crucial ingredient in the model of this chapter is costly buyer search. We know from Diamond (1971), that costly buyer search can lead to monopoly pricing. However, in our model we have neutralized the force of that argument by introducing informed consumers (exogenously as in Stahl, 1989, or endogenously via advertising as in this chapter). Indeed, given a finite number of sellers and advertising costs not too high, the Nash Equilibrium always exhibits price dispersion. What is remarkable is that as the number of sellers increases, pricing becomes asymptotically monopolistic. By analyzing the asymptotic behavior in some detail, we have learned that advertising per seller is curtailed with seller entry because the net marginal benefits to advertising efforts shrink to zero. At the same time, aggregate industry-wide advertising stays bounded above zero, which reinforces the shrinking benefits to advertising by an individual seller, because that seller gains nothing unless he is the lowest-priced seller about whom a typical buyer is informed. In other words, our model demonstrates how a seemingly more competitive environment (i.e. more sellers) can actually lead to less competitive behavior because the incentives for competitive pricing is significantly lessened by the larger number of competitors.

It is also insightful to compare these results with results from repeated games in which monopoly pricing can be maintained as a dynamic Nash Equilibrium with, for example, trigger strategies (e.g. Friedman, 1986). Essentially, the threat of reversion to competitive behavior is used to deter firms from deviating from tacit collusion. Our result has some of this flavor (in that the 'heat' of competition acts to deter competitive advertising and pricing), but there are more important differences. First, a grim trigger strategy supporting collusion can succeed in deterring all competitive instincts. However, in our model, only individual competitive behavior is deterred; the aggregate level of advertising remains positive, the lower bound on the price distribution declines, and the distribution of 'sales prices' remains dispersed. Second, the repeated game equilibrium can support joint profit maximization of the sellers, whereas in our model the sellers (in aggregate) 'waste' money on advertising. Third, tacit collusion in a repeated game seems less realistic with a very large number of firms (because of practical detection problems), but it is the large number case of our model that has near monopoly pricing. Thus, the forces leading to monopolistic pricing in our model are very different from the forces in a repeated game setting.

A major component of e-commerce that is missing from our model is the explicit existence of intermediaries who collect price information and sell it to consumers or who contract to shop for the consumer.

A major stride in analysing this important aspect of e-commerce has been recently made by Baye & Morgan (2000).

NOTES

1. In this case, α is precisely the probability of receiving an ad. However, in the general case, there may be word-of-mouth leakage to the buyers who do not receive ads, in which case α, which is the total fraction of buyers who become informed (directly and indirectly), could be greater than the probability of receiving an ad. With this interpretation, buyers who are 'uninformed' are truly uninformed both directly and indirectly prior to search, so it will be reasonable to assume that they distribute themselves uniformly among the sellers, in contrast to Butters' assumption.

2. The act of obtaining a price quotation is assumed to be a costly effort relative to the cost of the ultimate purchase transaction which is assumed to be zero. Thus, if a buyer receives an ad for a price she wants to accept, there is no additional transaction cost.

3. We are implicitly assuming that the expected value of a transaction with some seller is always at least c. For instance, the buyer may need to purchase a number of other commodities known to be available at competitive prices from all the sellers, so the expected value of a shopping trip is substantially larger than c independent of the separate expected surplus of the advertised item. Stahl (1994) considers a model in which uninformed buyers do not enter the market.

4. The support is the smallest closed set with probability one.

5. This 'no-real-search' implication is an artifact of the assumption of homogeneous search costs. Given one reservation price for all buyers, sellers will never choose higher prices, and hence, buyers never find it optimal to search further. Rob (1985) and Stahl (1996) consider models with heterogeneous search costs and find real search for low-search-cost buyers, but no search for a positive mass of high-search-cost buyers. Extending the present model to heterogeneous search costs is left for future study.

6. To see this, observe from Eq. (7), assuming $\alpha_B < 1$ that $\gamma'(\alpha_B) = R(B)(1 - y)$, where $y \equiv \gamma'(\alpha_P)/[(N-1)R(P)]$. Then, $r'(B) = R(B)\alpha'(B) - R'(B)(1 - \alpha_B)y$, so $L'(B) = R'(B)[1 + y(1 - \alpha_B)/\alpha_B] - L(B)\alpha'(B)/\alpha_B$, which is strictly positive for all B satisfying $L(B) = 0$. But since $L'(B) > 0$ for all roots of $L(\,)$, there can be only one root. If $\alpha_B = 1$ and Eq. (7) does not hold as an equality, then $L'(B) = R'(B) > 0$, so again B is unique.

7. Given strictly convex advertising costs, it is clear from Fig. 1 that $B_0 < P$, since $\alpha_P < \bar{\alpha} < 1$.

8. That $\gamma'(0)$ is strictly positive is crucial to our asymptotic results. It can be easily seen from Eq. (14) why $\gamma' = 0$ is a singular case.

9. A curious non-monotonic behavior is exhibited when $R(\hat{p})/2 < \gamma' < R(\hat{p})$. For example, suppose $\gamma = 0.75 \cdot R(\hat{p})$. By Theorem 1, pure monopoly pricing is the unique SNE for $N = 2$, 3 and 4. Then for all $N \geq 5$, there is no pure-strategy SNE, meaning that the mean price is less than $\hat{p}$. However, as $N \rightarrow \infty$, the mixed-strategy SNE converges back to the monopoly price.

10. The model of Stahl (1989) is similar under the interpretation that the proportion of zero-search-cost types is like having a group that has received ads from every seller and treating advertising as a prior decision. The result was that the SNE price

distribution converged to monopoly pricing because the rewards for price competition declined with seller entry. Now with endogenous advertising which turns out to be more intense at lower prices, the forces that drove the former result are even stronger: competition is too fierce, so sellers opt for their captive market share instead.

11. Note that this limit is *not* an equilibrium of the model with $c=0$, since clearly marginal-cost pricing is the unique equilibrium when $c=0$. Further, this result holds if and only if advertising costs are strictly convex, since as shown in Section IV.1, with linear advertising costs, $P=\hat{p}$, independent of c.

12. Grossman & Shapiro (1984) examine a Butters-type advertising model with differentiated products and also find the possibility of over-advertising. They do not, however, examine the effects of seller entry since they focus exclusively on a continuum model.

13. Recently, Robert & Stahl (1993) consider a similar model except that (i) each buyer demands one unit up to a choke price of $v>0$, and (ii) the search cost is a transportation cost and *not* a cost of obtaining a price quotation (such as a telephone call). The effect of entry in their model also leads to more monopolistic seller pricing, but the limit is not monopoly pricing. Unlike the present model with zero transportation costs and positive price quotation costs, when transportation costs are positive but obtaining price quotations is costless (above the transportation cost), there is a gap (of size c) in the support of the price distribution and no seller advertises a price above r-c. This forces more competitive pricing and keeps the average price bounded below the monopoly price. Robert (1989) analyzed a similar model to Robert and Stahl, except he assumed that initial marginal advertising costs $\gamma'(0)=0$. Also, he did not characterize the asymptotic behavior of the SNE price distribution, and he did not study the impact of entry on total surplus.

ACKNOWLEDGMENTS

The author thanks Michael Baye for encouragement and suggestions, and the Center for Economic Research, Tilburg University, The Netherlands for partial financial support.

REFERENCES

Allen, B., & Hellwig, M. (1986). Price-Setting Firms and Oligopolistic Foundations of Perfect Competition. *American Economic Review, 76*, 387–392.

Baye, M., & Morgan, J. (2000). Information Gatekeepers on the Internet and the Competitiveness of Homogeneous Product Markets. *American Economic Review*, forthcoming.

Butters, G. R. (1977). Equilibrium Distributions of Sales and Advertising Prices. *Review of Economic Studies*, *44*, 465–491.

Diamond, P. (1971). A Model of Price Adjustment. *Journal of Economic Theory, 3*, 156–168.

Freidman, J. W. (1986). *Game Theory with Applications to Economics*. Oxford.

Grossman, G, & Shapiro, C. (1984). Informative Advertising with Differentiated Products. *Review of Economic Studies*, *51*, 63–81.

MasColell, A. (1982). The Cournotian Foundations of Walrasian Equilibrium Theory An Exposition of Recent Theory. In: W. Hildenbrand (Ed.), *Advances in Economic Theory,*

Fourth World Congress of the Econometric Society (pp. 183–224). Cambridge University Press.

Rob, R. (1985). Equilibrium Price Dispersion. *Review of Economic Studies*, *52*, 457–504.

Robert, J. (1989). *Search and Price Advertising*. Department de Sciences Economiques, Universite de Montreal.

Robert, J., & Stahl, D. (1993). Informative Advertising in a Sequential Search Model. *Econometrica*, *61*, 657–686.

Rothschild, M. (1973). Models of Market Organization with Imperfect Information: A Survey. *Journal of Political Economy*, *81*, 1283–1308.

Stahl, D. O. (1996). Oligopolistic Pricing with Sequential Consumer Search and Heterogeneous Search Costs. *International Journal of Industrial Organization*, *14*, 243–268.

Stahl, D. O. (1989). Oligopolistic Pricing with Sequential Consumer Search. *American Economic Review*, *79*, 700–712.

Stahl, D. O. (1994). Oligopolistic Pricing and Advertising. *Journal of Economic Theory*, *64*, 162–177.

APPENDIX

Proof of Lemma 1

If there is an atom in F(p; r), there is a chance that two or more sellers may be charging the lowest price. In that event, we assume that the buyer randomizes between the low-priced sellers with each having an equal chance of being chosen.For example, if there is an atom of mass $f > 0$ at price $\tilde{p}$, and if a buyer is informed about a price $\tilde{p} \leq r$ from K sellers while the information about all the other sellers is that their prices are higher than $\tilde{p}$, then each of the K sellers have a 1/K change of selling to that buyer. The probability of this event is $[\alpha(\tilde{p})f]^K \cdot [1 - A(\tilde{p}; r)]^{N-K}$. Thus, in Eq. (3)

$$\varphi(\tilde{p}) = \sum_{K=0}^{N-1} \frac{(N-1)!}{(N-1-K)!(K^{+}1)!} [\alpha(\tilde{p})f]^K \cdot [1 - A(\tilde{p}; r)]^{N-1-K}. \tag{A1}$$

Suppose F(p; r) has an atom of mass $f > 0$ at $\tilde{p}$. Let $\alpha_j = \alpha(\tilde{p}; r)$, and consider $p_j = \tilde{p} - \varepsilon$ for $\varepsilon > 0$. Then, $\lim_{\varepsilon \to 0} \varphi(p_j = [1 - A(\tilde{p}; r) + \alpha(\tilde{p}; r)f]^{N-1} =$

$$\sum_{K=0}^{N-1} \frac{(N-1)!}{(N-1-K)!K!} [\alpha(\tilde{p}; r)f]^K \cdot [1 - A(\tilde{p}; r)]^{N-1-K}. \tag{A2}$$

Hence, using Eq. (3) and (A1–2), $\lim_{\varepsilon \to 0} [E\pi(p_j, \alpha_j - E\pi(\tilde{p}, \alpha_j] =$

$$R(\tilde{p})\alpha(\tilde{p}; r) \cdot \sum_{K=0}^{N-1} \frac{(N-1)!K}{(N-1-K)!(K+1)!} [\alpha(\tilde{p}; r)f]^K \cdot [1 - A(\tilde{p}; r)]^{N21-K}$$

which is strictly positive if $\alpha(\tilde{p}; r) > 0$, implying that undercutting $\tilde{p}$ is profitable: a contradiction. Therefore, if $f > 0$, then $\alpha(\tilde{p}; r) = 0$. Q.E.D.

Proof of Lemma 2

(a) If $\alpha = 1$ for all $p \leq P$, then we have classic Bertrand price competition so F is concentrated at 0, implying that $E\pi = -\gamma(1) < 0$. Since this cannot be part of a SNE, we must have $\bar{\alpha} < 1$, in which case $E\pi(r,0) = R(r)(1 - \bar{\alpha})^{N-1}/N > 0$. Therefore, $E\pi^* > 0$.

(b) (1) If $P > r$, then a sale could occur at P only if all other sellers have $p_k = P$, and market share would be f^{N-1}/N, where f is the atom at P. But since $P > r$, all buyers will have searched all sellers and be fully informed. Therefore, a deviation to $P - \varepsilon$ for some $\varepsilon > 0$ will capture the whole market. Given $E\pi^* > 0$, this is a profitable deviation. Hence, we must have $f = 0$, and $E\pi(P, \alpha) \leq 0$, which is inconsistent with (a). Therefore, $P \leq r$.

(2) Suppose $\hat{p} < P \leq r$. In this case, for $p_j \in (\hat{p}, P)$, sales revenues are decreasing and market share is decreasing in p, so $p_j = \hat{p}$ dominates P, a contradiction. Therefore, $P \leq \min\{r, \hat{p}\}$.

(3) Suppose $P < \min\{r, \hat{p}\}$. Then, for $p_j \in (P, \min\{r, \hat{p}\})$, $E\pi(p_j, \alpha_j) = R(p_j)[\alpha_j + (1 - \alpha_j)/N](1 - \bar{\alpha})^{N-1} - \gamma(\alpha_j)$, which is increasing in p_j – a contradiction. Therefore, $P = \min\{r, \hat{p}\}$. Q.E.D.

Proof of Lemma 3

Suppose to the contrary that there exists an open interval $(p', p'') \subset [B, P]$ that is not contained in the support of $F(p; r)$. Observe that $A(p; r)$ will be constant on this open interval, so

$$E\pi(p_j, \alpha_j) = R(p_j)\{\alpha_j[1 - A(p'; r)]^{N-1} + (1 - \alpha_j)(1 - \bar{\alpha})^{N-1}/N\} - \gamma(\alpha_j),$$

which is increasing in p_j, so p' cannot be in the support. Hence, we can extend the open interval in the direction of B, leading to the conclusion that B is not in the support – a contradiction. Q.E.D.

Proof of Lemma 4

Suppose to the contrary that there exists two prices $p' < p'' \leq \hat{p}$, both in the support of $F(p; r)$ such that $\alpha(p') = 0 = \alpha(p'')$. But then $E\pi(p',0) = R(p')(1 - \bar{\alpha})^{N-1}/N < R(p'')(1 - \bar{\alpha})^{N-1}/N = E\pi(p'',0)$, a contradiction. Q.E.D.

Proof of Lemma 5

Suppose there exists an atom at $p' < P$. By Lemma 1, $\alpha(p') = 0$, and by Lemma 4, there can be no atom at P. Since $\alpha(P)$ must be chosen optimally, it follows that $E\pi(P, \alpha(P)) \geq R(P)(1 - \bar{\alpha})^{N-1}/N$, which in turn is strictly greater than

$R(p')(1-\bar{\alpha})^{N-1}/N = E\pi(p',0)$, which contradicts the optimality of p'. Q.E.D.

Proof of Lemma 6

From Lemma 4, $\alpha(p) > 0$ for all $p < P$, so the l.h.s. of Eq. (7) must be non-negative for all $p < P$. Suppose the l.h.s. of Eq. (7) is negative at $p = P$ and $\alpha_P = 0$. Then since $\gamma(\ )$ is strictly convex and the l.h.s. of Eq. (7) is continuous in p and α_j, it follows that $\alpha(p) = 0$ for some interval below P, which contradicts Lemma 4. Therefore, $\partial E\pi(p, \alpha_j)/\partial\alpha_j \geq 0$ for all $p \in [B, P]$. Next, suppose the l.h.s. of Eq. (7) is strictly positive at P. Then, $\alpha(P) = 1$, and by Eq. (9), $\alpha(p) = 1$ for all $p \in [B, P]$, implying that $\bar{\alpha} = 1$. But then using Eq. (7), $\partial E\pi(P,1)/\partial\alpha_j = -\gamma(1) < 0$, a contradiction. [Also, $E\pi(P,1) = -\gamma(1) < 0$.] Q.E.D.

Proof of Lemma 7

Note that if $\gamma'(0) > [(N-1)/N]R(P)$, then Eq. (9) cannot be satisfied, so there is no solution for $\alpha(p)$ that is compatible with a SNE. If $\gamma'(0) = [(N-1)/N]R(P)$, then it must be that $\bar{\alpha} = 0$, in which case $E\pi(\cdot, 0)$ is strictly increasing up to $\hat{p}$; hence $P = \hat{p}$, which violates the premise that $\gamma'(0) < [(N-1)/N]R(\hat{p})$. Therefore, a SNE must have $\gamma'(0) < [(N-1)/N]R(P)$. The constraint on r then follows from Lemma 2. Q.E.D.

Proof of Theorem 1

(a) Given a pure-strategy SNE at price P, then by Eq. (1), $r > P$, so by Lemma 2 we must have $P = \min\{r, \hat{p}\} = \hat{p}$. Moreover, by Lemma 1, $\alpha(\hat{p}; r) = 0$, and, by Eq. (4), $\bar{\alpha} = A(\hat{p}; r) = 0$. Hence, $E\pi(p_j, \alpha_j) = R(p_j)\{\alpha_j + (1-\alpha_j)/N\} - \gamma(\alpha_j)$. Then $\partial E\pi/\partial p_j = R'(p_j)\{\alpha_j + (1-\alpha_j)/N\}$, which is increasing up to $\hat{p}$. Further, $\partial E\pi/\partial\alpha_j = [(N-1)/N]R(\hat{p}) - \gamma'(\alpha_j)$. Hence, $\alpha(\hat{p}; r) = 0$ is optimal iff $\gamma'(0) = [(N-1)/N]R(\hat{p})$.

(b) Suppose $\gamma'(0) \geq [(N-1)/N]R(\hat{p})$ and there exists a strictly mixed-strategy SNE. By Lemma 3, the support is a connected interval $[B, \hat{p}]$. Hence, first-order conditions must hold at $\hat{p}$. Differentiating Eq. (5) with respect to α_j and evaluating at $\hat{p}$ yields: $[(N-1)/N]R(\hat{p})(1-\bar{\alpha})^{N-1} = \gamma'[\alpha(\hat{p}; r)] \geq \gamma'(0)$, which violates the premise unless $\bar{\alpha} = 0 = \alpha(\hat{p}; r)$. But then, $\partial E\pi/\partial p_j = R'(p_j\{\alpha_j + (1-\alpha_j/N\}$, which is increasing up to $\hat{p}$. Thus, contradicting the supposition that $B < \hat{p}$. Q.E.D.

Proof of Theorem 2

First, consider the case with $r > R^{-1}[\gamma'(0)N/(N-1)]$. The discussion following Eq. (9) and Fig. 1 established that $\alpha_P \in [0, a_L]$. Since $a_L < 1$, by Eq. (9), $\alpha(p; r, \alpha_P) > \bar{\alpha}_P$ for all $p < P$. Observing that $\bar{\alpha}(a_L) = a_L$, if $\alpha_P = a_L$, then

$\alpha(p; r, a_L) > \bar{\alpha}(a_L)$ for all $p < P$. Hence, $F' < A'/\bar{\alpha}(a_L)$, and integrating: $F(p; r, a_L) < A(p; r, a_L)/\bar{\alpha}(a_L) = 1$. Therefore, there must be an atom at P. But by Lemma 1, α_P must be zero, which it is not. Therefore, $\alpha_P = a_L$ is not a SNE solution, and $F(p; r, a_L) < 1$. By continuity of $F(P; r, \cdot)$, there exists an $\varepsilon > 0$, such that $\alpha_P \in (a_L - \varepsilon, a_L)$ is also not a solution and $F(p; r, \alpha_P) < 1$. As α_P decreases from a_L towards zero, if there is an $\alpha_P \in (0, a_L)$ such that $F(p; r, \alpha_P) = 1$, then that α_P defines a SNE. If no such α_P exists, then $\alpha_P = 0$ defines the SNE with an atom at P, consistent with Lemma 4.

Second consider the case with $r = R^{-1}[\gamma'(0)N/(N-1)]$. If $P = \hat{p}$, which exceeds r, then the argument above applies. Suppose $P = r$. Then, from Eq. (9) it must be that $\alpha_P = \bar{\alpha} = 0$. But from Lemma 4, $\alpha(p) > 0$ for all $p < P$, which implies that $F(p) = 0$ for all $p < P$; that is, $B = P$. Also, from Eq. (6), $[1 - A(p; r, 0)]^{N-1} = \gamma'(0)\{1/R(p) + 1/[(N-1)R(P)]\} = N\gamma'(0)/[(N-1)R(P)] = 1$, where the strict inequality applies for all $p < P$, which implies that $B = P$. Hence, in this case Eqs (7–14) uniquely generate a pure-strategy SNE at $R^{-1}[\gamma'(0)N/(N-1)]$. [Of course, the given r is not consistent when $c > 0$, but we are not imposing consistency in this theorem.] Q.E.D.

Proof of Theorem 3

Because of the potential non-uniqueness of the α_P in Theorem 2, we will find a consistent α_P^* and r^* simultaneously. Let $P' \equiv R^{-1}[\gamma'(0)N/(N-1)]$, and take $\gamma'(0) < [(N-1)/N]R(\hat{p})$, so $P' < \hat{p}$, and no pure-strategy SNE exists. Consider the following mappings:

(i) $P_n = \min\{\hat{p}, \max\{P', P_{n-1} - H[P_{n-1}, c, F_0(\cdot; P_{n-1}, \alpha_{P,n-1})]\}\}$, and
(ii) $\alpha_{P,n} = \min\{\max\{\alpha_{P,n-1} + F_0(\cdot; P_{n-1}, \alpha_{P,n-1}) - 1, 0\}, a_L(P_{n-1}\}$.

[From the definition of a_L, it depends only on R(P).] Each mapping is continuous from a compact convex set into itself: $[P', \hat{p}]$ and [0, 1] respectively. Therefore, by Brower's Fixed Point Theorem there exists a fixed point (P^*, α_P^*). (a) Note from (i) that if $H[P^*, c, F_0(\cdot; P^*, \alpha_P^*)] > 0$, then $P^* = P'$; but by Theorem 2, $\alpha_P^* = 0$ and $F_0(\cdot; P^*, \alpha_P^*)$ has all mass concentrated at P', so $H[0, c, F_0(\cdot; 0, \alpha_P^*)] = -c$, a contradiction. Thus, we have $H[P^*, c, F_0(\cdot; P^*, \alpha_P^*)] \leq 0$ and $P^* > P'$; and observe that if $H[P^*, c, F_0(\cdot; P^*, \bar{\alpha}^*)] < 0$, then $P^* = \hat{p}$. (b) Note from (ii) that if $F_0(P^*; P^*, \alpha_P^*) > 1$, then $\alpha_P^* = a_L(P^*) > 0$; but this contradicts the fact that $F(P^*; P^*, a_L) < 1$ [from the proof of Theorem 2]. Thus, we have $F_0(P^*; P^*, \alpha_P^*) \leq 1$; and observe that if $F_0(P^*; P^*, \alpha_P^*) < 1$, then $\alpha_P^* = 0$ as required. If $P^* < \hat{p}$, so $H[P^*, c, F_0(\cdot; P^*, \alpha_P^*)] = 0$, then $r^* = P^*$ is a consistent buyer reservation price. If $H[P^*, c, F_0(\cdot; P^*, \alpha^*_P)] < 0$, then the consistent buyer reservation price r^* exceeds $P^* = \hat{p}$, and that is consistent. Q.E.D.

Proof of Lemma 8

That $\bar{\alpha}$ and α_P converge to 0 can be easily seen from Fig. 1 or Eq. (9). Then, using Eq. (9), $(1 - \bar{\alpha})^{N-1} \to \gamma'(0)/R(P)$. Observe that for sufficiently large N, $(1 - \bar{\alpha})^{N-1}$ is arbitrarily close to $\exp\{-N\bar{\alpha}\}$. Q.E.D.

Proof of Lemma 9

That $\alpha(p) \to 0$ follows because by Lemma 8 the right-hand-side of Eq. (8) converges to zero. For the second assertion, note from Eq. (8) that for $\alpha \approx 0$, $\gamma''\alpha^2 \approx [1 - R(P)/R(p)]\gamma'(0)/[NR(P)]$ for $p < P$. Therefore, $N\alpha(p) \to \infty$. Q.E.D.

Proof of Lemma 10

(a) Referring to Eq. (6), it will suffice to show that the right-hand side converges to $\gamma'(0)/R(p) > 0$. By Lemma 9, $\alpha(p) \to 0$, so for sufficiently large N, we can write $\gamma(\alpha) \approx \alpha\gamma'(0)$. Thus, the second term on the r.h.s. of Eq. (6) is approximately $[(\alpha - \alpha_P)\gamma'(0)]/[R(p)\alpha]$. In the first term, we first use Eq. (9) to substitute for $(1 - \bar{\alpha})^{N-1}$, and then using Lemma 9, all the terms converge to zero as $N \to \infty$ except for one which asymptotically approaches $(\alpha_P/\alpha) \cdot \gamma'(0)/R(p)$. Hence, together the r.h.s. of Eq. (6) asymptotically approaches $\gamma'(0)/R(p) > 0$ as claimed.

(b) Next observe that if there is a scalar $A > 0$ such that $A(p) = A$ for all N, then $[1 - A(p)]^{N-1}$ will converge to zero, a contradiction. Q.E.D.

Proof of Theorem 4

By definition, $\bar{\alpha} = \int_B^P \alpha(p)dF(p)$. Now multiply both sides by N. From Lemma 8, the left-hand side converges to $\ln[R(P)/\gamma'(0)]$. Given Lemma 8, if F(p) does not converge to zero for some $p < P$, then the right-hand side diverges to ∞, a contradiction. For the second assertion, use Lemmas 6 and 8 in Eq. (10) to get that in the limit $R(B) = \gamma'(0)$. Q.E.D.

Proof of Theorem 5

(a) Using Theorem 4 and Eq. (1), it follows that $H(r, c, F_0) \to -c$ as $N \to \infty$. Therefore, there is an N' such that for all $N > N'$, $H(r, c, F_0) < 0$ for all $r \in [0, P]$, so the reservation price $r^* > P$. Then, by Lemma 2, $P = \hat{p}$.

(b) By Theorem 4, $B \geq R^{-1}[\gamma'(0)]$, so R'/R is finite for all $p \in [B, P]$. Referring to Eq. (12), we see that first term in {} is $(1 - A)/(N - 1)$. Given Lemmas 6 and 8, the second term in {} asymptotically approaches $(1 - \alpha)/[(N - 1)^2\alpha]$. Recalling that $F'(p) = [\partial A/\partial p]/\alpha$, it follows that F' asymptotically approaches $[R'(p)/R(p)]\cdot\{N\alpha+1-\alpha\}/(N\alpha)^2$, which by Lemma 9 converges to zero for all $p<P$. Therefore, $\int_B^{P-} F'(p)dp \to 0$, as $N\to\infty$, implying that $F\{P\}>0$ for sufficiently large N. It follows from Lemma 5 and Eq. (9) that $\alpha_P^*=0$, and hence, there is a unique SNE for sufficiently large N. Q.E.D.

Proof of Theorem 6

As $\lambda\to 0$, Fig. 1 reveals that both a_L and a_H converge to one, so $\bar{\alpha}\to 1$ and $\gamma'(\alpha_P)\to 0$. Since $\bar{\alpha}\leq 1-F\{P\}$, it follows that $F\{P\}\to 0$. Using Eq. (9) in Eq. (8), the λ's can be eliminated by division, so the solution $\alpha(p; r, \alpha_P)$ is independent of λ. Next, using Eq. (9) in Eq. (6), as $\lambda\to 0$, the right-hand-side of Eq. (6) becomes arbitrarily close to zero; hence, $A(p; r, \alpha_P)$ becomes arbitrarily close to 1 for all p, implying that $\partial A(p; r, \alpha_P)/\partial p$ is arbitrarily close to zero for all $p>B$. Consequently, using Eq. (13–14) and the fact that $F\{P\}\to 0$, for any α_P and any $p>B$, $1-F(p; r, \alpha_P)$ is arbitrarily close to zero. Further, from Eq. (10), one can see that $B\to 0$. Therefore, $F(p; r, \alpha_P)\to 1$ for all $p>0$. Q.E.D.

Proof of Theorem 7

(a) By Lemma 7, $P=\hat{r}$. Suppose $P\to P'>\hat{r}$. If there is no atom at P', then clearly $B\to B'<P'$; if there is an atom at P', then $\alpha_{P'}=0$, so by Eq. (10) and the hypothesis that $P'>\hat{r}$, we have $B\to B'<P'$. Given $\gamma(\,)$ is strictly convex, it follows that $B_0'<P'$; hence, $H(P', F_0)=\int_{B'}^{P'} D(p)F_0(p)dp>0$, which implies that $r<P'$, a contradiction. Therefore, $P\to\hat{r}$. Using Eq. (8–9), it can be shown that $B\to\hat{r}$.

(b) From Fig. 1, it is clear that α_P and $\bar{\alpha}\to 0$, since $P\to\hat{r}$. Q.E.D.

STRATEGIC VERTICAL CONTRACTING WITH ENDOGENOUS NUMBER OF DOWNSTREAM DIVISIONS

Kamal Saggi and Nikolaos Vettas

ABSTRACT

We characterize vertical contracts in oligopolistic markets where each upstream firm may contract with multiple downstream firms (divisions). If the number of divisions is chosen before the terms of fee-and-royalty contracts, each upstream firm minimizes the number of its downstream firms. When the contracts are chosen sequentially, it is the leader that minimizes the number of divisions. Lastly, if the contracts are chosen before the number of divisions, or if the contracts involve either only fees or only royalties, then upstream firms do not have an incentive to minimize the number of their respective divisions.

I. INTRODUCTION

This chapter examines vertical contracts in oligopolistic markets, focusing on strategic considerations at both the 'upstream' and the 'downstream' stage. In particular, upstream firms make two key strategic decisions in our model: each upstream firm chooses the number of its independent downstream firms (or divisions) as well as the terms of their vertical contracts with those firms. We consider vertical contracts that specify a fixed fee and a per-unit payment which we call royalty (this can also be viewed as a wholesale price). Later in the

Industrial Organization, Volume 9, pages 101–131.

ISBN: 0-7623-0687-4

chapter, we also consider contracts that involve either only a fixed fee or only a royalty.

Past work that endogenizes the number of downstream firms, has shown that upstream firms may prefer to have a higher number of downstream firms for strategic reasons. In particular, Baye, Crocker & Ju (1996) demonstrate that, in a two-stage game where upstream firms choose the number of their downstream divisions and quantity competition follows in the downstream market, there is an incentive to increase the number of downstream divisions since this practice represents a commitment to more aggressive downstream behavior, a 'divide and conquer' strategy (see also Corchón, 1991; Polasky, 1992).[1] In earlier work (Saggi & Vettas, 1999) we have shown that allowing for contracts that involve both fees and per-unit payments dramatically alters the strategic incentives of upstream firms. In particular, when fee-and-royalty contracts are chosen after the number of downstream firms, each upstream firm chooses to have only *one* downstream firm. In the present chapter, we extend our previous analysis in important ways: here, we consider alternative timing of the strategic choices as well as alternative contracting arrangements.

Section II presents the basic model that corresponds to a three-stage game. Two upstream firms choose the number of their independent (i.e. maximizing own profit) downstream firms and then choose their royalty rates (and fees). In the last stage, downstream firms compete in quantities. Demand is parameterized in a way that allows us to consider how competition changes as we move from perfect substitutes to upstream monopoly. As a first step, in Section III we take the number of downstream firms as given and examine the last two stages of the game. The goal is to characterize contracts and competition with an arbitrary number of downstream firms per upstream firm. At this point, our analysis employs the concept of a *brand reaction curve* that indicates the aggregate output of all downstream firms that contract with the same upstream firm, taking as given the output of all rival downstream firms. When choosing royalty rates, an upstream firm would like its downstream firms to be more *passive* against one another but more *aggressive* against rival downstream firms. In equilibrium, the larger the number of its own downstream firms relative to that of rival firms and the higher the degree of product differentiation, the higher the royalty rate chosen by an upstream firm. More generally, higher competition between rival upstream firms and lower competition between downstream firms contracting with the same upstream firm tends to decrease royalty rates.[2]

To see the key tension captured by our model, it is useful to review previous work that examines two related scenarios that correspond to two extreme cases in our framework. Consider first an upstream monopoly that contracts with multiple oligopolistic downstream firms. In this situation, the upstream firm can extract the entire downstream monopoly profit by using a fixed fee and a royalty rate. The royalty forces the downstream firms to internalize the horizontal externality that exists among them and raises the market price from the oligopoly to the monopoly level, while the fee transfers the monopoly profit upstream. Essentially, the upstream firm shifts the downstream firms' reaction functions and makes them more passive against one another. This intuition is central in papers such as Dixit (1983) and Mathewson & Winter (1984).[3] Next, consider two upstream firms each contracting with a single downstream firm. In contrast to the previous case, now, under Cournot competition, each upstream firm would like its downstream firm to be more aggressive against the rival downstream firm. It follows that the equilibrium involves non-positive royalties.[4] This result is known from the 'strategic contracting and delegation' literature which includes, for example, Brander & Spencer (1985), Bonanno & Vickers (1988), Fershtman & Judd (1987), McGuire & Staelin (1983), Sklivas (1987) and Vickers (1985). The central insight of this research is that competing principals have a unilateral incentive to make their agents commit to more aggressive behavior (and, in our framework, this commitment could take place through the choice of lower royalties). When each upstream firm contracts with multiple downstream firms, forces related to both upstream as well as downstream competition are responsible for shaping vertical contracts.

Section IV reviews the main result from Saggi & Vettas (1999), since this is the starting point for this chapter. There, we showed that each upstream firm prefers to minimize the number of its own downstream firms. The logic of this result is as follows. In our framework, competition and marginal incentives in the downstream market can be controlled either through the choice of number of firms or through the choice of royalty rates. However, a larger number of downstream firms implies that, for any royalty rate, an upstream firm's brand reaction function becomes steeper so that the rival has a stronger incentive to behave strategically.

In Section V we consider the case where firms choose royalties (and fees) sequentially rather than simultaneously. We find that the result mentioned above has to be modified only slightly: upstream firms have a strict incentive to minimize the number of their downstream firms when, in the subsequent stage, they pick their royalties simultaneously or before their rival, whereas when they choose royalty rates after their rival their profit does not depend on the number of their downstream firms. We also find that, if both upstream firms have the same number of downstream firms, the leader (the upstream firm that chooses first) chooses a lower royalty rate while the follower chooses a higher royalty relative to their respective royalties under simultaneous choice.

It is crucial, however, whether the choice of royalties follows that of the number of firms or not. In Section VI we consider alternative scenarios such as fee-only contracts or royalties chosen before (or simultaneously with) the number of firms. Then, with some qualification, the logic of previous work (Baye, Crocker & Ju, 1996) applies and each firm prefers a higher number of firms. In other words, we find that as long as royalties are viewed as given at any level that is not 'too low' (not just at zero), the strategic incentive for commitment via a higher number of firms is present. This result requires products to be close enough substitutes: we also generalize previous work that considers only fees by allowing the degree of differentiation between products to vary (as products become more differentiated a lower number of downstream firms, instead of a higher, becomes profitable).

In addition to royalties not being fixed when the number of firms is chosen, it is also critical that contracts involve not only royalties but fees as well. Section VII explores the equilibrium when only royalties can be charged. Then the result that both upstream firms prefer to minimize the number of their downstream firms is no longer true. In fact, assuming that one upstream firm chooses a single downstream firm, the other upstream firm has an incentive to increase its number of firms without end. We also compare the royalty-only and the fee-and-royalty cases. For a fixed number of downstream firms, upstream firms may be better off with royalty-only contracts than with fee-and-royalty contracts, the intuition being that competition with fee-and-royalty contracts implies royalty rates that are too low from the viewpoint of joint profit maximization. If now firms could choose the type of their contracts, they may find themselves in a prisoners' dilemma situation (with both firms better-off under royalty-only contracts but having a unilateral incentive to pick fee-and-royalty contracts). This result is already present in previous work examining similar situations.[5] By endogenizing the number of downstream firms, our analysis shows that this prisoners' dilemma structure may not exist, since fee-and-royalty contracts imply an incentive to minimize the number of downstream firms. As noted above, this incentive is not present under royalty-only contracts.

As is clear from the above discussion, this chapter is related to a number of important literatures. In particular, it creates a link between work on strategic contracting and work exploring the choice of downstream divisions. In addition to the papers mentioned above, there are a number of other related papers and, while space does not allow us to review all of them, we discuss here the most relevant. Our chapter is related to Rey & Stiglitz (1995) who show that vertical restraints, which may affect intrabrand competition, can be used to reduce interbrand competition. They consider exclusive territories which, like picking a single downstream firm, eliminate downstream competition. However, the two analyses are very different. First, Rey & Stiglitz (1995) do not examine downstream oligopolistic competition within the same brand, they only compare perfect competition and exclusive territories. As a result, they do not analyze the role of the number of downstream competitors which is at the heart of our approach. Second, they consider downstream price competition and the strategic motives are very different. They show that with exclusive territories and prices being strategic complements one's rival retailers have an incentive to increase price (as a response to one's higher price and that of one's retailers). As a result, there is a lower perceived elasticity of demand and higher prices.[6] In contrast, our result that a lower number of downstream firms is desirable requires that both fees and royalties can be used; when only royalties can be used, the result does not hold and the strategic incentives are actually reversed. In other related work, Kühn (1997) shows that wholesale prices can be used as a commitment to relax competition between upstream firms. He considers fully nonlinear wholesale schedules and finds that (with constant marginal costs) the number of downstream firms is irrelevant for strategic contracting.[7] In contrast, our analysis, under fee-and-royalty contracts, assigns a critical role to the number of downstream firms. The difference is due to the fact that with fee-and-royalty contracts upstream firms can only control the position of the downstream firms' reaction functions but with fully nonlinear contracts the slope can be controlled as well. While fee-and-royalty contracts are, of course, restrictive in that the marginal price does not vary with quantity, they are very common in many markets.[8] For this reason, and also since much of the previous work to which we want to compare our results considers fee-and-royalty contracts, it is important to understand the strategic role of the number of competitors under such contracts.

It is important to discuss at this point why upstream firms may be concerned with affecting the downstream firms' behavior through the design of vertical contracts instead of directly determining the final product prices. The reason is that this practice may not be an available option. In particular, firms are not allowed to engage in such 'resale price maintenance', which is *per se* illegal under U.S. antitrust laws.[9] Of course, in some vertical structures firms attempt to indirectly determine prices through the use of 'suggested prices' or other means.[10] Downstream firms may try to avoid following such suggestions if they are not in their individual interest and may succeed in doing so if monitoring is costly.[11]

Our analysis focuses on the strategic role of the number of downstream divisions. Of course, in reality there are additional considerations (for example, related to the technology of production or distribution, the information structure, or legal restrictions) that may affect the choice of vertical structures. We abstract from a number of issues that are, without doubt, important for vertical contracts but have been analyzed elsewhere. In particular, we do not examine the implications of uncertainty (for example, a royalty may be desirable for risk-sharing or incentives reasons even if it is not desirable for strategic reasons). Allowing for other such factors may, of course, modify the results.[12]

II. THE BASIC MODEL

There are two upstream firms, A and B, each with $n_i \geq 1$ downstream firms, $i = A, B$.[13] To simplify the analysis, we assume that production cost is zero both upstream and downstream. To capture the idea that the product sold by the downstream firms contracting with one upstream firm may be differentiated from the product sold by firms contracting with a different upstream, we assume that the market demand function is as follows:

$$p_i = a - Q_i - sQ_j, \qquad i, j = A, B, \tag{1}$$

where

$$Q_i = \sum_{k=1}^{n_i} q_i^k, \qquad i = A, B$$

denotes the aggregate output of all the downstream firms selling product i (each downstream firm contracting with upstream firm i produces quantity q_i^k, $k = 1, \ldots, n_i$ and sells at a price p_i). The parameter $s \in [0, 1]$ in the demand function measures the degree of substitution between the two products.

Our basic model is a three-stage game. First, the two upstream firms, A and B, simultaneously choose the number of their downstream firms, $n_i \geq 1$, $i = A, B$. Then, the two upstream firms (simultaneously) make a take-it-or-leave-it offer to each of their downstream firms that specifies a pair (f_i, r_i), where f_i is a fixed fee (independent of sales level) and r_i is a royalty rate per unit sold. Thus, the total payment (to an upstream firm) of a firm that produces q units equals

$$f_i + r_i q.$$

The outside option of each downstream firm is normalized to zero and we assume that the same contract is offered to all downstream firms associated with a given upstream firm. In the third stage of the game, firms that accept the contract compete in quantities in the product market. We study the subgame perfect equilibrium of this game.

Before proceeding further, a few remarks are in order. First, note that our formulation admits as special cases situations that have been studied in the literature. If $n_A = n_B = 1$ (each upstream firm has only one downstream firm) we have the case of pure interbrand competition. This has been the focus of much of the strategic contracting literature. If, on the other hand, we have $s = 0$ (or $n_j = 0$) then there is only intrabrand competition. Furthermore, the second stage of the game can be slightly modified to isolate the role royalties and fees play in transferring profits (from downstream to upstream firms). In particular, later in the chapter we consider scenarios where firms are able to use only one of the two instruments. The case where only fixed fees are used corresponds to the literature where the only strategic variable is the number of downstream firms, as in Baye, Crocker & Ju (1996).

Second, the timing adopted here (i.e. upstream firms choose the number of downstream firms before royalties) reflects a situation where it is more costly to change the number of downstream firms than to change royalties (or wholesale prices) and appears to be a natural assumption in many markets. Later in the chapter we discuss in detail the implications of alternative timing assumptions. In particular, we consider the case where the number of downstream firms is chosen after the royalty rates as well as the case where the two decisions are made simultaneously (see Section VI).

Finally, note that we assume quantity competition in the downstream market. Clearly, it matters whether the strategic variables in the downstream market are prices or quantities, since downstream reaction functions are increasing in case of price competition and decreasing in case of quantity competition.[14] In our framework, quantity competition allows us to capture the idea that an upstream firm wants its downstream firms to be passive against its own firms but aggressive against rivals. Since this trade-off has important implications for the design of vertical contracts, it seems appropriate to proceed with the assumption of quantity competition in the downstream market.

III. EQUILIBRIUM CONTRACTS

We first analyze the choice of contracts and downstream quantity competition for a given number of downstream firms, n_A and n_B. In addition to being a necessary first step in the derivation of equilibrium, this analysis is also of independent interest. For example, in some markets, upstream firms may be unable to fully control the number of their downstream firms and this number may reflect a variety of factors. While introducing additional factors that influence the choice of number of downstream firms is outside the scope of the present model, it is useful to study competition when there exist multiple upstream firms each contracting with multiple downstream firms.

III.1. Derivation of the Equilibrium Royalties and Quantities Using 'Brand Reaction Functions'

Profit, net of royalty payments, for a downstream firm m that accepts a contract involving a royalty rate r_i is:

$$(a - Q_i - sQ_j - r_i)q_i^m, \qquad i, j = A, B. \tag{2}$$

where again Q_i denotes the aggregate quantity produced by all downstream firms that contract with firm i, $i = A, B$. The reaction function of downstream firm m is

$$q_i^m(Q_{i,-m}, Q_j; r_i) = \frac{1}{2}(a - Q_{i,-m} - sQ_j - r_i), \tag{3}$$

where $Q_{i,-m} \equiv \Sigma_{k=1,k\neq m}^{n_i} q_i^k$ denotes the aggregate quantity produced by all other downstream firms contracting with firm i except for firm m. At this point, the direct way forward is to calculate the equilibrium output levels (as functions of the number of firms and royalty rates),

$$q_i = \frac{a - r_i - n_j(a(s-1) + r_i - sr_j)}{1 + n_i + n_j + n_i n_j(1 - s^2)}, \tag{4}$$

and the corresponding aggregate downstream profits and then calculate the equilibrium royalty rates (see Saggi & Vettas, 1999, for details). However, here we follow a different approach that uses the concept of a *brand reaction function*. This alternative approach provides important insights concerning the strategic incentives that are exploited throughout the chapter.

Consider again the reaction function of a given downstream firm m, as given by (3). This represents, of course, the optimal output for firm m taking the output levels of *all* other firms as given. Note that each of these reaction functions has slope (relative to Q_j) equal to $s/2$. Adding (3) over m and rearranging, we obtain the brand reaction function for i:

$$Q_i(Q_j) = \frac{n_i}{n_i + 1}(a - r_i - sQ_j), \qquad i, j = 1, 2 \tag{5}$$

This function indicates the aggregate output of all of firm i's downstream firms given the output of all of firm j's downstream firms.

Remark 1. *The brand reaction function for firm i has slope $sn_i/(n_i + 1)$. Thus, for a given output Q_j of firm j, the higher is n_i the higher the aggregate output*

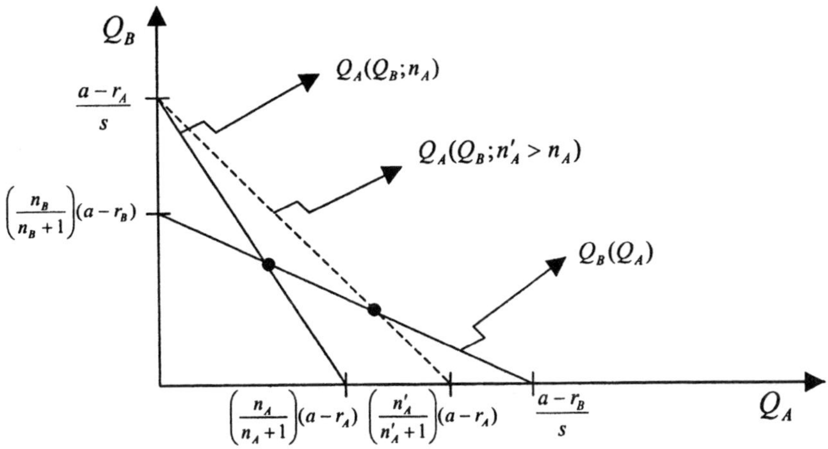

Fig. 1. Brand reaction curves.

of firm i's downstream firms. In addition, decreasing the royalty r_i generates a parallel upward shift of firm i's brand reaction function.

Figure 1 illustrates how firm *A*'s brand reaction function changes when the number of *A*'s downstream firms increases from n_A to n'_A. As n_A increases, firm *A*'s reaction function rotates and its slope $sn_A/(n_A+1)$ increases (note that with Q_B graphed against Q_A, the reaction function becomes flatter). The intuition here is that downstream firms do not internalize the effect of their output decisions on other downstream firms, including the ones selling the same brand. As n_A increases, the effect of this externality becomes stronger and downstream firms behave more aggressively. Note that this effect underlies the result that (when royalties are not used) a larger number of downstream firms offers a stronger strategic position (e.g. as in Baye, Crocker & Ju, 1996).

Returning to the construction of the equilibrium, it is clear that firm *i*, would like its downstream firms to produce an output level that maximizes their *total* profit. Thus, while choosing its royalty rate, r_i, firm *i* acts like a Stackelberg leader along *j*'s brand reaction function (taking r_j as given). In other words, it wants its downstream firms to collectively produce output Q_i that solves

$$\max_{Q_i} \{[a - Q_i - sQ_j(Q_i)]Q_i\}. \tag{6}$$

or, substituting firm *j*'s optimal response,

$$\max_{Q_i} \left\{ [a - Q_i - \frac{sn_j}{n_j + 1} (a - r_j - sQ_i)]Q_i \right\}.$$

The first-order condition (with respect to Q_i) can be written as:[15]

$$a - \frac{sn_j}{n_j + 1} (a - r_j) = \frac{2[(1 - s^2)n_j + 1]}{n_j + 1} Q_i$$

Solving for Q_i, we obtain the Stackelberg leader's output level:

$$Q_i^{SL} = \frac{[a + s(r_j - a)]n_j + a}{2[(1 - s^2)n_j + 1]}. \tag{7}$$

Substituting into firm j's reaction function, we obtain the Stackelberg follower output level:

$$Q_j^{SF} = \frac{n_j}{n_j + 1} [a - r_j - sQ_i^{SL}]$$

$$= \frac{n_j}{n_j + 1} \left\{ \frac{[(2 - s - s^2)n_j + 2 - s]a - [(2 + s - 2s^2)n_j + 2]r_j}{2[(1 - s^2)n_j + 1]} \right\}. \tag{8}$$

Now, for a given r_j, firm i wants to choose a royalty rate r_i that will make its brand reaction curve pass through the point (Q_i^{SL}, Q_j^{SF}). In other words, it chooses r_i such that

$$Q_i^{SL} = \frac{n_i}{n_i + 1} [a - r_i - sQ_j^{SF}]. \tag{9}$$

Substituting (7) and (8) into (9) and solving for r_i we obtain the reaction functions (10) of the upstream firms in terms of royalties:

$$r_i(r_j; n_i, n_j) = \frac{[n_j((s^2 - 1)n_i + 1) - n_i + 1][n_j(a(s - 1) - sr_j) - a]}{2(n_j + 1)[(1 - s^2)n_j + 1]n_i}, \qquad i, j = A, B. \tag{10}$$

Then, solving the system of the two royalty reaction functions, we find the equilibrium royalty rates:

$$r_i^* = r(n_i, n_j) = \frac{a[2 - s + n_i(2 - s^2 - s)][n_i - 1 - n_j(s^2 n_i - n_i + 1)]}{n_i[n_i n_j(s^4 - 5s^2 + 4) + (n_i + n_j)(4 - 3s^2) + 4 - s^2]}, \ i, j = A, B. \tag{11}$$

Before examining the equilibrium royalties in detail, it is useful to discuss the monotonicity of the royalty reaction functions. It is easy to see that these can be *either increasing or decreasing, depending on the numbers of downstream firms*. Moreover, it is possible that one of these functions is decreasing while the other is increasing. Specifically, standard calculations show that:

Remark 2. *(Monotonicity of royalty reaction functions): The royalty r_i is decreasing in r_j if and only if*

$$n_i < n_j[(s^2 - 1)n_i + 1] + 1. \tag{12}$$

The equilibrium royalty captures a central tension in the analysis (see Saggi & Vettas, 1999, for a detailed discussion). Increasing the royalty rates makes downstream firms more passive. This is desirable in order to control competition with other downstream firms contracting with the *same* upstream firm but undesirable from the point of view of competing with firms contracting with a *rival* upstream firm. Both of these effects need to be considered in order to determine whether positive royalties are used in equilibrium.

III.2. Monotonicity of Equilibrium Contracts and Limit Behavior

An analysis of the equilibrium contracts can be found in Saggi & Vettas (1999) so we only report here results that are needed for our subsequent analysis. First consider the two extreme cases. When $s = 0$, thc royalties are

$$r_i^*(s=0) = a(n_i - 1)/2n_i, \qquad i = A, B, \tag{13}$$

and the downstream market is driven to the monopoly price ($a/2$). When $s = 1$, the quantity for each downstream firm associated with upstream firm i can be calculated as:

$$q_i = \frac{a - r_i - n_j(r_i - r_j)}{n_i + n_j + 1}. \tag{14}$$

Further, if $s = 1$, the royalty reaction functions become

$$r_i(r_j;\, n_i,\, n_j) = \frac{(n_i - n_j - 1)(r_j n_j + a)}{2(n_j + 1)n_i} \tag{15}$$

and the equilibrium royalties are

$$r_i^*(s=1) = \frac{a(n_i - n_j - 1)}{n_i(n_A + n_B + 3)}, \qquad i, j = A, B \tag{16}$$

so that firm A employs a positive royalty if and only if $n_A > n_B + 1$. Furthermore, when $n_A > n_B + 1$ firm B does not employ a positive royalty. In particular, when $n_i = n_j = 1$ and $s = 1$, the royalties are

$$r_i^*(n_i = n_j = 1;\, s = 1) = -\,a/5 < 0, \qquad i = A, B. \tag{17}$$

If negative royalties (that is, subsidies) are not possible, the equilibrium involves zero royalties.[16]

We now examine how equilibrium royalties vary with the number of downstream firms. First, we find that a firm's royalty decreases with the number of rival downstream firms, regardless of the magnitude of s. To see this, it is enough to differentiate r_i^* from (11) with respect to n_j. We obtain

$$\frac{\partial r_i^*}{\partial n_j} = \frac{-\,2as^2(n_i+1)[(s^2+s-2)n_i+s-2][(s^2-1)n_i-1]}{n_i[(s^4-5s^2+4)n_in_j-(3s^2-4)(n_i+n_j)-s^2+4]^2} < 0. \tag{18}$$

Regarding the monotonicity with respect to own firms, the result is not unambiguous. While numerical examples indicate that royalties typically increase in the number of own firms, this result is not generally true. In particular, when $s = 1$, from (16) we obtain

$$\frac{\partial r_i^*}{\partial n_i} = \frac{a[n_j(n_j+2n_i+4)+n_i(2-n_i)+3]}{n_i^2(n_i+n_j+3)^2}. \tag{19}$$

It is easy to see, for example, that when $n_j = 1$ this expression is positive for $n_i = 5$ and negative for $n_i = 6$. Figure 2 illustrates that, as n_i increases, r_i is initially negative, increases to positive levels and then decreases converging to zero from above. We can summarize these results as follows.

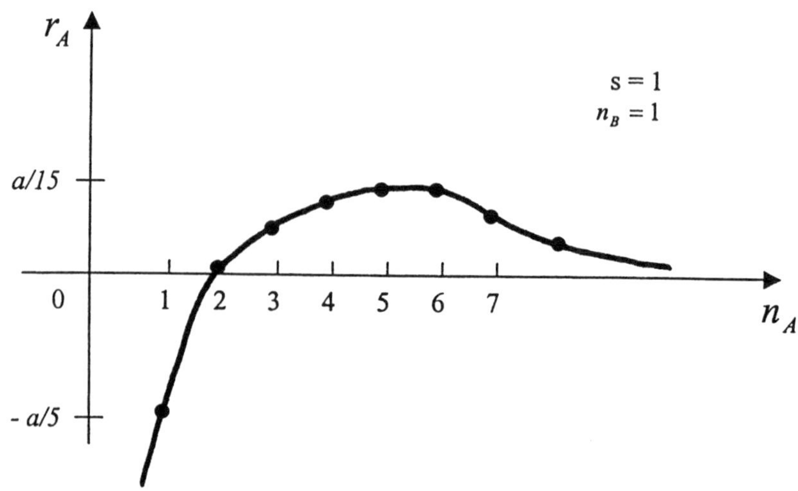

Fig. 2. Equilibrium royalty as a function of own number of firms.

Remark 3. *The equilibrium royalty of firm i (i) decreases as the number of rival downstream firms (n_j) increases, and (ii) can either increase or decrease in the number of own firms (n_i).*

Given the above discussion, it is interesting to examine the limit behavior of equilibrium royalty rates and profits. As Fig. 2 shows, when $s=1$, $r_i^* \to 0$ as $n_i \to \infty$. More generally, from (11) we find that, as the number of own downstream firms increases, the equilibrium profit converges to

$$\lim_{n_i \to \infty} \pi_i^* = \frac{a^2(s^2+s-2)^2[1+(1-s^2)n_j](n_j+1)}{[n_j(s^4-5s^2+4)+4-3s^2]^2}.$$

Note that, in the limit, profit is positive if $s<1$. If $s=1$ both firm i's royalty and profit tend to zero as $n_i \to \infty$.

IV. SIMULTANEOUS CHOICE OF THE NUMBER OF DOWNSTREAM FIRMS

Thus far, we have taken the number of downstream firms as fixed. In this section we examine the choices of n_A and n_B by the upstream firms (given that equilibrium choices of royalties and quantities will follow). Substituting the equilibrium royalty rates (11) into the downstream quantities and prices, we obtain the upstream profits. Then, each firm i chooses its number of downstream firms to maximize its upstream profit and it can be shown that:

Proposition 1. *(Saggi & Vettas, 1999) Suppose that the upstream firms choose first the number of downstream firms and then their fee-and-royalty contracts. Then, if $s>0$, in equilibrium we have $n_A^*=n_B^*=1$, so that each upstream firm chooses to have only one downstream firm. Moreover, it is strictly optimal for firm i to have only one downstream firm, regardless of the number of downstream firms chosen by firm j.*

Note that, if there is no substitution among the two products ($s=0$), we know from the analysis in the previous section that any number of downstream firms n_i^* is optimal for firm i, $i=A, B$ (and i can always achieve the monopoly profit).

The above result is the core of the analysis in Saggi & Vettas (1999) and the reader is referred to this paper for a more complete discussion of this result. The main effect at work is as follows. A higher number of downstream competitors and a lower royalty rate can be both used by an upstream firm to commit to more aggressive behavior in the downstream market. However, when choosing its number of downstream firms, an upstream firm must consider the effect on the rival's royalty choice. In particular, a larger number of downstream firms implies that, for any royalty rate, the rival has a stronger incentive to behave strategically (by choosing a lower royalty rate). Since it is desirable to have higher rival royalty rates, each firm minimizes its number of downstream competitors (and then behaves aggressively in terms of royalties given the number of downstream firms).

A simple example helps illustrate the effect of increasing the number of downstream firms. Suppose that $s=1$ and $n_B=1$. Table 1a reports how the equilibrium changes as n_A increases from 1 to 2, and Fig. 3a provides a diagrammatic illustration.

V. SEQUENTIAL CHOICES OF ROYALTIES

We now consider a modification of the above model, where the two firms choose their royalty rates *sequentially*. This case is of interest since some firms,

Table 1. Numerical example: n_A increases from 1 to 2 (with $n_B=1$ and $s=1$)

n_A	n_B	r_A	r_B	q_A	$Q_A=n_Aq_A$	$q_B=Q_B$	p	π^A	π^B
1	1	$-\frac{a}{5}$	$-\frac{a}{5}$	$\frac{2a}{5}$	$\frac{2a}{5}$	$\frac{2a}{5}$	$\frac{a}{5}$	$\frac{2a^2}{25}$	$\frac{2a^2}{25}$
2	1	0	$-\frac{a}{3}$	$\frac{a}{6}$	$\frac{a}{3}$	$\frac{a}{2}$	$\frac{a}{6}$	$\frac{a^2}{18}$	$\frac{a^2}{12}$

1a: Fee-and-Royalty Contracts

n_A	n_B	r_A	r_B	q_A	$Q_A=n_Aq_A$	$q_B=Q_B$	p	π^A	π^B
1	1	—	—	$\frac{a}{3}$	$\frac{a}{3}$	$\frac{a}{3}$	$\frac{a}{3}$	$\frac{a^2}{9}$	$\frac{a^2}{9}$
2	1	—	—	$\frac{a}{4}$	$\frac{a}{2}$	$\frac{a}{4}$	$\frac{a}{4}$	$\frac{a^2}{8}$	$\frac{a^2}{4}$

1b: Fee-Only Contracts

n_A	n_B	r_A	r_B	q_A	$Q_A=n_Aq_A$	$q_B=Q_B$	p	π^A	π^B
1	1	$\frac{a}{3}$	$\frac{a}{3}$	$\frac{2a}{9}$	$\frac{2a}{9}$	$\frac{2a}{9}$	$\frac{5a}{9}$	$\frac{2a^2}{27}$	$\frac{2a^2}{27}$
2	1	$\frac{7a}{22}$	$\frac{3a}{11}$	$\frac{7}{44}a$	$\frac{7}{22}a$	$\frac{9}{44}a$	$\frac{21}{44}a$	$\frac{49}{484}a^2$	$\frac{27}{484}a^2$

1c: Royalty-Only Contracts

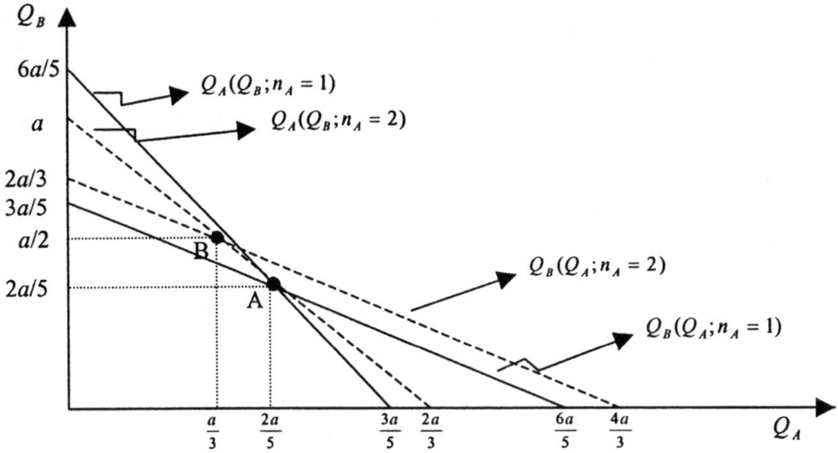

(a)

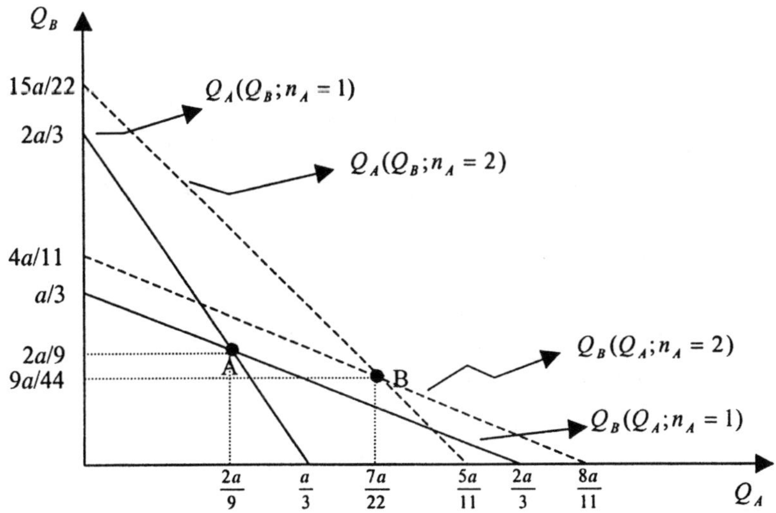

(b)

Fig. 3. (a) Fee-and-royalty contracts: equilibria when n_A increases from 1 to 2 (with $n_B = 1$ and $s = 1$); (b) Royalty-only contracts: equilibria when n_A increases from 1 to 2 (with $n_B = 1$ and $s = 1$).

because of their stronger position in the market or for other reasons, may have the ability to commit to their vertical contracts before their rivals choose theirs. Moreover, the study of this case allows us to better understand the strategic relation between the choices of royalty rates and the number of downstream firms.

To simplify the exposition, we assume that $s=1$. Given the number of firms and the royalty rates, the output of a downstream firm associated with upstream firm i is again given by (14). As before, the total profit of firm A (after it collects both fees and royalties) if it chooses royalty r_A and B chooses r_B is $\pi^A(r_A, r_B)=pn_Aq_A=(a-n_Aq_A-n_Bq_B)n_Aq_A$ and maximization gives rise to the reaction functions (15).

Now, instead of solving the system of these two reaction functions, as we did before when royalties were chosen simultaneously by the two firms, we have to allow for the fact that the first mover (say firm A) anticipates how B's royalty choice depends on its own choice. Thus, firm A solves

$$\max_{r_A} \pi^A(r_A, r_B(r_A)). \tag{20}$$

Substitution of (15) into the profit expressions yields

$$\pi^A(r_A, r_B(r_A))=\pi^A\left(r_A, \frac{(n_B-1-n_A)(n_Ar_A+a)}{2n_B+2n_An_B}\right)=\frac{n_A(-n_Ar_A-2r_A+a)(n_Ar_A+a)}{4(n_A+1)^2}. \tag{21}$$

Note that n_B drops out from the expression: firm A's profit when it is a leader with respect to the choice of royalties does not depend on the number of B's downstream firms. Maximization implies that the leader's royalty is:

$$r_A^L=\arg\max_{r_A} \pi^A(r_A, r_B(r_A))=-\frac{a}{n_A(n_A+2)}. \tag{22}$$

Substituting into B's reaction function we obtain the follower's royalty

$$r_B^F=r_B(r_A^L)=a\frac{n_B-1-n_A}{2n_B(n_A+2)}. \tag{23}$$

Clearly, we have:

Remark 4. *The leader's royalty, r_A^L, is (i) negative and (ii) increasing in the number of own firms, n_A.*

It is also easy to see that $\partial r_B^F/\partial n_A=-a(1+n_B)/2n_B(n_A+2)^2<0$ and thus:

Remark 5. *The follower's royalty, r_B^F decreases in the number of firms of the leader.*

Now we can calculate the equilibrium profit for the leader, $\pi^A(r_A^L, r_B^F)$. Substituting we obtain:

$$\pi^A(r_A^L, r_B^F) = \frac{a^2}{4(n_A+2)}. \tag{24}$$

Similarly, we can calculate the follower's profit

$$\pi^B(r_A^L, r_B^F) = \frac{a^2(n_A+1)}{4(n_A+2)^2}. \tag{25}$$

Thus, we have:

Proposition 2. *The Stackelberg leader's profit is decreasing in the number of its own firms. The Stackelberg follower's profit is also decreasing in the number of downstream firms of the leader. Both the Stackelberg leader's and the Stackelberg follower's profit does not depend on the number of the follower's firms.*

The proof is obvious for the leader from (24). For the follower, from (25) we have $\partial\pi^B(r_A^L, r_B^F)/\partial n_A = -a^2 n_A/4(2+n_A)^3 < 0$.

In particular, note that the Stackelberg leader's profit goes to zero as the number of its own firms goes to infinity.[17] The analysis of the sequential royalty choice, together with the simultaneous royalty case analyzed earlier, imply the following result.

Proposition 3. *An upstream firm is strictly worse-off with a larger number of downstream firms (of its own) when both fixed fees and royalty rates are used, regardless of whether the firm chooses its royalty rate simultaneously with or before its upstream rival's royalty rate. An upstream firm's profit does not depend on its number of downstream firms when its royalty rate is chosen after its upstream rival's royalty rate.*

We can also compare the royalty rates under sequential choices to the rates from the simultaneous choices, r^*, given by (16).

Proposition 4. *Compared to the simultaneous choice of royalties and for a given number of downstream firms, the follower (firm B) always charges a higher royalty. The leader (firm A) charges a higher (lower) royalty if and only if n_B is greater (smaller) than $n_A + 1$.*

To prove this result observe that the leader's profit is $\pi^A(r_A, r_B(r_A))$ and

$$\left.\frac{d\pi^A}{dr_A}\right|_{r_A=r_A^L} = \frac{\partial\pi^A}{\partial r_A} + \frac{\partial\pi^A}{\partial r_B}\frac{\partial r_B}{\partial r_A}. \tag{26}$$

In a Nash equilibrium (in royalties) each firm takes the other firm's royalty rate as given ($\partial r_B/\partial r_A = 0$) and so we have $\partial\pi^A/\partial r_A = 0$. As long as π^A has a unique maximum (this holds with linear demand), this implies that $r_A^L > r^* \Leftrightarrow \partial r_B/\partial r_A > 0$. However, from (12) we know that $\partial r_B/\partial r_A > 0 \Leftrightarrow n_B > n_A + 1$. It follows that $r_A^L > r^* \Leftrightarrow n_B > n_A + 1$. Now compare r_B^F to r^*. From above, when $\partial r_B/\partial r_A > 0$ then $r_A^L > r^*$. But, then, $r_B^F = r_B(r_A^L) > r_B(r^*) = r^*$. When, on the other hand, $\partial r_B/\partial r_A < 0$ then $r_A^L < r^*$. But, in this case we also have $r_B^F = r_B(r_A^L) > r_B(r^*) = r^*$. So we always have $r_B^F > r^*$. Finally, note that when $n_B = n_A + 1$ we have $\partial r_B/\partial r_A = 0$ and so $r_A^L = r^* = r_B^F$.[18]

A special case of interest is the following:

Remark 6. *When $n_B = n_A$ (and in particular when each upstream has only one downstream), under sequential choice of royalties, the leader chooses a lower royalty rate while the follower chooses a higher royalty relative to their respective royalties under simultaneous choice.*

VI. ROYALTIES BEFORE THE NUMBER OF DOWNSTREAM COMPETITORS

To better understand the strategic incentives present and, in particular, the role royalties play in influencing the choice of number of downstream firms, we now examine some alternative contracting arrangements. In particular, we examine situations where the choice of royalty rates does not follow that of the number of downstream firms. This could be either because royalties cannot be used at all, or because royalties must be chosen before or at the same time as the number of competitors.

Suppose first that royalties cannot be used and that only fixed fees can be used to transfer profit upstream. In the absence of royalties, the only strategic choice of upstream firms is the number of downstream firms. When $s = 1$, this case corresponds to the literature on strategic choice of downstream divisions discussed above (e.g. Baye, Crocker & Ju, 1996) and is presented here for comparison purposes. Naturally, our conclusion is the same as in that work, that there is an incentive to have a higher number of downstream units (and, without some cost for adding downstream firms, an equilibrium fails to exist). Thus the strategic incentives are reversed when royalties cannot be used. We further examine the fee-only case when $s < 1$. Not surprisingly, the result that a larger number of downstream firms is preferred does not hold as the degree of differentiation between the two products increases.

The fee-only case can be viewed as having royalties fixed at rate zero. We generalize the result from this case and show that the strategic incentive to have a higher number of downstream firms (and the non-existence of equilibrium) is present when royalties are fixed at any level that is not 'too low'. In particular, *for given royalty rates* that are not so low that the relevant brand reaction function is beyond the Stackelberg point, each firm has an incentive to act strategically and have more downstream firms than its rival. Thus, when our basic model is changed so that royalties are chosen first and the number of firms second or, alternatively, so that fee-and-royalty contracts and numbers of firms are chosen simultaneously, we find that, subject to the important qualification noted above, the strategic incentives are reversed. The formal analysis and results follow.

VI.1. Only fees

At the fee stage, both firms simply extract the downstream profits of their respective downstream firms (with per firm profit equal to $p_i q_i$). The expressions for downstream output choices are as before (see (4)) except that we need to substitute $r_i = 0$. Consider the choice of number of firms. Notice that now the upstream firms cannot control the marginal incentives of their downstream firms. Taking n_j as given, upstream firm i chooses n_i to maximize $\pi_i^{r=0} = n_i p_i q_i$. The first order condition is:

$$\frac{\partial \pi_i^{r=0}}{\partial n_i} = \frac{a^2(1+n_j - sn_j)^2(1+n_j - n_i - n_i n_j + s^2 n_i n_j)}{(1+n_j+n_i+n_i n_j - s^2 n_i n_j)^3}.$$

Setting this term equal to zero gives the reaction function of firm i

$$n_i = \frac{1+n_j}{1+(1-s^2)n_j}, \qquad i, j = A, B. \tag{27}$$

In particular, when $s=1$ the reaction function of firm i is

$$n_i = n_j + 1, \qquad i, j = A, B. \tag{28}$$

Clearly, the two reaction functions are upward sloping and parallel to each other: each firm wants to have one more downstream firm than its rival.[19] It follows that there is no equilibrium when products are perfectly homogenous ($s=1$). To illustrate the consequences of increasing the number of downstream firms when only fees can be used, consider again a simple example we discussed in the fee and royalty case. Suppose, with $s=1$ and $n_B=1$, that n_A increases from 1 to 2. The results are reported in Table 1b. It is easy to see that as its number of firms increases, A's total profit increases (while that of firm B decreases). Thus, firm A has an incentive to increase the number of its firms from $n_A = 1$ to $n_A = 2$.

Consider now the case where $s=0$: the two products are perfectly differentiated. Then, as can be seen from (27), the optimal choice of each firm is to pick a single downstream firm in order to extract monopoly profit. This is intuitive: under upstream monopoly for both products, when only fees can be used, each monopolist would prefer a single downstream since increasing the number of downstream firms simply decreases total downstream profit. When $0<s<1$, solving the system given by (27) we find that there is a symmetric equilibrium where the optimal number of firms of each type equals

$$n^* = \frac{1}{\sqrt{1-s^2}}. \tag{29}$$

As is obvious, as s approaches zero, n^* goes to 1 whereas as s approaches 1, n^* goes to infinity. In general, as products become more homogenous, each firm wants to increase its number of downstream firms. As noted above, when $s=1$, our analysis here is almost identical to Baye, Crocker & Ju (1996) where the only difference is that they specify a cost for each additional downstream firm and thus find an equilibrium with a finite number of divisions.

VI.2. Choosing Number of Downstream Competitors After or at the Same Time as Royalties

Suppose now that firms choose royalty rates and numbers of downstream firms simultaneously. To simplify the exposition, we present the results for the case of $s=1$.[20] Each firm solves the following problem:

$$\max_{r_i, n_i} n_i p_i(r_i, r_j) q_i(r_i, r_j).$$

It will be shown that for royalties that are not 'too low', the incentive to have a higher number of firms than the rival (that we saw before) is also present in this case. Thus to have an equilibrium it is required that the royalties specified be low enough. To see this point, note that to have an equilibrium the number of firms, n_i, must be optimal for firm i given the two royalty rates and the rival firm's number of firms. Thus, the first order condition of firm i with respect to n_i must hold.[21] Solving the first order condition, we find that, given r_i, r_j, and n_j, the optimal value of n_i is

$$n_i^* = \frac{(n_j+1)(a+r_j n_j)}{a+n_j r_j - 2n_j r_i - 2r_i}. \tag{30}$$

Using the above equation, it is easy to check that *when royalties are non-negative* we have $n_i^* > n_j + 1$ and thus

$$n_i^* > n_j.$$

But, clearly, this condition cannot hold simultaneously for both firms (it cannot be that $n_i > n_j$ and $n_j > n_i$). Further, note that (30) must be also satisfied in an equilibrium of the game where royalties are chosen before the number of firms. Therefore:

Remark 7. *When royalties must be non-negative, an equilibrium in the number of downstream firms cannot exist (we could only have an equilibrium if we allow for an infinite number of downstream firms). Similarly, there cannot exist an equilibrium with non-negative royalties when royalties and the number of firms are chosen simultaneously.*

Thus, with non-negative royalties when contracts are chosen at the same time as the number of firms, or when royalties are chosen before the number of firms, each upstream firm has an incentive to increase the number of its downstream firms (and this results in the non-existence of equilibrium). More generally, we show that the preceding strategic incentive is present not only for zero royalties but also for any fixed royalty rates that are not too low. Specifically, the key observation is that, unless r_i is so low that (given r_j and n_j) firm i's brand reaction function is already beyond the Stackelberg point, i has an incentive to choose $n_i^* > n_j$.

To see more precisely how the royalty rates affect the choice of the number of firms, we proceed as follows. To simplify the exposition, focus on symmetric equilibria. So suppose that $r_i = r_j = r$. With equal royalties, and using (30), (31) becomes $n_i^* = (n_j + 1)(a + rn_j)/(a - rn_j - 2r) > n_j$ which is true if and only if $r > -a/n_j(3 + 2n_j)$ (and $r < a/(n_j + 2)$). The key point here is that firm i wants to have a higher (respectively: lower, equal) number of downstream firms than firm j if and only if the royalty r is above (below, equal to) $-a/n_j(3 + 2n_j)$. What is the intuition for this result? As discussed above, firm i wants to make its brand reaction function pass through the Stackelberg point on j's brand reaction function. In particular, (10) gives the royalty that makes the reaction function go through the Stackelberg point. Imposing symmetry and solving (10) for r we find the same critical level as above, that when $r = -a/n_j(3 + 2n_j)$ the reaction function passes exactly through the Stackelberg point when $n_i^* = n_j = n$. Thus, if $r > -a/n_j(3 + 2n_j)$ firm i needs a higher than n_j number of firms to make its reaction curve go through the Stackelberg point. If, on the other hand, $r < -a/n_j(3 + 2n_j)$, to make its reaction curve go through the Stackelberg point, firm i chooses $n_i^* < n_j$. Finally, when $r = -a/n_j(3 + 2n_j)$ the optimal is $n_i^* = n_j$. This reasoning allows us to construct symmetric equilibria as follows. Consider (30) and impose symmetry also with respect to the number of firms to obtain $n = (n + 1)(a + rn)/(a - rn - 2r)$ or

$$rn(3 + 2n) + a = 0. \tag{32}$$

Having both firms choose a royalty rate and number of firms that satisfy (32) constitutes an equilibrium. Note that, as discussed above, (32) requires that the royalty rates are negative. In particular the royalty rates have to satisfy $r = -a/n(3+2n) < 0$. For example $n_A = n_B = 1$ and $r_A = r_B = -a/5$ is an equilibrium of the game where royalties and the number of firms are chosen simultaneously. Asymmetric equilibria could be also constructed but to save space we do not discuss these.

We summarize as follows.

Proposition 5. *Suppose that $s = 1$. When royalties and the number of downstream firms are chosen at the same time, there is no equilibrium with nonnegative royalties (since, with such royalties, each firm has an incentive to have a higher number of firms than its rival) but there is a multiplicity of equilibria with negative royalties. When royalties are chosen before the number of downstream firms, if nonnegative royalties are chosen in the subsequent stage each firm would like to have a higher number of downstream firms than its rival and, thus, there is no equilibrium.*

VII. ROYALTY-ONLY CONTRACTS AND COMPARISON

The final step in our analysis to allow firms to use only royalties. Thus, now, in the second stage we consider linear contracts. The main result here is that removing the firms' ability to transfer part of the profit upstream in the form of a fixed fee changes the strategic incentives and each firm prefers to increase its number of downstream firms, as long as the rival number of firms are not too large.

VII.1. Equilibrium with Royalty-Only Contracts

First, both firms simultaneously choose their number of downstream firms. Next, they choose their royalty rates and finally downstream firms compete in the product market. At the royalty stage, n_i and n_j are already determined, and taking r_j as given, firm i solves the problem

$$\max_{r_i} n_i r_i q_i(r_i, r_j) \tag{33}$$

where the output decisions are as before (see (4)). Note that the objective function indicates that only a part of downstream profit can be transferred upstream via the royalties (while under fee-and-royalty contracts the entire profit can be transferred). The first order conditions for the choice of royalties are given by

$$\frac{n_i(a - 2r_i - asn_j + an_j - 2n_j r_i + sr_j n_j)}{1 + n_i + n_j - n_i n_j + s^2 n_i n_j} = 0, \qquad i, j = A, B.$$

Solving these first order conditions yields the royalty rates

$$r_i^{f=0} = \frac{a[2(1+n_i)+n_j(2-s)+(2-s-s^2)n_in_j]}{4(n_i+1)+n_j(4n_i-s^2n_i+4)}, \qquad i,j=A,B.$$

Next consider the choice of number of firms. Each upstream firm solves the problem

$$\max_{n_i} \{n_i r_i^{f=0} q_i(r_i^{f=0}, r_j^{f=0})\}.$$

After appropriate substitutions and when $s=1$, the total upstream profit (maximand) can be rewritten as

$$\pi_i^{f=0} = \frac{a^2 n_i(n_j+1)(2n_i+n_j+2)^2}{[4n_i+4n_j+3n_in_j+4]^2[n_i+n_j+1]} \tag{34}$$

While the analytical expressions for the above problem are rather complicated, it is easy to show that there exists no symmetric equilibrium in this game. We have $\partial\pi_i^{f=0}/\partial n_i = a^2(n_j+1)(2n_i+n_j+2)[n_i(10n_jn_i+8n_i-3n_j^3+11n_j^2+30n_j+16)+4n_j^3+16n_j^2+20n_j+8]/[4n_i+4n_j+3n_in_j+4]^3[n_i+n_j+1]^2$. It is easy to check from this expression that when n_j is small, $\partial\pi_i^{f=0}/\partial n_i>0$ for all n_i and that when n_j is big enough, $\partial\pi_i^{f=0}/\partial n_i<0$ for all $n_i>1$.[22] This may suggest that there exists a symmetric equilibrium, in which each upstream firm selects a 'medium' number of downstream firms. It turns out that this is not the case. To see this, impose symmetry on the above first order condition and write

$$\left.\frac{\partial\pi_i^{f=0}}{\partial n_i}\right|_{n_i=n_j=n} = \frac{-a^2(n+1)(n^3-9n^2-12n-4)}{(2n+1)^2(n+2)^3(3n+2)}.$$

While the above first order condition admits a positive root at approximately $n=10.21$, inspection of the second order condition reveals that this root picks out a local minimum for both firms. In fact, it is straightforward to show that at $n_j=10.21$, firm i can benefit from increasing its number of firms. Thus, there is no positive root of the above first order condition at which profits are maximized.

Proposition 6. *Suppose that s = 1 and that contracts can specify only royalties. Then if an upstream firm chooses to have only one downstream firm, the other upstream firm's best response would be to have an infinite number of downstream firms. Furthermore, there exists no symmetric equilibrium in which both firms pick a finite number of downstream firms.*

To illustrate the effect of an increase in the number of downstream firms when fees cannot be used, consider again the simple example we discussed above.

Suppose that $s = 1$, $n_B = 1$ and n_A increases from 1 to 2. The results of this change are reported in Table 1c and illustrated graphically at Fig. 3b. We find that, in contrast to the fee-and-royalty case, now as n_A increases both firms royalty rates decrease (where firm *A*'s royalty decreases less than that of *B*). Further, firm *A*'s total profit increases while that of firm *B*'s decreases. As a result, in this case, firm *A* would like to increase the number of its firms from $n_A = 1$ to $n_A = 2$.

The intuition is as follows. When firms can only choose royalty rates, the incentive to increase the number of downstream firms in order to assume a more aggressive posture against the rival is *reinforced* by the incentive to induce competitive behavior in the downstream market by increasing the number of downstream firms, as long as the number of rival firms are small (which is the case in this example).[23] Consequently, firm *A* wants to proliferate the number of its downstream firms without end, so long as firm *B* does not have too many firms.

The above proposition does not rule out asymmetric equilibria. In fact, we can show that when firm *A* chooses an infinite number of downstream firms it is optimal for firm *B* to have one downstream firm and, conversely, when *B* chooses to have one downstream firm it is optimal for *A* to have and infinite number. Since this is not the main focus of the paper, we do not expand more on this issue. The important point we want to emphasize here is that under royalty-only contracts, if one firm picks one or only a few downstream firms, then its rival wants to proliferate its downstream firms without end.[24]

VII.2. Comparison

Now we turn to a comparison of the cases of royalty-only and fee-and royalty contracts. The comparison for a given number of downstream firms, can be summarized as follows. While royalty-only contracts allow downstream firms to keep part of their profits, upstream firms (as well as downstream) may be better off with royalty-only contracts. The reason is that, under fee-and-royalty contracts, royalties are too low from the viewpoint of maximization of joint upstream profits and making fees infeasible makes firms choose higher royalty rates. Indeed, under some conditions, upstream firms may be in a prisoners' dilemma when choosing the type of their contracts since both could be better-off if they could commit to royalty-only contracts, but it is a dominant strategy for each to choose a fee-and-royalty contract. This observation has been already made in the literature, in very similar frameworks.[25] The new point here is that when the number of downstream firms is endogenous (and precedes the choice of contracts) then a restriction to royalty-only contracts is not expected to yield higher profit for upstream firms. In other words, the prisoners's dilemma structure vanishes when the number of firms is endogenized. The reason is that, as shown above, while under royalty only contracts, firms prefer to increase the number of downstream firms, with fee-and-royalty contracts they prefer to have only one.

Some formal results follow. First, we compare royalty-only and fee-and-royalty contracts when the number of downstream firms is given. We show that the two upstream firms may make themselves better-off by 'tying their hands': if both firms charge only royalties, and no fixed fees, their profits may increase. To see the intuition, observe that, under fee-and-royalty contracts, in equilibrium competition induces firms to choose royalty rates that are too low. By way of illustration, if we suppose $s=1$ and $n_i = n_j(=n)$ then the (symmetric) royalties that maximize joint upstream profits are $a(2n-1)/4n$ which are higher than the equilibrium contracts $-a/n(2n+3)$ from (16).[26] Removing fees from the set of strategic variables allows the firms to jointly commit to less aggressive behavior (higher royalties) and this may lead to higher profits (in some, but not all, cases).

This intuition is confirmed as follows. We compare maximized profits of the upstream firms under the two different contracts. Consider the case when $s=1$, when downstream competition is at its strongest. Straightforward but lengthy calculations show that the ratio of maximized profits under the two-instrument contract to profits under the royalty-only contract is given by

$$\rho \equiv \frac{\Pi_i^{f,r}}{\Pi_i^{f=0}}$$

$$= \frac{[2-s+n_i(2-s^2-s)][(n_j(s^2-1)-1][4(n_i+1)+n_j(4n_i-s^2n_i+4)]^2[\Phi]}{n_i[n_j(s^2n_i+sn_i-2n_i+s-2)-2(n_i+1)]^2[\Psi]}$$

where

$$\Phi \equiv n_j[s^2n_i - 1 - n_i] - [1+n_i]$$

and

$$\Psi \equiv [n_j[s^4n_i - 5s^2n_i + 4n_i - 3s^2 + 4] + n_i[4-3s^2] + 4 - s^2]^2.$$

To gain some insight into the above ratio of profits, suppose $n_A = n_B = n$ and $s=1$. In this case, ρ becomes

$$\rho = \frac{(1+2n)}{n}\left[\frac{n+2}{2n+3}\right]^2.$$

It is easy to show that ρ is decreasing in n and we can state the following result.

Remark 8. *Firms may enjoy higher profits under royalty-only contracts than under fee-and-royalty contracts. If* $n_A = n_B = n$ *and* $s = 1$*, while the two-instrument contract yields higher profits when* $n = 1$*, the royalty-only contract dominates when* n *is equal to 2 or higher. As* s *becomes smaller, the two-instrument contract becomes more profitable relative to a royalty-only contract, since controlling downstream competition becomes less important. In the limit when* $s = 0$*, firms always make higher profits with the two-instrument contract.*

Thus, under some conditions, the upstream (as well as the downstream) firms may be better-off in a world where royalty-only contracts can be used as compared to a world where both fees and royalties can be used. This observation suggests a game where upstream firms first choose the type of their contracts (royalty-only or fee-and-royalty) and then compete in these contracts. It is straightforward to show that, although both firms could be better off if they choose royalty-only contracts, each has a dominant strategy (in the first stage) to choose a two-instrument contract. Thus, the two upstream firms may find themselves in a prisoners' dilemma when choosing the nature of their vertical contracts. Note that, previous work (Gal-Or, 1991a; Rey & Stiglitz, 1995) has demonstrated this prisoners' dilemma feature of contracts' choices under downstream price competition. We show here that this result also holds under quantity competition but it is crucial that each upstream firm has more than one downstream firm. In particular, it is required that the number of downstream firms per upstream firm is two or higher; if they each just have one downstream firm then this result does not hold.

The above result, however, keeps the number of downstream firms fixed. By making this number an endogenous choice of the upstream firms, we see that the picture may change dramatically and the prisoners' dilemma feature described above may not be present. The reason should be clear by now. Suppose that we consider the same enlarged game where firms can choose their contract type, but now also allow them to choose the number of their downstream firms before they compete in contracts. From the above analysis we know that in equilibrium each upstream firm chooses a single downstream under fee-and-royalty contracts, whereas when only royalties can be used and a competitor has a low number of downstream firms there is an incentive to increase indefinitely the number of downstream firms. As a result, when the number of firms is endogenized, it is problematic to state that competition in two-instrument contracts leads to lower profit than competition in royalty-only contracts and the strategic incentives are modified.

VIII. CONCLUSION

In this chapter we have examined strategic interactions in vertically related markets when both upstream and downstream competition is oligopolistic. The strategic variables that each upstream firm controls are the terms of its vertical contracts as well as the number of its own downstream firms. We use the concept of a brand reaction function to illustrate how these two strategic variables affect competition in the downstream market.

When fee-and-royalty contracts are chosen after the number of downstream divisions, each upstream firm prefers to minimize the number of its downstream firms. We show how this result is modified when royalties are chosen sequentially: the leader in terms of royalties has a strict incentive to minimize the number of its downstream firms, while the follower's profit does not depend on the number of its downstream firms. We then show that it is crucial that royalties be chosen after the number of firms: if they are chosen before or simultaneously with the number of firms, then (under some conditions) each upstream firm has an incentive to increase the number its downstream firms. Similarly, when contracts cannot involve fees but only royalties, it is no longer true that the firms have an incentive to minimize the number of their downstream firms. Finally, we discuss the firms' incentives to use royalty-only contracts relative to fee-and-royalty contracts and how this comparison is related to the number of downstream firms.

Our model shares with most of the work in this area the feature that upstream firms can commit to their contracts and that these contracts are observable. The implications of relaxing the assumption that upstream firms can commit to contracts have been explored in other work.[27] Note also that observability of contracts is an important assumption in our framework: its main implication is that, at the level of downstream competition, all firms know the terms of all contracts. Uncertainty about contracts would represent uncertainty about the true reaction functions of competitors. To highlight the strategic motives, we have kept the model as simple as possible. Clearly, a number of factors that play an important role in vertically related markets (such as imperfect information), are not captured in this model. Such factors could lead to different conclusions regarding equilibrium contracts. Further, we have employed a simple parameterization with linear demand (and costs). This formulation allows us to highlight the strategic motives in a clear way and we expect the main implications to be more generally true. Nevertheless, alternative formulations may require modification of our results.

NOTES

1. This point is also related to the idea that horizontal mergers may not be profitable for strategic reasons (see Salant, Switzer & Reynolds, 1983).

2. To illustrate this implication, consider a software or microchip producer who supplies multiple computer manufacturers and faces little competition for his product. Our analysis implies that, to the extent that the supplier has the bargaining power, he would tend to offer contracts that involve relatively high royalties, with the goal of reducing the competition between computer manufacturers. In contrast, to shift to a different industry, consider a fast-food company which faces severe competition from other fast-food companies and has franchising agreements with a number of (downstream) sellers. In this case, we expect that the franchising agreements will tend not to put much weight on royalties, because the fast-food company wants to strengthen the strategic position of its franchisees relative to rival franchisees. This would be especially true in the case of a newcomer into the industry which has only a small number of franchises and is competing with established firms that have a larger number of franchisees. Empirically, while fees and royalties are both widely used, there are indeed differences with some contracts specifying relatively higher royalties and other higher fees.

3. See Katz (1989) for a survey on vertical contracts; in particular, for downstream price competition as a negative externality, see pp. 678–679.

4. Negative royalties correspond to wholesale unit prices lower than the upstream cost.

5. In particular, Gal-Or (1991) shows that, under price competition in the downstream market, upstream firms may benefit from a commitment to not using a fee or other 'vertical restraints'. Under these circumstances, firms choose higher wholesale prices, which facilitates sustaining higher retail prices. In our model, with quantity competition between firms contracting with different upstream firms, these incentives are reversed, and thus our analysis is different. Whereas our results also show that firms may benefit when fees cannot be used, in our case this result relies on the existence of multiple downstream firms. Also, in Gal-Or (1991) each upstream firm contracts with only one downstream, whereas we examine competition between (multiple) downstream firms. See also Rey & Stiglitz (1995, Proposition 3) for a related result.

6. See also Slade (1998) for related empirical work that emphasizes the interaction between exclusive-dealing clauses and strategic factors.

7. With increasing marginal costs, however, an increase in the number of retailers reduces equilibrium prices (see pp. 55–56).

8. In general, the class of contracts that can be written should depend on the informational structure. See Rey & Tirole (1986) for informational assumptions underlying two-part tariffs. An upstream firm may be able to only observe the amount of the product supplied to a downstream but not the actual quantity sold to final consumers. Then, if marginal prices are not equal for all quantities, downstream firms may set up a secondary market for the product.

9. See Ippolito (1991) for an analysis of evidence from litigation.

10. The Monopolies and Mergers Commission in the U.K. has recently decided to ban the use of 'recommended retail prices' by electric-appliance makers. See *The Economist*, May 31, 1997, p. 53.

11. In general, the ability to set prices may be left in the hands of downstream firms for the additional reason that they have better information about demand.

12. Several other important factors that affect contract design have been examined in the literature. They include adverse selection, risk-aversion, and incentives (moral hazard). See, for example, Mathewson & Winter (1985) for a theoretical analysis from an agency viewpoint, and Lafontaine (1992) for an overview of theoretical results as well as an empirical study. In particular, one of the findings in this last paper, that empirically fees and royalties are not negatively related, is inconsistent with many other models but is consistent with our framework: higher royalties make the downstream market less competitive and thus the fees that can be extracted are not necessarily lower. See also Lafontaine & Slade (1997) for a discussion of empirical regularities in vertical contracts.

13. The case of more than two upstream firms represents a straightforward extension of the model.

14. A price competition formulation would imply higher royalties than our formulation.

15. It is easy to check that the second-order conditions for a maximum are satisfied.

16. Note that total output is decreasing (and market prices are increasing) in royalty rates and therefore upstream firms benefit if they are not allowed to use negative royalties. While competition forces each firm to be aggressive and subsidize its downstream firm, both firms would be better off with a zero royalty than with the equilibrium (negative) rate. Specifically, compared to a zero rate, a royalty equal to $-a/5$ decreases the market price form $a/3$ to $a/5$ and each upstream firm's total profit from $a^2/9$ to $2a^2/25$.

17. In contrast, recall that when an upstream firm faces no competition ($s=0$) then, even with an infinite number of downstream firms, the upstream firm can achieve the monopoly profit.

18. These inequalities can be also easily checked with direct calculations using (22), (23) and (16).

19. Note that this is the same expression reported in Corchón (1991) and Polasky (1992).

20. As above, when $s \rightarrow 0$ the incentive to have a smaller number of downstream firms dominates.

21. We have $\partial\pi_i^{r,n}/\partial n_i = -(a - r_i - n_j r_i + n_j r_j)(a n_i - 2 r_i n_i - 2 n_i n_j r_i + n_i n_j r_j - a - a n_j - n_j r_j - r_j n_j^2)/(1 + n_i + n_j^2)$.

22. Note that in the expression for $\partial\pi_i^{f=0}/\partial n_i$ the denominator is positive, as well as all terms of the numerator with the possible exception of $(4-3n_i)n_j^3$. For n_j large enough, this term dominates and $\partial\pi_i^{f=0}/\partial n_i$ has the same sign as $(4-3n_i)$. It is easy to show that, as $n_j \rightarrow \infty$, it is optimal to have $n_i = 1$ (as this gives higher profit than $n_i = 2$ and $\partial\pi_i^{f=0}/\partial n_i < 0$ for $n_i \geq 2$).

23. We can show that $\partial(\partial\pi_i^{f=0}/\partial n_i)/\partial n_j < 0$, i.e. the marginal incentive to increase your own firms is strictly decreasing in the number of firms of your rival.

24. The possibility of asymmetric equilibria in which both firms pick a finite number of firms cannot be ruled out, though we have been unable to find such equilibria.

25. See, in particular, Gal-Or (1991) and Rey & Stiglitz (1995).

26. Of course, when $s=0$, there is no downstream competition across the firm types so that the royalties charged by upstream firms also maximize joint profits.

27. See, for example, Fershtman & Kalai (1997) on unobservable strategic contracting. A related but different issue is whether the commitment problem faced by an upstream firm that licenses to multiple downstream firms can be solved using nondiscrimination clauses. See, for example, DeGraba & Postlewaite (1992) who find that such clauses are effective with linear prices and McAfee & Schwartz (1994) who find that they are not effective under two-part tariffs (both papers consider a monopolist upstream).

ACKNOWLEDGMENTS

We wish to thank Jim Anton, Jack Hughes, Kai-Uwe Kühn, Ping Lin, Margaret Slade, and Juuso Välimäki for helpful comments and discussions, and Wen Zhou for excellent research assistance. We are responsible for all errors and omissions.

REFERENCES

Baye, M., Crocker, K., & Ju, J. (1996). Divisionalization, franchising, and divestiture incentives in oligopoly. *American Economic Review*, *86*, 223–236.

Bonanno, G., & Vickers, J. (1988). Vertical separation. *Journal of Industrial Economics*, *36*, 257–266.

Brander, J., & Spencer, B. (1985). Export subsidies and international market share rivalry. *Journal of International Economics*, *18*, 83–100.

Corchón, L. (1991). Oligopolistic competition among groups. *Economics Letters*, *36*, 1–3.

DeGraba, P., & Postlewaite, A. (1992). Exclusivity clauses and best price policies in input markets. *Journal of Economics and Management Strategy*, *1*, 423–454.

Dixit, A. (1983). Vertical integration in a monopolistically competitive industry. *International Journal of Industrial Organization*, *1*, 63–78.

Fershtman, C., &. Judd, K (1987). Equilibrium incentives in oligopoly. *American Economic Review*, *77*, 927–940.

Fershtman, C., & Kalai, E. (1997). Unobserved delegation. *International Economic Review*, *38*, 763–774.

Gal-Or, E. (1991). Duopolistic vertical restraints. *European Economic Review*, *35*, 1237–1253.

Ippolito, P. (1991). Resale price maintenance, Empirical evidence from litigation. *Journal of Law and Economics*, *34*, 263–294.

Katz, M. (1989). Vertical contractual relations. In: R. Schmalensee & R. Willig (Eds), *Handbook of Industrial Organization* (Vol. 1, Chap. 11). North-Holland.

Kühn, K. (1997). Nonlinear pricing in vertically related duopolies. *Rand Journal of Economics*, *28*, 37–62.

Lafontaine, F. (1992). Agency theory and franchising: Some empirical results. *Rand Journal of Economics*, *23*, 263–283.

Lafontaine, F., & Slade, M. (1997). Retail contracting: Theory and practice. *Journal of Industrial Economics*, *45*, 1–25.

Mathewson, G., &. Winter, R (1984). An economic theory of vertical restraints. *Rand Journal of Economics*, *15*, 27–38.

Mathewson, G., & Winter, R. (1985). The economics of franchise contracts. *Journal of Law and Economics*, *28*, 503–526.

McAfee, R., & Schwartz, M. (1994). Opportunism in multilateral contracting: Nondiscrimination, exclusivity, and uniformity. *American Economic Review*, *84*, 210–230.

McGuire, T., & Staelin, R. (1983). An industry equilibrium analysis of downstream vertical integration. *Marketing Science*, *2*, 161–191.

Polasky, S. (1992). Divide and conquer: On the profitability of forming independent rival divisions. *Economic Letters*, *40*, 365–371.

Rey, P., & Stiglitz, J. (1995). The role of exclusive territories in producers' competition. *Rand Journal of Economics*, *26*, 431–451.

Rey, P., & Tirole, J. (1986). The logic of vertical restraints. *American Economic Review*, *76*, 921–939.

Saggi, K., & Vettas, N. (1999). On intrabrand and interbrand competition: The strategic role of fees and royalties. *European Economic Review*, forthcoming.

Salant, S., Switzer, S., & Reynolds, R. (1983). Losses from horizontal merger: The effects of an exogenous change in industry structure on Cournot-Nash equilibrium. *Quarterly Journal of Economics*, *48*, 185–199.

Sklivas, S. (1987). The strategic choice of managerial incentives. *Rand Journal of Economics*, *18*, 452–458.

Slade, M. (1998). Beer and the tie: Did divestiture of brewer-owned public houses lead to higher beer prices? *Economic Journal*, *108*, 565–602.

Vickers, J. (1985). Delegation and the theory of the firm. *Economic Journal*, Supplement 95, 138–147.

WHY PEOPLE BUY LIABILITY INSURANCE UNDER THE RULE OF SIMPLE NEGLIGENCE

Keith J. Crocker and Neil Doherty

ABSTRACT

If people take due care, they will not be found negligent under the Rule of Simple Negligence. Moreover, under this rule, investment in due care is privately and socially optimal. These results jointly imply that liability insurance should not exist. But it does. One explanation relies on errors in the administration of tort law or in the imperfect observation of care. We offer an alternative explanation. Our model is one of symmetric uncertainty where establishing due care requires information on the individual's risk type, which is a random variable. Even though we assume that information on type is costless, we find that, by pooling across risk classes, liability insurance permits risk averse individuals to avoid the classification lottery that arises from establishing type. Insurance also has subtle effects on an injurer's private costs, which depend on the concavity or convexity of the demand for precaution. Given parameter variability across individuals, one would expect to observe a segmented market with some insuring while others choose to become informed and take due care. Moreover, we show that the voluntary purchase of liability insurance is Pareto improving.

Industrial Organization, Volume 9, pages 133–148.

ISBN: 0-7623-0687-4

I. INTRODUCTION

Why does anyone bother to purchase liability insurance? One might think that the purpose would be to protect an individual from the financial losses associated with the tort liability resulting when one's negligent actions inadvertently cause injury to another. But, under the commonly applied Rule of Simple Negligence, the taking of due care completely shields an injurer from liability, thereby obviating the need for the protection afforded by liability insurance. As a consequence of this result, which requires both knowledge of the relevant negligence standard by injurers as well as perfect accuracy by the courts in the application of that standard, the traditional justification for liability insurance has relied on the existence of an unresolved randomness in the application of the negligence rule.[1] Such uncertainty exposes a risk averse injurer to the risk of choosing a level of care insufficient to escape tort liability, thereby generating a demand for liability insurance. In contrast, this chapter examines an environment where the courts apply the negligence standard without error, and in which injurers can become costlessly informed about that standard. Nonetheless, we find that potential injurers may choose to remain uninformed and purchase liability insurance to protect themselves from the risk of being found negligent.

Justifying the demand for liability insurance as protection from mistakes in tort administration is somewhat tenuous. It is undoubtedly true that lay jurors do commit errors in applying sometimes complex behavioral standards to a set of facts. Yet, the same rationale cannot be so easily applied in jurisdictions such as Great Britain, where liability insurance thrives despite the fact that civil cases are decided by merit appointed judges. Nor does a second type of mistake, the imperfect observation of care by the courts, provide a completely satisfactory explanation for liability insurance. While this could be an issue in the case of automobile insurance, where reconstructing an injurer's behavior may be problematic, such reasoning is less convincing when applied to some other types of liability insurance. For example, in cases of medical malpractice, doctors generally follow closely documented procedures which should, in principle, avoid this kind of uncertainty. Yet, such controls and record keeping do not appear to have dampened the demand for medical malpractice insurance.

This chapter explores another explanation for liability insurance: that injurers have incomplete knowledge of their risk type and, therefore, of the relevant negligence standard. As a result, injurers are initially uninformed about the cost of establishing due care, so that the purchase of liability insurance jointly hedges the risk associated with the random cost of investing in care, which we call *classification risk*, as well as the risk that an uninformed investment in care may be insufficient to escape negligence.

We consider an environment with risk averse injurers who are initially uninformed about the appropriate level of precaution required to escape a finding of negligence. The choice facing the injurer is whether to obtain costless, and perfect, information about the negligence standard and then take due care to avoid liability or, alternatively, to remain uninformed and purchase liability insurance instead. Throughout, we assume that an individual's chosen level of precaution is freely observable to courts and to insurance firms, and that courts apply the appropriate negligence standard ex post without error. Even so, we find that liability insurance may be the preferred option because of the lottery over precautionary expenditures that injurers face if they choose to become informed and take due care.[2] Thus, the demand for liability insurance arises not to protect victims who may be injured by accidents, but, rather, to shield potential injurers from the classification risk they face in reaching a decision regarding the appropriate amount of precaution to take. Since the appropriate choice for a particular injurer depends on risk aversion, the variance of the classification lottery, and the shape of the precautionary technology, one would expect to observe in practice a market with some injurers purchasing insurance while others become informed and avoid tort liability.

The chapter proceeds as follows. The next section describes the environment of the model, and examines the benefits to the injurer of becoming informed and taking due care or, alternatively, of remaining uninformed and purchasing liability insurance. Section III derives conditions under which the injurer would prefer the insurance alternative, and demonstrates that, in such a situation, the purchase of insurance results in a Pareto improvement over the alternative of becoming informed and taking due care. The final section contains concluding remarks.

II. THE MODEL

We consider an environment with risk averse potential injurers who possess the von Neumann-Morganstern utility function $U(W)$, where W denotes an injurer's wealth. An injurer has the initial wealth $\bar{W}$, and may cause an accident resulting in a monetary loss αL with probability $p(x)$, where x is the level of precaution taken by the injurer. We assume that the safety technology satisfies $p(0) = 0$, $p(x) > 0$, $p'(x) < 0$ and $p''(x) > 0$, and that the per-unit cost to the injurer of taking precaution is c. Finally, there is assumed to be two types of injurers in the model indexed by α. Specifically, $\alpha \in \{\alpha_1, \alpha_2\}$ where $1 > \alpha_1 > \alpha_2 > 0$, and the proportion of α_1-types is given by the parameter λ.

Before proceeding, it will be useful to define formally the rule of simple negligence and the standard of due care. This standard, which is generally referred to as the 'Hand Rule',[3] defines due care as the level which minimizes the total social costs of accidents and precaution, which may be written as

$$T(\alpha, x) \equiv \alpha p(x)L + cx. \tag{1}$$

The costs are minimized by the level of precaution, $X(\alpha)$, which solves the first-order condition $\alpha p'(X(\alpha))L + c = 0$.[4] In the following discussion, we shall use $X(\alpha)$ to denote the level of due care associated with a type α individual. We now state the negligence standard which prevails under the Rule of Simple Negligence.

Definition. Under the Rule of Simple Negligence, an injurer is negligent, and therefore liable for accident damages αL, if, and only if, $x < X(\alpha)$.

The sequence of events in the model is described by the time line depicted in Fig. 1. We assume that injurers, insurance firms and the courts are initially symmetrically uninformed about an individual's type, α. The choice facing the injurer is whether to remain uninformed and purchase liability insurance, or, alternatively, to forego insurance and become informed of one's type, followed by due care and immunity from tort liability.[5] After this decision, the agent selects a utility-maximizing level of precaution, which is followed by the potential accident state. If an accident occurs, the court proceedings costlessly reveal an agent's type to all actors, and the Rule of Simple Negligence is applied to determine liability.

Before proceeding, we present two results that will be useful in the analysis that follows.

Lemma 1. Let $X(\alpha)$ denote the due care standard of precaution for a type α injurer. Then

(i) $T((\alpha, X(\alpha))$ is a strictly concave function of α; and
(ii) X is a concave (convex) function if $-2(p'')^2 + p'''p' < (>) 0$.

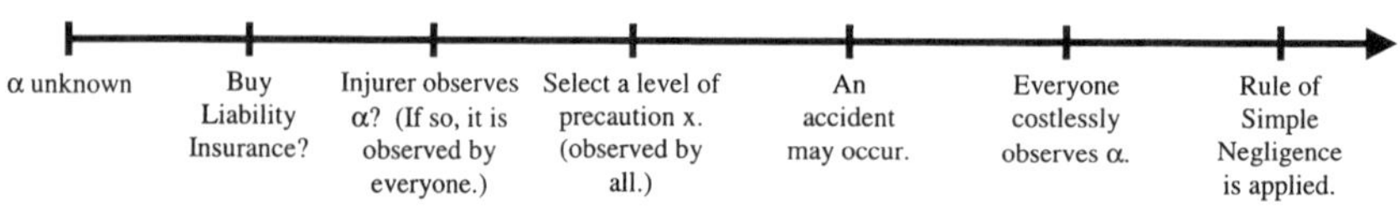

Fig. 1.

Proof. (i) follows directly from the definition of X and an application of the envelope theorem.

To demonstrate (ii), define the function

$$G(x, \alpha) \equiv p'(x)\alpha L + c$$

so that X is implicitly defined by the relationship $G(X, \alpha) = 0$. Then,

$$X'(\alpha) \equiv -\frac{G_\alpha}{G_X} = -\frac{p'}{p''\alpha}.$$

Also,

$$X''(\alpha) \equiv -\frac{-G_{\alpha x}G_\alpha - G_\alpha\left(G_{\alpha x} - \dfrac{G_{xx}G_\alpha}{G_x}\right)}{G_x^2} = \left[\frac{-p'L^2}{G_x^2}\right]\cdot\left[-2p'' + \frac{p'''p'}{p''}\right].$$

from which (ii) follows immediately after noting $p' < 0$. Q.E.D.

The concavity of T guarantees that the expected total costs of accidents and precaution are always minimized by observing type prior to the selection of precaution. In contrast, the demand for precaution, X, may be either concave or convex, depending on the higher order derivatives of p. For example, precaution technologies of the form

$$p_1(x) \equiv K - \frac{c}{L}\frac{(n-1)}{n}(1+x)^{n-1/n},$$

where K is a constant, generate the precautionary demand $X_1(\alpha) \equiv \alpha^n - 1$, which is concave (convex) for $n < 1$ $(n > 1)$. Similarly, the precaution technology

$$p_2(x) \equiv K - \frac{c}{L}\log(1+x)$$

results in the linear demand $X_2(\alpha) \equiv \alpha - 1$.

We now turn to an examination of the choices facing a potential injurer.

Information and Due Care

An injurer who observes her type prior to selecting a level of precaution, and who anticipates an application of the Rule of Simple Negligence by the courts, will choose the level of due care, $X(\alpha)$, which eliminates tort liability.[6] After risk type is ascertained and due care is taken, all uncertainty is removed and there would be no demand for liability insurance. There is, however, ex ante uncertainty by the injurer regarding the outcome of the classification lottery, α. Thus, the expected utility of a consumer who adopts this strategy may be written as

$$V_{DC} \equiv U(\bar{W} - cX(\alpha_1)) + (1 - \lambda)U(\bar{W} - cX(\alpha_2)) \tag{2}$$

since λ $(1 - \lambda)$ is the probability that the injurer discovers that she is a type α_1 (α_2).

We may write equation (2) as

$$V_{DC} = U(\bar{W} - c(\lambda X(\alpha_1) + (1 - \lambda)X(\alpha_2)) - R_1) \tag{3}$$

where R_1 is defined to be the risk-premium resulting from the concavity of U.

Furthermore, in the same fashion we may rewrite (3) as

$$V_{DC} = U(\bar{W} - c[X(\bar{\alpha}) - R_2] - R_1) \tag{4}$$

where R_2 is the risk premium associated with the shape of the function X and $\bar{\alpha} \equiv \lambda\alpha_1 + (1\text{-}\lambda)\alpha_2$ is the average type in the population. From Pratt (1969, p. 125), it follows that

$$R_1 \approx -\sigma_X^2 \frac{U''}{U'}; \text{ and} \tag{5}$$

$$R_2 \approx -\sigma_\alpha^2 \frac{X''}{X'} \tag{6}$$

where σ_x^2 and σ_α^2 are the variances of X and α, respectively.

Remain Uninformed and Purchase Liability Insurance

Alternatively, an injurer could choose to remain uninformed and purchase insurance against the tort liability. Insurance firms can observe whether an injurer has chosen to become informed and the level of precaution actually chosen. Thus, competitive insurers would charge uninformed individuals actuarially fair premiums, and injurers would purchase full insurance. Since, under the negligence standard, the degree of exposure to liability by the insurance firm depends on the level of precaution chosen by the insured, the actuarially fair premiums are

$$\pi \equiv \begin{cases} \bar{\alpha}p(x)L & \text{for} \quad x < X(\alpha_2) \\ \lambda\alpha_1 p(x)L & \text{for} \quad X(\alpha_2) \leq x < X(\alpha_1) \\ 0 & \text{for} \quad x > X(\alpha_1). \end{cases} \tag{7}$$

Given this premium schedule, the insured's problem can be written as

$$\underset{x}{\mathrm{Max}}\ U(\bar{W} - \pi - cx), \tag{8}$$

the solution to which is characterized by the following lemma.

Lemma 2. If the injurer purchases insurance, the utility maximizing level of precaution, denoted $\hat{x}$, satisfies

(i) $\hat{x} = X(\alpha_2)$ if $\lambda \le \alpha_2/\alpha_1$, and
(ii) $\hat{x} = X(\lambda\alpha_1)$ if $\lambda > \alpha_2/\alpha_1$.

Proof. A solution of the second type is depicted in Fig. 2, and is characterized by a tangency of an injurer's indifference curve and the premium schedule, so that $\lambda\alpha_1 p'(x)L = -c$. In order for this solution to obtain, we must have

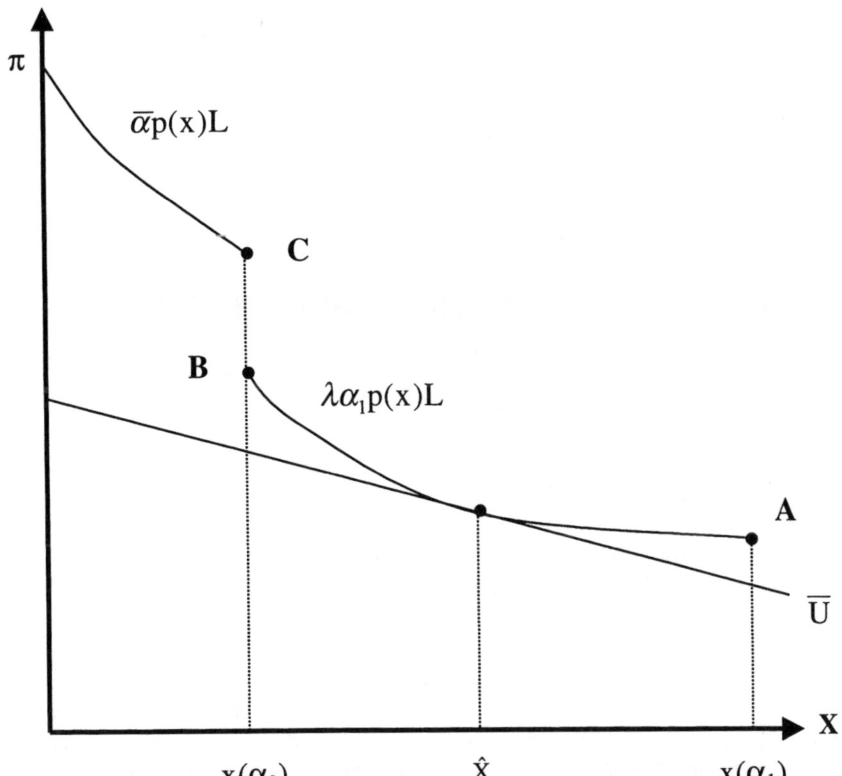

Fig. 2.

$$\lambda\alpha_1 p'(X(\alpha_2))L < -c < \lambda\alpha_1 p'(X(\alpha_1))L$$

where the term on the left- (right-) hand side is the slope of the premium schedule at point B (A). The right hand inequality follows immediately from the fact that $X(\alpha_1)$ solves $\alpha_1 p'(X(\alpha_1))L = -c$. Similarly, the left hand inequality holds when $\lambda \leq \alpha_2/\alpha_1$ since $\alpha_2 p'(X(\alpha_2)) = -c$.

Finally, when $\lambda \leq \alpha_2/\alpha_1$, the utility-maximizing choice of x along the $\lambda\alpha_1 p(x)L$ premium schedule is $X(\alpha_2)$. To conclude the proof, it is sufficient to show that the slope of $\bar{\alpha}p(x)L$ is always less than $-c$ so that an injurer never chooses $x < X(\alpha_2)$. Since $\bar{\alpha}\, p'(x)L$ is an increasing function of x, to demonstrate this result it is sufficient to show that the slope at $X(\alpha_2)$ is less than $-c$, which follows immediately from the observation that

$$\bar{\alpha}p'(X(\alpha_2))L = -\frac{\bar{\alpha}c}{\alpha_2} < -c.$$

Q.E.D.

Thus, the utility of an injurer who purchases liability insurance may be written as

$$V_{LI} \equiv U(\bar{W} - p(\hat{x})\lambda\alpha_1 L - c\hat{x}),$$

and, in the analysis that follows, we will for convenience concentrate on the case where $\lambda > \alpha_2/\alpha_1$ so that $\hat{x} = \equiv X(\lambda\alpha_1)$.

III. THE DEMAND FOR LIABILITY INSURANCE

We now turn to an examination of the circumstances under which the purchase of liability insurance is preferred by the injurer to the alternative of becoming informed and taking due care.

Equations (4) and (9) imply that $V_{LI} - V_{DC}$ may be written as

$$U(\bar{W} - p(\hat{x})\lambda\alpha_1 L - c\hat{x}) - U(\bar{W} - cX(\bar{\alpha}) + cR_2 - R_1), \tag{10}$$

where $\hat{x} \equiv X(\lambda\alpha_1)$ from Lemma 2. Moreover, from Lemma 1, $X' > 0$, so that $\hat{x} \equiv X(\lambda\alpha_1) < \bar{x} \equiv X(\lambda\alpha_1 + (1-\lambda)\alpha_2)$. It follows immediately that a sufficient condition for liability insurance to be preferred to due care by the injurer is

$$R_1 - cR_2 - p(\hat{x})\lambda\alpha_1 L > 0. \tag{11}$$

The attractiveness of liability insurance depends on the degree of risk aversion through the risk premium R_1, on the shape of the demand for precaution function X which determines the sign and magnitude of R_2, and on the size of the liability insurance premium $p(\hat{x})\lambda\alpha_1 L$. In the special case where $\alpha_2 = 0$, it follows that $\hat{x} = X(\bar{\alpha})$, so that (11) would be a necessary condition as well.

In the case of a risk neutral injurer ($R_1 = 0$), it might appear that a sufficiently convex X function, which would imply $R_2 < 0$, could satisfy (11) and, hence, make liability insurance the preferred option.

We now demonstrate that this cannot be the case.

Lemma 3. A risk neutral injurer will not purchase liability insurance.

Proof. We shall prove the result by contradiction. Suppose that a risk neutral injurer preferred insurance to due care. Then

$$\lambda cX(\alpha_1) + (1-\lambda)cX(\alpha_2) > p(\hat{x})\lambda\alpha_1 L + c\hat{x} \tag{12}$$

which implies

$$\begin{aligned} &\lambda[p(X(\alpha_1))\alpha_1 L + cX(\alpha_1)] + (1-\lambda)[p(X(\alpha_2))\alpha_2 L + cX(\alpha_2)] \\ &\quad > p(\hat{x})\lambda\alpha_1 L + c\hat{x} + \lambda p(X(\alpha_1))\alpha_1 L + (1-\lambda)p(X(\alpha_2))\alpha_2 L. \end{aligned} \tag{13}$$

Equation (13) may be written as

$$\lambda T(\alpha_1, X(\alpha_1)) + (1-\lambda)T(\alpha_2, X(\alpha_2)) > \lambda T(\alpha_1, \hat{x}) + \Delta(\lambda) + (1-\lambda)c\hat{x} \tag{14}$$

Where $\Delta(\lambda) \equiv \lambda p(X(\alpha_1))\alpha_1 L + (1-\lambda)p(X(\alpha_2))\alpha_2 L$. Since $\hat{x} \geq x_2$, it follows that

$$\Delta(\lambda) - (1-\lambda)p(\hat{x})\alpha_2 L = \lambda p(X(\alpha_1 L + (1-\lambda)\alpha_2 L(pX(\alpha_2) - p(\hat{x})) > 0. \tag{15}$$

Consequently,

$$\Delta(\lambda) + (1-\lambda)c\hat{x} > (1-\lambda)[p(\hat{x})\alpha_2 L + c\hat{x}] = (1-\lambda)T(\alpha_2, \hat{x}) \tag{16}$$

which, in conjunction with (14) implies

$$\lambda T(\alpha_1, X(\alpha_1)) + (1-\lambda)T(\alpha_2 X(\alpha_2)) > \lambda T(\alpha_1, \hat{x}) + (1-\lambda)(T(\alpha_2, \hat{x}). \tag{17}$$

But, since $X(\alpha_i)$ minimizes $T(\alpha_i, x)$, this statement is a contradiction. Thus, (12) cannot hold, and a risk neutral injurer would never purchase liability insurance. Q.E.D.

Alternatively, a risk averse injurer may prefer liability insurance to due care as long as the aversion to risk generates a risk premium sufficiently large to overcome the higher cost of liability insurance. To see this, consider the special case where $\alpha_2 = 0$, so that (11) is both necessary and sufficient for insurance to be preferred by the injurer. Using Pratt's (1974) approximation for the risk premium and inequality (11), the injurer is indifferent between due care and insurance when

$$\sigma_x^2\rho_u - c\sigma_\alpha^2\rho_x - p(\hat{x})\lambda\alpha_1 L \approx 0, \tag{18}$$

where $\rho_u \equiv -U''/U'$ and $\rho_x \equiv -X''/X'$.

Consider, for the moment, the case in which $\rho_x = 0$. If we were to increase α_1, while adjusting λ downward to hold $\lambda\alpha$ constant, $\hat{x}$ and, therefore, the liability insurance premium, would remain unchanged, yet the variance of the classification lottery over α would increase. Since X is a monotonically increasing function of α, this would increase σ_x, and reduce the degree of risk aversion, ρ_u, required to make the injurer prefer liability insurance. The indifference curve between due care and insurance for the case in which the demand for precaution is linear is depicted in Fig. 3 by the solid line.

Besides determining the value of the classification lottery, the values of λ, α_1 and α_2 determine the size of the externality that arises from the injurer's behavior. With insurance, injurers determine the level of precaution $\hat{x}$ without knowledge of risk type α_i where $X(\alpha_2) \leq \hat{x} < X(\alpha_1)$. Consequently, should the individual cause an accident and be found to be of type α_1, she will be found to be negligent and the insurer will have to pay damages under the policy. Thus, while all cost imposed on injured third parties is externalized under due care, with insurance, the injurer internalizes to the extent that she might be an α_1-type. The higher the value $\lambda\alpha_1$ the greater the internal cost, the greater the insurance premium and, ceteris paribus, the less attractive is insurance relative to due care.

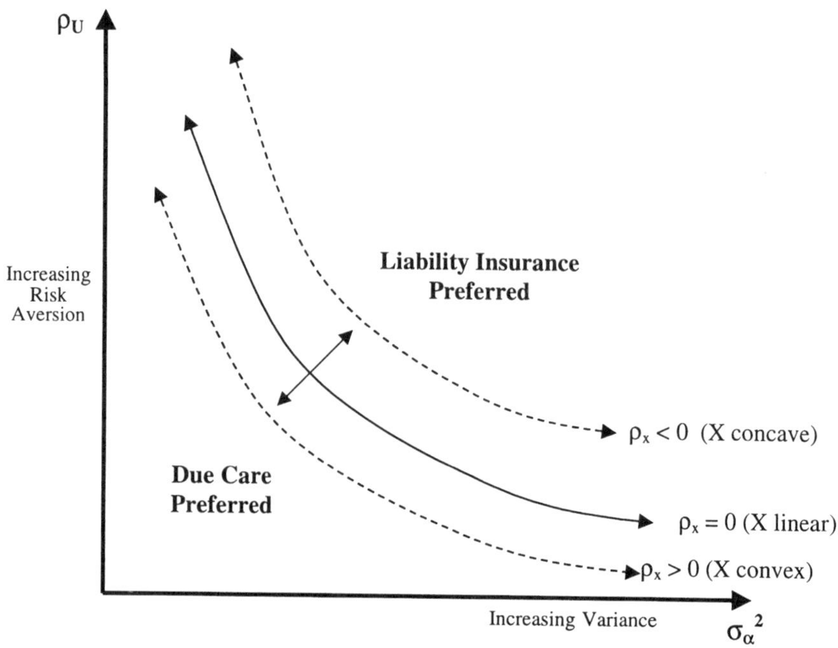

Fig. 3.

The attractiveness of liability insurance turns not only on the degree of risk aversion and the size of the tort externality, but upon the shape of the demand for precaution as well. Concavity of X ($\rho_x < 0$) guarantees that the expected costs of precaution are lower when an injurer obtains information on type prior to taking precaution than when precaution is chosen based on the injurer being an average type. Linearity ($\rho_x = 0$) implies these costs are equal, while convexity ($\rho_x > 0$) would favor a choice of precaution prior to observing type. These effects are shown in Fig. 3 by shifts of the indifference curve away from the origin in the case of concave precautionary demand, and toward the origin when demand for precaution is convex. We now summarize these results.

Theorem. Risk averse injurers may purchase liability insurance instead of becoming informed about type and taking due care. Higher levels of injurer risk aversion favor liability insurance, while concavity of precautionary demand makes due care more attractive.

As a result, the demand for liability insurance should arise from risk averse injurers who face a high variance classification lottery, but for whom the probability of being classified adversely (an α_1-type) is low, so that λ is small and the liability insurance premium (and thus the tort internality) is modest. Indeed, it is precisely for such low-probability, high variance outcomes that individuals tend to buy liability insurance.

IV. THE EFFICIENCY OF VOLUNTARY INSURANCE

Finally, a decision to purchase liability insurance by the injurer will result in a Pareto improvement relative to the alternative regime in which the injurer opts to become informed and then exercises due care. Clearly, the injurer must be made better off, for otherwise she would have not chosen to purchase the insurance. Moreover, it is easy to show that the injured party prefers the insurance outcome. In the absence of liability insurance, the injurer takes due care $X(\alpha_i)$ in state α_i, so that a risk-averse victim would purchase full insurance against injury at a premium of $p(X(\alpha_i))\alpha_i L$. When the injurer purchases insurance, however, the level of care chosen, $\hat{x}$, is insufficient to escape liability in the α_1 state, so that the victim is fully compensated for losses in that case and does not need the protection afforded by the insurance. Indeed, only in the α_2 state is the injurer not negligent, and the victim can purchase insurance against injury at a state-contingent premium of $p(\hat{x})\alpha_2 L$, which is less than that of the no-insurance case because $\hat{x} > X(\alpha_2)$. Thus, the victim is better off in each possible state, so that the voluntary purchase of liability insurance must be efficient.

Even though Pareto-improving, the voluntary purchase of liability insurance does not minimize the total costs of accident and precaution. The decision to purchase insurance entails a trade-off between the information required to select more efficient (cost-minimizing) levels of precaution, on the one hand, and the injurer's aversion to the risk associated with the classification lottery, on the other.

There is an inherent tension between these goals which arises because of a missing insurance market.

There are two lotteries in the model: the injurer faces the randomness of classification while the victim is confronted with the insurable financial loss associated with an accident. Were classification insurance available, the injurer could shed the risk associated with the first lottery, and then become informed about type and take the (non-negligent) cost-minimizing level of precaution. Victims would then purchase first-party insurance against accidents to mitigate the randomness associated with the second lottery. With insurance against both classification risk and accidental injury, there is no friction between risk aversion and the selection of precaution by an injurer.

The problem is that the precautionary decision often entails the selection of an action by the injurer which may be inherently uninsurable, such as shoveling the snow off a sidewalk or clearing an oil spill off of a factory floor. Because of the difficulty involved with insuring such activities, the market for classification insurance is imperfect at best. The result is a distinctly second-best outcome, where its insurability of accident losses is distorted, through the mechanism of liability insurance, to deal with the (inherently uninsurable) classification risk. As a consequence, classification risk is mitigated, but at the cost of inefficient levels of precaution.

IV. CONCLUSIONS

Typically, individuals are not well informed of the risks they impose on others, of the liability rules that might internalize those costs to the injurer, and of the possibilities for mitigating those risks. Imperfect information poses a choice: either become informed about these issues and invest in due care to escape potential liability, or buy liability insurance. Of course, if information were very costly, people would choose to become insured. But, even if information were free, we find that a demand for insurance will exist. Insurance is a substitute for perfect information and investing in due care.

The choice between investing in information and taking due care or buying insurance involves a balance of three effects. First, the prospective cost of due care is viewed as random by the initially uninformed injurer. Information on risk type might reveal the cost of investing in due care to be large or small. By permitting the investment in care to precede observation of type, liability insurance hedges the injurer's classification risk, a strategy which is of greater value when the individual is highly risk averse or when the classification lottery has high variance.

Second, the choice between due care and insurance changes the size of the externality created by the injurer's activity. With due care, the injurer escapes liability and all cost of injury is borne by the victim. With insurance, the decision is undertaken without knowledge of risk type and there is no guarantee that the chosen level of care will satisfy the due care standard that will be established subsequently for the injurer's type. Thus, under liability insurance part of the cost of an accident is internalized by the injurer. It is this internalized cost that is the insurable exposure and which forms the basis for the premium. Clearly, this internalized cost generates a bias against insurance demand.

The third component of the demand for insurance arises from the properties of the precaution technology. With full information, different levels of investment in care will be chosen by injurers of different type. These levels of precaution satisfy the due care standard and are socially efficient insofar as they minimize the joint costs to society of accidents and care. Without prior information on type, as in the case where the injurer opts to remain uninformed and purchase liability insurance instead, a common choice of care will be selected by all potential injurers. While such a choice cannot minimize the costs of accident and precaution ex post, it may nevertheless involve a private ex ante cost saving to the injurer if the precaution function is convex. In contrast, if this function is concave, the expected costs of care from those choosing to become informed will tend to be higher than for those choosing the insurance alternative.

These various effects define a demand function by injurers for liability insurance. Liability insurance will be in high demand for the more risk averse injurers and for those facing a high classification variance and a convex precaution technology. Finally, in identifying this demand model, we have resisted the temptation to consider the cost of information. If information on establishing risk type and the required level of due care is costly, it is trivial to show that some will forego this investment and buy insurance. With such transaction costs, the demand for insurance will be more robust than we have indicated. We have shown, however, that such costs are not necessary to generate a demand for liability insurance, and that insurance can be rationalized from the more subtle interaction of risk aversion, classification risk and the properties of the precaution technology.

NOTES

1. See Shavell (1987) for a discussion, as well as a demonstration of the simple negligence result discussed above. Sarath (1991) discusses how uncertain legal standards generate a demand for liability insurance and this theme is developed by Bajtelsmit & Thistle (1994) who argue that legal uncertainty is a necessary, but not sufficient, condition for liability insurance to exist. We will show that legal uncertainty is not a necessary condition.

2. The assumption of costless information is conservative. Our purpose is to explain liability insurance. Clearly, if injurers bear a cost of becoming informed about the relevant negligence standard (say, by consulting an attorney or other agent), or about their investment opportunities in care (by conduction a risk analysis), then liability insurance becomes even more attractive.

3. This rule was first enunciated by Justice Hand in *United States v. Carroll Towing Co.*, 159 F. 2d 169 (2d Cir. 1947).

4. The derivative conditions assumed for the precautionary technology, p, guarantee that the relevant second-order conditions are satisfied.

5. It is easy to show that these decisions are mutually exclusive. If an individual observes type and takes due care, there is no demand for liability insurance since the injurer is shielded from tort liability. Alternatively, once liability insurance is purchased, there will be no gain to observing type, since the insurance protects the individual from liability. Also, we assume throughout that the insurance firms cannot independently discover an agent's α.

6. This result is well known. See, for example, the discussion in Shavell (1987).

REFERENCES

Bajtelsmit, V., & Thistle, P. (1994). Uncertain Legal Standards and the Market for Insurance, paper presented to the American Risk and Insurance Association, Toronto.

Calfee, J., & Craswell, R. (1984). Some Effects of Uncertainty on Compliance with Legal Standards. *Virginia Law Review*, *70*, 965–1003.

Diamond, P. (1974). Single Activity Accidents. *Journal of Legal Studies*, *3*, 107–164.

Pratt, J. W. (1964). Risk Aversion in the Small and in the Large. *Econometrica*, *32*, 122–136.

Sarath, B. (1991). Uncertain Litigation and Liability Insurance. *Rand Journal of Economics*, *2*, 218–231.

Shavell, S. (1985). Uncertainty over Causation and the Determination of Civil Liability. *Journal of Law and Economics*, *28*, 587–609.

Shavell, S. (1987). *Economic Analysis of Accident Law*. Harvard University Press, Cambridge, Mass.

APPENDIX

In the body of the chapter, we assumed that the insurer knew which consumers had taken become informed and taken due care and who remained uninformed about tort liability. The insurer offered a policy to those who were uninformed which was priced on a weighted average of the expected losses of the high risk, α_1, and low risk, α_2, types. Consequently, this policy might prove attractive to the type 1 injurer. This was not a problem in section II, since the insurer knew who was informed and would simply refuse to sell the policy to a known type 1. Now suppose that this assumption no longer holds; the insurer cannot observe information status. The known type 1 injurer might well be tempted to buy the insurance policy instead of taking due care. And, the insurer cannot exclude this purchase since it cannot distinguish between informed and uninformed consumers. In short, there is an adverse selection problem. In this section we will identify the necessary equilibrium conditions for purchase of insurance when insurers cannot monitor information status of injurers.

A type i injurer who discovers his type and takes due care, will have wealth

$$W_i = \bar{W} - cX(\alpha_i)$$

Now, if insurance is designed for the uninformed at a premium π_u, the wealth of the consumers in the loss and no loss states will be:

$$W_N \equiv \bar{W} - \pi_u - c\hat{x}; \qquad W_L \equiv \bar{W} - \pi_u - c\hat{x} - L + I$$

Writing $W_i \equiv \bar{W} - cX(\alpha_i)$ as the utility of the informed type 'i' injurer who takes due care for his risk type, the following self selection constraint is needed to prevent informed type 1's masquerading as uninformed and purchasing liability insurance:

$$U(W_1) \geq (1 - \alpha_1 p(\hat{x}))U(W_N) + \alpha_1 p(\hat{x})U(W_L) \qquad SSC(1)$$

We assume that SSC(1) is binding. We assume that the self selection constraint for type 2 injurers, is not binding. This is trivial; if they buy insurance they avoid risk but at a premium that subsidizes the type 1's whereas if they simply take due care they avoid risk without paying this subsidy. Thus:

$$U(W_2) > (1 - \alpha_2 p(\hat{x}))U(W_N) + \alpha_2 p(\hat{x})U(W_L) \qquad SSC(2)$$

Now consider whether an uniformed injurer would choose to become informed and take the level of due care appropriate for his type, or buy insurance. For the time being, we retain the assumption that the cost of becoming informed is zero. The expected utility gain is $\Delta V = V_{DC} - V_{LI}$:

$$\Delta V = \lambda U(W_1) + (1 - \lambda)U(W_2) - [(1 - \bar{\alpha}p(\hat{x})\ U(W_N) + \bar{\alpha}p(\hat{x})U(W_L)].$$

Substituting SSC(1) and adding and subtracting $(1-\lambda)[(1-\alpha_2 p(\hat{x}))U(W_N) + \alpha_2 p(\hat{x})U(W_L)]$, gives

$$\Delta V = \lambda[(1-\alpha_1 p(\hat{x})U(W_N) + \alpha_1 p(\hat{x})U(W_L)] - [(1-\bar{\alpha}p(\hat{x})U(W_N) + \bar{\alpha}p(\hat{x})U(W_L)] + (1-\lambda)[(1-\alpha_2 p(\hat{x})\, U(W_N) + \alpha_2 p(\hat{x})U(W_L)] + (1\text{-}\lambda)[U(W_2) - (1-\alpha_2 p(\hat{x})U(W_N) - \alpha_2 p(\hat{x})U(W_L)].$$

Now, since $\bar{\alpha} = \lambda\alpha_1 + (1-\lambda)\alpha_2$, the first three terms on the right hand side of ΔV cancel out leaving only the final term. However, the final term is simply $(1-\lambda)$ times the non-binding self selection constraint SSC(2). Consequently, ΔV must be positive. Under these circumstances, all injurers would choose to become informed and take the appropriate level of due for their risk type. No one would buy insurance. The reason insurance is unattractive is as follows. Since the insurer is unable to distinguish uninformed from informed high risk types, remaining uninfomed and buying an insurance policy entails a signalling cost, one must signal that one is not an informed high risk type. This signalling cost effectively eliminates the downside risk from becoming informed. The inability of the insurer to monitor information status, implies that it must offer a menu in which the informed are never worse off than if they remained uninformed. Thus, the value of the information lottery must be positive and nobody would choose to buy insurance.

Given the inability to monitor, would people ever choose to buy insurance? The answer depends on the cost of information. If information is sufficiently costly, it is straightforward to show that the value of ΔV will switch sign to become negative. This can be seen by inserting an information cost k into ΔV as follows then simply substituting as before.

$$\Delta V = \lambda U(W_1 - k) + (1-\lambda)U(W_2 - k) - [(1-\bar{\alpha}p(\hat{x})U(W_N) + \bar{\alpha}p(\hat{x})U(W_L)]$$

We can now state the result of this Appendix:

Theorem. If insurers cannot monitor the information status of injurers, then a necessary condition for injurers to buy insurance, is that it is costly for injurers to acquire information on risk type and on due care standards.

BEHAVIOR-BASED PRICE DISCRIMINATION AND CONSUMER SWITCHING

Maria Arbatskaya

ABSTRACT

A common practice among utility companies is to offer discounts to consumers who use a rival's services in an attempt to induce them to switch suppliers. This chapter examines a two-period model of price competition on a Hotelling line that captures this type of price discrimination. In the first period, firms have no information about individual consumers' preferences and, therefore, they post a single price. In the second period, each firm gains information about consumers' first-period purchase decisions. We show that firms have an incentive to use this information to price discriminate. A firm charges a lower price to its rival's customers ('pays consumers to switch') whenever the firm is not too disadvantaged with respect to its marginal cost. Even when consumers' switching costs are non-trivial, a re-segmentation of the market prevails in the unique subgame-perfect equilibrium to the game. An analysis of the impact of this kind of price discrimination on consumer surplus, firm profits, and social welfare is also presented.

I. INTRODUCTION

In the United States and the United Kingdom, among other countries, deregulation of energy markets is on the way.[1] Though savings to consumers

Industrial Organization, Volume 9, pages 149–171.

ISBN: 0-7623-0687-4

from switching can be considerable,[2] many consumers are reluctant to switch their suppliers. In fact, the U.K. Trade Secretary admitted that despite average household savings of up to 65 pounds and 20 pounds a year on gas and electricity bills, more than 80% of domestic users remained with their original gas and electricity providers by the end of November, 1999. The small amount of switching is a puzzling observation, which has important implications for current and future industry deregulations. It became a concern to industry regulators and prompted investigations into consumer switching behavior. While empirical studies document consumer switching and try to explain it,[3] most of the theoretical literature on switching costs has produced models in which switching does not occur in an equilibrium (see the literature review in Klemperer (1995)). Caminal & Matutes (1990) and Chen (1997) are among few exceptions. This chapter shows that switching is to be expected in the segment pricing model, and it further analyzes the direction and the amount of switching that takes place.

'Paying consumers to switch' is a common practice in many markets. The most well-known example is that of AT&T, which offered discounts on long-distance calls to MCI's customers in an effort to persuade them to change their service provider. Credit card companies offer teaser interest rates on new accounts. Energy consumers receive offers of up to $50 for signing up with a marketer in U.S. electricity and natural gas markets in the states open to competition. On the other hand, retailers give consumers coupons on purchases of the same product and mail-order companies send flyers with rebates to their customers ('paying consumers not to switch').[4] These examples illustrate a particular kind of price discrimination studied in the chapter – discrimination based on the consumer purchase history.

In a small community, the information about consumer purchase history is likely to be available to shopkeepers. Supermarkets and department stores do not usually keep track of the consumption pattern of each customer. Still, they can target coupons to their current customers or to a rival's current customers.[5] In this chapter, we only assume that a firm knows who the consumer's current supplier is and can price discriminate based on this information. The results apply best to repeated-purchase markets in which the fact that a consumer does not purchase from a firm means that the consumer buys from a rival firm.[6] In the duopolistic markets we consider, sellers would know exactly where the consumer purchases the product.

Even when information is readily available to firms, their ability to price discriminate between groups of consumers may be limited by the existence of switching costs. At the extreme, when consumers have prohibitively high switching costs, firms enjoy a local monopoly power. It is commonly believed that higher switching costs are associated with lower consumer surplus since they discourage consumers from switching suppliers, grant sellers higher monopoly power, and thus lead to higher prices. This may not be true in a dynamic model: competition for market shares in the first period of the game can be intensified to the extent that an increase in switching costs benefits consumers overall, even though they would be worse off in the second period due to higher switching costs and therefore higher second-period prices (Klemperer (1987a)).[7]

The literature on third-degree price discrimination have suggested that the ability to price discriminate may decrease the profits of firms competing in imperfectly competitive markets.[8] The reason is that firms try to steal each other's customers. When firms can discriminate, they charge higher prices in their strong market (with relatively low elasticity of demand) and lower prices in their weak market. When one firm's weak market is another firm's strong market, this leads to 'all-out competition' and lower profits for each firm. This intuition is confirmed in the current framework of the price discrimination on the Hotelling line.

We consider two firms located at the ends of the Hotelling line. They sell a differentiated product and compete in prices for two periods.[9] In the first period, firms set uniform prices and generate shares of consumers. In the second period, firms can price discriminate between their own and their rival's consumers and consumers can switch at a cost to the alternative seller. We assume that consumers have fixed preferences (they do not change their location on the Hotelling line) and are not acting strategically in their first-period choice of supplier. The latter assumption simplifies the analysis and allows us to obtain a unique subgame-perfect equilibrium (SPE) of the game. Consumers may act in their short-term interest due to the lack of information.[10]

This dynamic segment pricing model allows us to show that competition in the first period will be intensified by the fact that firms with larger first-period shares have a larger pool of loyal consumers which can be exploited in the second period. Higher switching costs further promote fierce competition for market shares in the first period, while price discrimination works in the opposite direction since it tends to lower second-period profits. The overall effect on total profits and consumer surplus depends on the relative strength of the two effects.

The analysis of the dynamic segment pricing model is divided into two parts. We first study segment pricing in a market with a fixed initial division of consumers between the firms, which in itself is an interesting model to consider. We find conditions on the initial market share of firm 1 which lead to a positive amount of switching and to two-sided switching. In the case of two-sided switching, the amount of switching is computed. It is also shown that profits are generally non-monotonic with respect to the extent of a firm's initial presence in the market. That is, a firm may actually prefer to have a smaller initial market share. Intuitively, the firm lets its rival have a larger pool of consumers to benefit from relaxed second-period competition.

In the second part, we endogenize the initial market division by considering the two-period model with uniform pricing in the first period and segment pricing in the second period. We find that some switching always occurs when switching costs are not too high and that whenever firms' marginal costs are not too asymmetric, a two-sided redistribution of the market occurs in the two-period model. In contrast to the segment pricing model, no switching ever takes place with uniform pricing in both periods. Any amount of switching is inefficient in the model with fixed consumer preferences, equal production costs, and perfectly inelastic demand. We find that when firms have similar marginal and consumer switching costs the ability of firms to price discriminate reduces the industry profits and social welfare, although it increases consumer surplus.

The next section provides an overview of the segment pricing model. Section III outlines the solution to the two-period uniform pricing model. In Section IV, the equilibrium pricing policies are computed in the case when price discrimination is based on an initial segmentation of the market. A two-period game that endogenizes the initial division of the market is solved in Section V. There we also compare the outcomes in segment pricing and uniform pricing models with respect to the realized profits, consumer surplus, and social welfare. Concluding remarks follow in Section VI.

II. SEGMENT PRICING: AN OVERVIEW

Consider two firms located at the ends of a Hotelling line with consumers uniformly distributed between the locations. Firms move simultaneously in both periods by choosing a uniform price in the first period and possibly discriminatory prices in the second period. The difference between the two periods is in the segmentation of the market: only in the second period can the firms separate their own and their rival's first-period customers. Marginal costs and exogenous switching costs (called switching costs hereafter) are known to all parties.[11]

In the first period, firm i charges a uniform price p_i to all consumers to maximize its total profits in two periods

$$\Pi_i = (p_i - c_i)\alpha_i(p_1, p_2) + \pi_i^*(\alpha_i(p_1, p_2)) \tag{1}$$

where c_i is the marginal cost of firm i, α_i is firm i's first-period measure of consumers, and π_i^* is the second-period profit as a function of the firm's initial market share, given the optimal choice of second-period prices, $i=1, 2$. It is assumed that there is no discounting between the periods. In the second period, firm i sets prices p_{ii} and p_{ij} to its own and its rival's first-period customers to maximize its second-period profits

$$\pi_i = \alpha_{ii}(p_{ii} - c_i) + \alpha_{ji}(p_{ij} - c_i)$$

where $i \neq j = 1, 2$; p_{ij} is the price charged by firm i to firm j's first-period consumers, and α_{ij} is the share of consumers purchasing from firm i in the first period and from firm j in the second period, $i, j = 1, 2$.

Throughout the chapter we assume that consumers have the same baseline valuation of the product $V > 0$ and their willingness to pay differs only because of the differences in their location, x, and the associated traveling costs. This reservation price V is high enough to guarantee that the market is fully covered so that in each period consumers buy a unit of the good from one or the other seller. Transportation costs are linear and equal to $t > 0$ per unit of distance. The distance between the firms and the range of consumer locations equal to $\theta > 0$. Denote by $s_i \geq 0$ the cost to a consumer of switching to firm i and by S the sum of the switching costs, $S = s_1 + s_2$. Assume that switching costs are not too high, so that condition

$$S < 2t\theta \tag{A1}$$

is satisfied.

Firm 1's consumer payoffs in the first period are $(V - p_1 - tx)$ and in the second period $(V - p_{21} - t(\theta - x) - s_2)$ if the consumer switches the supplier and $(V - p_{11} - tx)$ otherwise. Firm 2's consumer payoffs are similarly defined. Given the dynamic nature of the game, we solve it backwardly. In the second period, firm i sets a price p_{ii} to its own first-period customers and price p_{ij} to its rival's customers. In choosing first-period prices, p_i, firms account for direct and strategic effects of their choices on their total profits.

We rule out the possibility of strategic behavior by buyers, assuming that in each period they maximize their current utility from consumption. It seems quite plausible that due to the lack of information consumers are not able to foresee the outcome of the second-period price competition between the firms. Other interpretations of this assumption are provided in the introduction. If consumers were to solve an intertemporal consumption problem, they could choose to buy from a less-preferred supplier in the first period so as to switch to their favorite supplier in the second period and enjoy a price discount that this supplier would give its rival's first-period customers. Rational consumers are not as sensitive as myopic consumers to first-period price cuts since a price cut by a firm means that the firm will have a larger first-period market share and hence it will charge a higher price to its own consumers in the second period.

By maximizing their overall utility, rational consumers try to avoid the situation in which they are locked in and exploited by the firm. Let u_{ij} be the utility the consumer located at x obtains from buying firm i's product in the first period and firm j's product in the second period, $i, j = 1, 2$. It is clear that the utility levels of the switchers do not depend on their locations since their traveling cost is $t\theta$ regardless of the location. For two-sided switching to occur, the switchers' utility levels have to be equal, $u_{12} = u_{21}$. This indifference requires one to set an ad hoc tie-breaking rule which lead to complications we choose to avoid by assuming that consumers maximize their static payoffs.

III. TWO-PERIOD UNIFORM PRICING

We consider a two-period uniform pricing model as a benchmark for comparison with the segment pricing model. It is assumed that firms simultaneously set prices in two periods and are restricted to charge the same price to all consumers. We show that under uniform pricing, two-sided redistribution of the market is not possible for any initial division of the market.

Denote the market share of firm i in period t, $\alpha_{i,t}$, and its uniform price, $p_{i,t}$, $i, t = 1, 2$. In the second period, consumers are divided into firm 1 and firm 2's customers and incur a non-negative cost of switching when changing their supplier. The marginal first-period consumer is located at $x = \alpha_{1,1}$. Suppose in the second period two-sided switching were possible. Then, there would be two marginal consumers (consumers just indifferent between buying from firm 1 and firm 2) located at x and y: $x < \alpha_{1,1} < y$ in firm 1 and firm 2's market segments. This is not possible. Firm 1's marginal consumer located at x is indifferent between buying at firm 1 and switching to firm 2 and hence a consumer located to the right of $\alpha_{1,1}$ strictly prefers to buy at firm 2 due to savings on traveling and switching costs. Hence, there cannot exist a firm 2's marginal consumer located at $y > \alpha_{1,1}$.[12]

Next we show that even one-sided switching cannot arise in the second period of the two-period model. Without loss of generality, suppose that some consumers switch from firm 2 to firm 1 in the second period. The difference between the competition in two periods is in the existence of the switching costs in the second period. For switching from firm 2 to firm 1 to occur, the marginal consumer in period two has to be located to the right of the marginal consumer in period one, $\alpha_{1,1} < \alpha_{1,2}$. This is not possible given that firm 1's customers have to incur an additional switching cost in period two.[13]

Therefore, no switching occurs when firms charge uniform prices in two periods and the two-period uniform-price competition reduces to a simple repetition of the static price competition. Firm i's profits are $\pi_i = (p_i - c_i)\alpha_i(p_1, p_2)$, where $\alpha_i(p_1, p_2) = \frac{1}{2t}(p_j - p_i + t\theta)$ is firm i's market share. In the unique Nash equilibrium to the game, firm i charges price $p_i^U = \frac{2}{3}c_i + \frac{1}{3}c_j + t\theta$, and it enjoys profits $\pi_i = \frac{1}{18t}(3t\theta - c_i + c_j)^2$ and a market share $\alpha_i^U = \frac{\theta}{2} + \frac{1}{6t}(c_j - c_i)$, $i \neq j = 1, 2$. The two-period total industry profits are $\Pi^U = \frac{2}{9t}(c_1 - c_2)^2 + 2t\theta^2$.

One-period consumer surplus under uniform pricing is the utility from consuming the product net of price and traveling expenditures,

$$CS_{t=1}^U = \int_{z=0}^{\alpha_1^U} (V - p_1^U - tz)dz + \int_{z=0}^{\alpha_2^U} (V - p_2^U - tz)dz$$

Consumer surplus in the two-period game is

$$CS^U = 2\theta V - \theta(c_2 + c_1) + \frac{1}{18t}(c_2 - c_1)^2 - \frac{5}{2}t\theta^2$$

and social welfare, $W^U = CS^U + \Pi^U$, is

$$W^U = 2\theta V - \theta(c_2 + c_1) + \frac{5}{18t}(c_2 - c_1)^2 - \frac{1}{2}t\theta^2$$

Note that consumer surplus, industry profits, and social welfare are increasing with the difference in the firms' marginal cost. The reason is that production efficiency improves as one of the firms becomes relatively more efficient, keeping the mean marginal cost constant. Switching costs in no way affects the equilibrium outcome under uniform pricing.

IV. SEGMENT PRICING IN MATURE MARKETS

In this section we assume that firm i (i = 1, 2) has a fixed initial consumer segment of measure α_i and that the firms can price discriminate between their own and the rival's customers, $\alpha_1 + \alpha_2 = \theta$. Using the heterogeneous product model with consumer switching costs we find the equilibrium to the pricing game, provide conditions for two-sided and one-sided switching, derive the upper bound for the amount of switching, and discuss some comparative statics results.

As shown in Section III, uniform pricing in mature markets necessarily leads to one-sided switching, if any switching at all. Price-discrimination, however, can result in two-sided consumer switching, i.e. some consumers switch from firm 1 to firm 2 while some other consumers switch from firm 2 to firm 1. The equilibrium market shares can be found by calculating the locations of the consumers just indifferent between staying with their initial supplier and switching to the other firm.

Equilibrium

By x denote the location of firm 1's marginal consumer. Since the consumer obtains $V - p_{21} - t(\theta - x) - s_2$ from switching to firm 2, and $V - p_{11} - tx$ from staying with the original supplier,

$$x = \frac{1}{2t}(p_{21} - p_{11} + t\theta + s_2) \tag{2}$$

Similarly, firm 2's marginal consumer derives the benefit of $V - p_{12} - ty - s_1$ from firm 1's product and $V - p_{22} - t(\theta - y)$ from firm 2's product. By y denote the location of firm 2's marginal consumer,

$$y = \frac{1}{2t}(p_{22} - p_{12} + t\theta - s_1) \tag{3}$$

Assume the locations of the marginal consumers are such that inequality $x < \alpha_1 < y$ holds, as in Fig. 1 (Case 1). This condition guarantees two-sided switching and we focus on this case in the chapter.

The firm profits are

$$\begin{cases} \pi_1 = x(p_{11} - c_1) + (y - \alpha_1)(p_{12} - c_1) \\ \pi_2 = (\alpha_1 - x)(p_{21} - c_2) + (\theta - y)(p_{22} - c_2) \end{cases} \tag{4}$$

where the locations of the marginal consumers are defined by equations (2) and (3).

From the linear system of four first-order conditions (π_i with respect to p_{ij}, $i, j = 1, 2$) we obtain the firms' (upward-sloping) best response functions. The resulting firms' prices and market shares are listed in columns 1 and 2 of Table 1. The second-order sufficient condition is satisfied since $\partial^2\pi_i/\partial p_{ij}^2 = -1/t < 0$, $i, j = 1, 2$.

Contrary to the independence result in Chen (1997), equilibrium prices depend on the firm's initial market share. The difference arises due to the heterogeneity of consumer preferences in our model. As the firm's market share expands, the composition of the firm's consumer base changes since the additional consumers are located further from the firm and closer to its rival. It

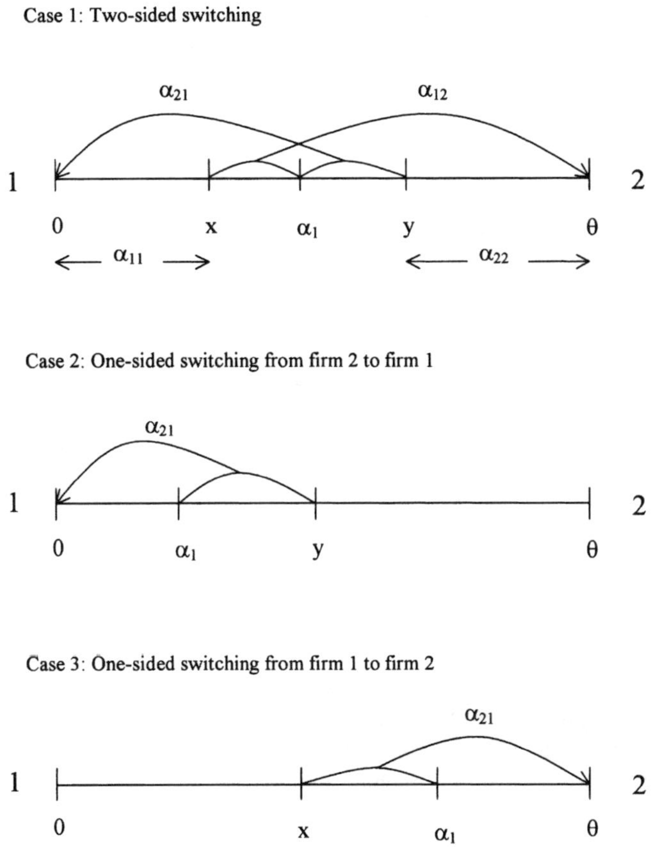

Fig. 1. Switching Patterns in Mature Markets.

Notes: For intermediate levels of firm 1's initial market share, α_1, two-sided switching occurs in the market (Case 1). For small α_1, consumers only switch from firm 2 to firm 1 (Case 2). Finally, for large α_1, consumers only switch from firm 1 to firm 2 (Case 3); α_{ij} is the share of consumers buying from firm i in period 1 and from firm j in period 2, $i, j = 1, 2$.

becomes more costly for the firm to keep its consumers from switching (in terms of the losses it incurs on the loyal consumers). The incentive to exploit the loyal customers dominates and the firm's price increases in the firm's initial market share. The price the firm charges its rival's customers decreases with the firm's initial market share, as the firm compensates the consumers for higher

Table 1. Prices and Market Shares in Mature Markets ($i \neq j = 1, 2$).

(1) Prices	(2) Market Shares
$p_{ii} = \frac{1}{3}(2c_i + c_j + 2\alpha_i t + t\theta + s_j)$	$\alpha_{ii} = \frac{1}{6t}(c_j - c_i + 2\alpha_i t + t\theta + s_j)$
$p_{ij} = \frac{1}{3}(2c_i + c_j + 3t\theta - 4\alpha_i t - s_i)$	$\alpha_{ji} = \frac{1}{6t}(c_j - c_i + 3t\theta - 4\alpha_i t - s_i)$

traveling costs. As a result, firm i tends to price discriminate[14] more against its own customers as α_i increases.

Consumer Switching

Let $\delta = c_1 - c_2$ denote the difference in the firms' marginal costs. The comparison of the locations of the marginal consumers $x = \frac{1}{6t}(-\delta + 2\alpha_1 t + t\theta + s_2)$ and $y = \frac{1}{6t}(-\delta + 2\alpha_1 t + 3t\theta - s_1)$ implies that $x < y$ under assumption (A1).

Proposition 1. Assumption A1 is a necessary and sufficient condition for a positive amount of switching in the price-discriminating equilibrium.

Proof. To show that some consumers will switch their suppliers in the equilibrium, we have to demonstrate that $\alpha_{12} + \alpha_{21} > 0$. This inequality holds for switching costs which are not too large, $s_1 + s_2 < 2t\theta$. ■

In other words, the simple mean of the firms' switching costs should not exceed the cost of traveling along the Hotelling line from one firm to the other. Condition (A1) guarantees that there is a re-segmentation of the market, with some consumers switching their suppliers.

Three possibilities arise with respect to the location of the initial market boundary α_1. For intermediate levels of α_1 the market boundary falls in-between the equilibrium locations of the marginal consumers (Case 1). For small firm 1's initial market shares, the market boundary is to the left of x and y (Case 2). Finally, for large α_1, the market boundary is to the right of x and y (Case 3). See Fig. 1. Define the lower and the upper critical values on firm 1's initial market share, $\underline{\alpha}_1 = \frac{1}{4}\theta + \frac{1}{4t}(-\delta + s_2)$ and $\overline{\alpha}_1 = \frac{3}{4}\theta - \frac{1}{4t}(\delta + s_1)$. Note that given (A1), $\underline{\alpha}_1 < \overline{\alpha}_1$. The corresponding critical values for firm 2 are $\underline{\alpha}_2 = \theta - \overline{\alpha}_1$ and $\overline{\alpha}_2 = \theta - \underline{\alpha}_1$.

Proposition 2 characterizes the conditions on the parameters of the model which lead to one-sided and two-sided exchange of customers between the firms in the equilibrium.

Proposition 2. Two-sided switching occurs when $\alpha_1 \in (\underline{\alpha}_1, \overline{\alpha}_1)$. Consumers only switch from firm 2 to firm 1 when $\alpha_1 \leq \underline{\alpha}_1$. Consumers only switch from firm 1 to firm 2 when $\alpha_1 \geq \overline{\alpha}_1$.

Proof. Case 1: two-sided switching. This case arises when $x < \alpha_1 < y$. Condition $x < \alpha_1$ is equivalent to $\alpha_1 > \underline{\alpha}_1$ and condition $\alpha_1 < y$ is equivalent to $\alpha_1 < \overline{\alpha}_1$.

Case 2: one-sided switching, from firm 2 to firm 1. This case arises when $\alpha_1 \leq x < y$ which is equivalent to $\alpha_1 \leq \underline{\alpha}_1$.

Case 3: one-sided switching, from firm 1 to firm 2. This case arises when $x < y \leq \alpha_1$ which is equivalent to $\alpha_1 \geq \overline{\alpha}_1$. ■

The results obtained in Proposition 2 are quite intuitive. When firm 1 has a sufficiently large initial market share, price competition results in a reduction in the firm's market share. On the other hand, when firm 1's market segment is sufficiently small, the firm ends up with a higher share of the market after some of the rival's consumers switch to the firm. For the intermediate values of firm 1's market segment, re-distribution of the market is two-sided, with an ambiguous net effect. For example, if $t = \theta = 1$, $c_1 = c_2 = c$ and $s_1 = s_2 = s$, then $\underline{\alpha}_1 = \frac{1}{4}(s+1)$ and $\overline{\alpha}_1 = \frac{1}{4}(3-s)$. That is, two-sided re-segmentation occurs if the initial market shares are not too asymmetric, i.e. for $\alpha_1 \in (\underline{\alpha}_1, \overline{\alpha}_1) \subseteq (0.25, 0.75)$.

To determine an upper bound on the amount of switching, one has to compute the measure of the switchers, $\alpha_{12} + \alpha_{21} = \frac{1}{6t}(2t\theta - S)$. This leads to the claim in Proposition 3.

Proposition 3. The measure of switchers in the market is

$$\alpha_{12} + \alpha_{21} = \frac{1}{3}\theta - \frac{1}{6t}S \tag{5}$$

There is an upper bound of 1/3 on the fraction of switchers in the equilibrium.

The proportion of consumers who switch their suppliers decreases with the sum of switching costs, S, and increases with the traveling costs, t. The reason for the latter result is that with higher traveling costs consumers are less inclined to visit a more distant firm for the discount it provides. Interestingly, the consumer propensity to switch does not depend on the initial division of the market and the marginal cost asymmetries between the firms, as long as $\alpha_1 \in (\underline{\alpha}_1, \overline{\alpha}_1)$. The result of Proposition 3, of course, depends on the assumption of the uniform distribution of consumer locations. However, an appropriate bound can be derived for other distributions as well.

Two-sided re-segmentation of the market will be in favor of firm i when $\alpha_{ji} > \alpha_{ij}$. This inequality is a restriction on firm i's initial market share

$$\alpha_i < \frac{1}{2}\theta + \frac{1}{4t}(c_j - c_i) + \frac{1}{8t}(s_j - s_i) \tag{6}$$

Firm i has to be relatively small to have its market share increased in the equilibrium as a result of price discrimination and consumer switching.

Next we explore the effects of marginal costs, switching costs, and the initial market shares on the firm profits, starting with the effect of the initial market share, α_i.

Initial-Market-Share Effect

From profits in equation (4) and the equilibrium prices in Table 1, it follows that firm i's profits are a convex quadratic function of α_i which attains its minimum at the critical value

$$\alpha_i^c = \frac{1}{2}\theta - \frac{1}{10t}(c_i - c_j + 2s_i + s_j) \tag{7}$$

It is interesting to explore how a firm's profits change with the extent of the firm's initial presence in the market. Notice that $d\pi_i/d\alpha_i \geq 0$ iff $\alpha_i \geq \alpha_i^c$, where α_i^c is the critical level of i's initial market share, $i = 1, 2$. This means that the profits of firm i, π_i, are U-shaped with respect to α_i, $\alpha_i \in (\underline{\alpha}_i, \overline{\alpha}_i)$ (See Fig. 2). In the neighborhood of the critical value α_i^c profits are higher if firm i were smaller or bigger.

For example, for $t = \theta = 1$, $c_1 = c_2$, and $s_1 = s_2 = 0$, we obtain $(\underline{\alpha}_i, \overline{\alpha}_i) = (0.25, 0.75)$ and $\alpha_i^c = 0.5$. Firm 1's profits

$$\pi_1 = \frac{5}{9}(2\alpha_1^2 - 2\alpha_1 + 1)$$

are U-shaped with the minimum attained at $\alpha_1 = 0.5$. For symmetric firms, the profits are at the lowest level when the market is initially divided equally between the firms. Proposition 4 summarizes the main finding.

Proposition 4. Firm profits are generally non-monotonic and U-shaped in its initial share of the market.

Firm i's profits are decreasing at small values and increasing at high values of the firm's initial market share, $\alpha_i \in (\underline{\alpha}_i, \overline{\alpha}_i)$. If firm i's market share is sufficiently low, $\alpha_i < \alpha_i^c$, where α_i^c is defined in equation (7), the firm would prefer to have an even smaller market share in the first period. The reason is

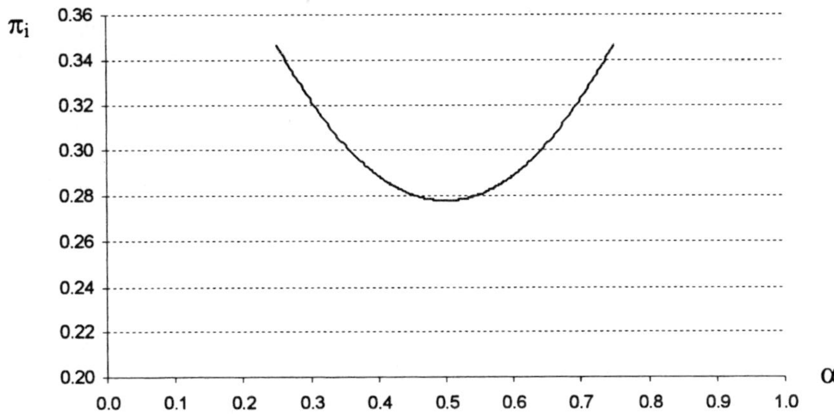

Fig. 2. Firm Profits As a Function of its Initial Market Share.

Notes: Firm *i*'s profits are a U-shaped function of its initial share of the market, $\alpha_i \in (\underline{\alpha}_i, \overline{\alpha}_i)$. For the assumed parameter values, $t=\theta=1$, $c_1=c_2$, and $s_1=s_2=0$, we obtain $(\underline{\alpha}_i, \overline{\alpha}_i)=(0.25,0.75)$ and $\alpha_i^c=0.5$.

that a relatively small firm would prefer its rival to have an even higher market share so that the rival prices less aggressively in the second period (the firm prefers its rival to be a 'fat cat'). This strategic effect can dominate the direct effect of a higher market share – the ability to exploit lock-in consumers. This is in contrast to the usual finding in the literature that firms have a severe fight for market shares.

Price Comparisons

The traditional result in the literature on third-degree price discrimination is that the non-discriminatory price lies between lower prices charged to switchers and higher prices charged to loyal customers. We can re-write prices as $p_{ii}=p_i^U-\frac{2}{3}\alpha_j t+\frac{1}{3}s_j$ and $p_{ij}=p_i^U-\frac{4}{3}\alpha_i t-\frac{1}{3}s_i$, $i\neq j=1,2$. The comparison of firm *i*'s uniform and discriminatory prices reveals that the price the firm sets to its rival's customers is always lower than the uniform price, $p_{ij}<p_i^U$. The relationship between the firm's uniform price and its price to its own customers is ambiguous. Firm *i*'s price in its own market segment is lower than the firm's uniform price, $p_{ii}<p_i^U$ iff $\alpha_i<\theta-\frac{1}{2t}s_j$. That is, firm *i*'s loyal consumers benefit from lower discriminatory prices when firm *i*'s share is not too high. When a firm has less than a half of the market it will not charge its own consumers more than the uniform price, given that $s_j < t\theta$. For sufficiently small switching costs, all-out competition ensues, with all prices being lower than the uniform price, $p_{ij} < p_i^U$, $i, j = 1, 2$.

The difference between the price firm i charges its own consumers and the rival's consumers is $p_{ii} - p_{ij} = 2t(\alpha_i - \frac{1}{3}\theta) + \frac{1}{3}S$. Firm i discriminates in favor of its own customers (PCNTS) when $p_{ii} < p_{ij}$. This inequality is equivalent to condition $\alpha_i < \frac{1}{3}\theta - \frac{1}{6t}S$, stating that firm i's market share is relatively low. Both firms cannot PCNTS since firms cannot each have a market share of less than a third of the market.

The difference between the prices charged by different sellers in the same market (price premium of firm i) is a function of the difference in the firms' marginal costs and the rival's switching costs and market share, $p_{ii} - p_{ji} = \frac{1}{3}(c_i - c_j) + \frac{2}{3}t\alpha_j + \frac{2}{3}s_j$. Whenever firm i is not too disadvantaged in its marginal cost, firm i's own consumers receive firm i's product at a premium.

Comparative Statics

Some comparative statics results for firms' equilibrium prices and profits are

(i) $\dfrac{\partial p_{ii}}{\partial s_i} = 0$ and $\dfrac{\partial p_{ij}}{\partial s_j} = 0$

(ii) $\dfrac{\partial p_{ii}}{\partial s_j} = 1/3$ and $\dfrac{\partial p_{ij}}{\partial s_i} = -1/3$

(iii) $\dfrac{\partial p_{ii}}{\partial \alpha_i} = 2t/3$ and $\dfrac{\partial p_{ij}}{\partial \alpha_i} = -4t/3$

(iv) $\dfrac{d\pi_i}{ds_i} < 0$

A firm's price to its own consumers, p_{ii}, does not depend on s_i and increases in s_j, the cost of switching to the rival. A firm's price to its rival's consumers, p_{ij}, does not depend on s_j and decreases in s_i. As the cost of switching to the firm increases the firm partially compensates its rival's consumers for switching. According to (iv), a firm obtains lower profits when it becomes more costly for consumers to switch to the firm. Statement (iv) is equivalent to inequality $\alpha_i < \bar{\alpha}_i \equiv \frac{3}{4}\theta - \frac{1}{4t}(c_i - c_j + s_i)$, which holds when two-sided switching occurs.

We could not find parameters for which a firm obtains higher profits when it becomes more costly for consumers to switch to the rival. That is, in simulations we always arrived at $d\pi_i/ds_j > 0$, which is true iff $\alpha_i > -\frac{1}{2}\theta + \frac{1}{2t}(c_i - c_j - s_2)$. Simulations also resulted in $d\pi_i/dc_i < 0$, which is true iff $\alpha_i < 2\theta - \frac{1}{t}(c_i - c_j) + \frac{1}{2t}(s_j - s_i)$. Since $d\pi_i/dc_i = -d\pi_i/dc_j$, the same condition guarantees that the firm profits increase in the marginal cost of its rival, $d\pi_i/dc_j > 0$, $i \neq j = 1, 2$.

V. TWO-PERIOD SEGMENT PRICING

In the previous analysis we treated the initial market shares of the firms as being exogenously given. In contrast, this section allows these shares to be endogenously determined in the first period. We look for a subgame-perfect equilibrium to the two-period game in which firms employ uniform pricing strategies in the first period and segment pricing policies in the second period. That is, in the second period firms can price discriminate between their own and their competitor's first-period customers. We assume that consumer behavior (the choice of which store to visit) is based only on the current prices and not on the future prices of the firms.

Equilibrium

In the first period, firm i charges a uniform price p_i to all consumers to maximize its total profits in two periods

$$\Pi_i = (p_i - c_i)\alpha_i(p_1, p_2) + \pi_i^*(\alpha_i(p_1, p_2))$$

where the first-period market share of firm i is $\alpha_i = \frac{1}{2}\theta + \frac{1}{2t}(p_j - p_i)$, $i \neq j = 1, 2$. The first-order conditions

$$\frac{\partial \Pi_i}{\partial p_i} = \left(\alpha_i + (p_i - c_i)\frac{\partial \alpha_i}{\partial p_i}\right) + \frac{d\pi_i^*}{d\alpha_i}\frac{\partial \alpha_i}{\partial p_i} = 0 \tag{8}$$

can be solved to yield the equilibrium first-period prices p_i, $i = 1, 2$. The term in the brackets in equation (8) is the direct effect and the second term

$$\frac{d\pi_i^*}{d\alpha_i} = \frac{\partial \pi_1}{\partial \alpha_1} + \frac{\partial \pi_1}{\partial p_{21}}\frac{\partial p_{21}}{\partial \alpha_1} + \frac{\partial \pi_1}{\partial p_{22}}\frac{\partial p_{22}}{\partial \alpha_1}$$

is the indirect (strategic) effect of a price change on the firm profits (by the envelope theorem, $\partial \pi_1/\partial p_{11} = \partial \pi_1/\partial p_{12} = 0$). The firms' best response functions and prices are summarized in Table 2.

Table 2. Best Response Functions and Prices in the First Period ($i \neq j = 1, 2$).

Best Response Functions $p_i(p_j)$	Prices p_i
$p_i = -\frac{1}{8}p_j + \frac{7}{8}c_i + \frac{1}{4}c_j - \frac{1}{2}s_i - \frac{1}{4}s_j + \frac{9}{8}t\theta$	$p_i = \frac{6}{7}c_i + \frac{1}{7}c_j - \frac{10}{21}s_i - \frac{4}{21}s_j + t\theta$

Perhaps surprisingly, the best response functions are downward-sloping. Firms consider first-period prices as strategic substitutes, $\partial^2\Pi_i/\partial p_i \partial p_j = -\frac{1}{18t} < 0$. If firm i were to ignore the strategic effect that its choice of the first-period price has on the market share α_i and, consequently, on the second-period price competition, it would follow an ordinary upward-sloping one-period best response $p_i = \frac{1}{2} p_j + \frac{1}{2} c_i + \frac{1}{2} t\theta$. The equilibrium first-period prices for non-strategic firms would be $p_i^U = t\theta + \frac{2}{3} c_i + \frac{1}{3} c_j$, $i \neq j = 1, 2$. Since in the first-order condition (8) $\partial\alpha_i/\partial p_i = -\frac{1}{2t} < 0$ but term $d\pi_i^*/d\alpha_i$ has an ambiguous sign, an unambiguous comparison between the strategic and non-strategic prices cannot be made. Whether a firm's prices are higher or lower due to the strategic effect of the first-period market share depends on the parameters of the model. The strategic effect is negative and hence the firm strategically lowers its first-period prices iff $c_i - c_j < \frac{4}{3} s_i - \frac{1}{6} s_j$. Firm i is more likely to reduce its first-period prices when the firm has a relatively low marginal cost. Intuitively, the firm reduces its first-period price to obtain a larger market share when competition is expected to yield higher second-period profits to the firm and when it is more difficult for the firm to attract its rival's customers and to keep its own customers.

Firm i enjoys a market share

$$\alpha_i = \frac{\theta}{2} + \frac{1}{14t}(2(s_i - s_j) + 5(c_j - c_i)) \tag{9}$$

which is positive provided

$$c_i - c_j < \frac{7}{5} t\theta + \frac{2}{5}(s_i - s_j)$$

For firms with similar marginal costs and consumer switching costs, each firm necessarily serves some consumers in the first period. We restrict our attention to the cases where both firms are present in the market in the first period.

From the solution to the one-period game with fixed initial shares presented in Section IV it follows that prices and market shares are as presented in Table 3.

Consumer Switching

Since the amount of consumer switching does not depend on the initial market shares, the results on the intensity of consumer switching established in Section IV hold. At least some consumers switch their suppliers since $\alpha_{12} + \alpha_{21} = \frac{1}{3}\theta - \frac{1}{6t} S > 0$, which is satisfied given condition (A1). Two-sided switching occurs when $\alpha_1 \in (\underline{\alpha}_1, \overline{\alpha}_1)$ where $\underline{\alpha}_1 = \frac{1}{4}\theta + \frac{1}{4t}(-\delta + s_2)$,

Table 3. Prices and Market Shares in the Second Period ($i \neq j = 1,2$).

(1) Prices	(2) Market Shares
$p_{ii} = \frac{3}{7} c_i + \frac{4}{7} c_j + \frac{2}{21} s_i + \frac{5}{21} s_j + \frac{2}{3} t\theta$	$\alpha_{ii} = \frac{1}{42t} (12c_j - 12c_i + 2s_i + 5s_j + 14t\theta)$
$p_{ij} = \frac{8}{7} c_i - \frac{1}{7} c_j - \frac{11}{21} s_i + \frac{4}{21} s_j + \frac{1}{3} t\theta$	$\alpha_{ji} = \frac{1}{42t} (-3c_j + 3c_i - 11s_i + 4s_j + 7t\theta)$

$\overline{\alpha}_1 = \frac{3}{4}\theta - \frac{1}{4t}(\delta + s_1)$, and $\delta = c_1 - c_2$. Let $\Delta_i = \frac{7}{3} t\theta + \frac{4}{3} s_i - \frac{11}{3} s_j$, $i \neq j = 1, 2$. Two-sided re-segmentation of the market always occurs in the two-period game with relatively symmetric marginal and switching costs. This leads us to the following proposition.

Proposition 5. In the equilibrium to the two-period game, two-sided consumer switching occurs iff the difference in firms' marginal costs is not too large, $c_i - c_j < \Delta_i$ for $i = 1, 2$.

Proof. Inequality $\alpha_1 > \underline{\alpha}_1$ is equivalent to $c_1 - c_2 < \Delta_1$ and inequality $\alpha_1 < \overline{\alpha}_1$ is equivalent to $c_2 - c_1 < \Delta_2$. ■

For example, when firms have the same consumer switching costs $s_2 = s_1 = s$ and are not too asymmetric in their marginal costs, $\delta \in (-\frac{7}{3}(t\theta - s), \frac{7}{3}(t\theta - s))$, two-sided switching emerges in the equilibrium.

The net re-distribution of the market is in favor of firm 2, $\alpha_{12} > \alpha_{21}$, iff

$$\frac{2}{5}(c_1 - c_2) < s_1 - s_2$$

Firm 1's market share is reduced in the equilibrium if firm 1's disadvantage in marginal cost is low enough, compared to firm 1's disadvantage in consumer switching costs.

Price Comparisons

Switching costs have the expected effects on firms' prices. An increase in consumers' cost of switching to firm j makes it more attractive for firm i to capture a higher market share in the first period. Firm i decreases its first-period price, $\partial p_i / \partial s_j < 0$, and increases its second-period prices to both consumer groups, $\partial p_{ik} / \partial s_j > 0$, $i, j, k = 1, 2$ and $i \neq j$. An increase in consumers' cost of switching to firm i makes it more difficult for firm i to attract its rival's customers. To capture a larger initial consumer base, the firm reduces its equilibrium price in the first period, $\partial p_i / \partial s_i < 0$, and increases the extent of price discrimination in the second period, charging an even higher price to its first-period customers, $\partial p_{ii} / \partial s_i > 0$, and a lower price to potential switchers, $\partial p_{ij} / \partial s_i < 0$.

The difference between the equilibrium prices firm i charges its own consumers and its rival's consumers is $p_{ii} - p_{ij} = -\frac{5}{7}(c_i - c_j) + \frac{13}{21} s_i + \frac{1}{21} s_j + \frac{1}{3} t\theta$.

At higher consumer switching and traveling costs and for relatively low marginal cost, a firm is more likely to pay consumers to switch. For example, firms with sufficiently symmetric marginal costs always PCTS. Firm i discriminates in favor of its own customers (PCNTS, $p_{ii} < p_{ij}$) iff

$$c_i - c_j > \frac{1}{15}(13 s_i + s_j + 7t\theta) > 0$$

which is satisfied when firm i has a sufficiently higher marginal cost than firm j.

Profits, Consumer Surplus, and Social Welfare

We end this section by comparing consumer surplus, industry profits, and social welfare in the segment pricing and uniform pricing models. The two-period profits of firm i are defined in equation (1). To compare industry profits in discriminatory equilibrium, $\Pi = \Pi_1 + \Pi_2$, with the profits in the two-period uniform pricing model, Π^U, define $\Delta\Pi = \Pi^U - \Pi$. This difference in profits is a concave quadratic function of the difference in the firms' marginal costs of production, $\delta = c_1 - c_2$. For firms not too asymmetric in marginal and consumer switching costs, price discrimination results in lower industry profits. Numerical simulations confirm that industry profits are lower under segment pricing, although one firm can possibly gain in the price-discriminating equilibrium.

The consumers obtain the benefit V from the product in two periods, have to pay a price for the product, and additionally incur switching and transportation costs. Consumer surplus in the two-period segment pricing model can be written as

$$CS = 2\theta V - (\alpha_{11} p_{11} + \alpha_{12}(p_{21} + s_2) + \alpha_{21}(p_{12} + s_1) + \alpha_{22} p_{22}) - (\alpha_1 p_1 + \alpha_2 p_2)$$

$$- \left(\frac{1}{2}\alpha_{11}^2 + \frac{1}{2}\alpha_{22}^2 + \alpha_{12}(\alpha_2 + \alpha_{12}/2) + \alpha_{21}(\alpha_1 + \alpha_{21}/2) \right) t - \left(\frac{1}{2}\alpha_1^2 + \frac{1}{2}\alpha_2^2 \right) t$$

The change in consumer surplus, $\Delta CS = CS - CS^U$, is a convex quadratic function of δ. Simulations show that, overall, consumers benefit from segment pricing despite the extra cost of switching that the consumers have to incur.

Social welfare is defined as a sum of consumer surplus and industry profits, $W = CS + \Pi$. Switching is inefficient in the model with fixed consumer preferences, equal production costs, and perfectly inelastic demand. Therefore, one would expect lower social welfare under segment pricing. Indeed, we find that when firms have similar marginal and consumer switching costs the ability of firms to price discriminate reduces social welfare. However, when the firms are sufficiently asymmetric, welfare can be increased due to the effect segment pricing has on the distribution of production between the firms. The change in social welfare, $\Delta W = W - W^U$, is a concave quadratic function of δ. For example, in case $s_1 = s_2 = 0$, the condition $|\delta| < 7t\theta/\sqrt{115}$ guarantees that price discrimination hurts firms, benefits consumers, and reduces the social welfare.

VI. CONCLUSION

Some analysts argue that the marketing strategy of offering discounts to switchers was a mistake on the part of AT&T. The company stopped offering \$100 inducements to switchers in 1997. At the same time, in deregulated energy markets consumers are currently offered \$50 inducements to switch. Also, natural gas tends to be cheaper at firms traditionally supplying electricity while electricity is cheaper at traditional gas suppliers (*Financial Times*, London, August 28, 1999, p. 6). In the framework of the two-period pricing model considered in this chapter, paying consumers to switch can arise in the equilibrium. Moreover, it is not possible for both firms to offer lower prices to their own customers, paying consumers not to switch (see Shaffer & Zhang (1998) for the same finding in their model). However, we find that firms engaged in price discrimination tend to obtain lower profits than they would receive if price discrimination was not possible, especially when the firms are relatively symmetric. Consumers gain from segment pricing despite the extra cost of switching that the consumers have to incur. Social welfare diminishes for firms with similar marginal and consumer switching costs.

As to the amount of switching, re-segmentation of the market between the first and second periods prevails even when consumers' switching costs are non-trivial. In the two-period pricing game, two-sided switching occurs when firms are not too asymmetric in their marginal costs. In the case of two-sided switching (and for consumers uniformly distributed along the Hotelling line) less than one third of consumers switch their suppliers and the amount of switching is decreasing with the sum of exogenous switching costs.

We also arrived at some surprising results. In contrast to Chen's (1997) independence result, equilibrium prices depend on the firm's initial market share. Interestingly, firm profits are generally U-shaped in its initial share of the market. This can be attributed to the strategic effect that a rival's first-period market share has on the rival's second-period pricing strategy. By considering the dynamic model with segment pricing and switching costs, we arrive at a rich set of comparative statics results. Most intriguingly, firms treat their first-period prices as strategic substitutes.

Our analysis can be extended to more than two firms competing for more than two periods (as in Taylor (1999)). An extension of our model to a larger number of periods is computationally difficult but interesting. In the environment where consumers can switch back and forth, the benefits from attracting switchers diminish. As a market opens up to competition, PCTS can be initially a more profitable strategy than later on. Once the first round of switching occurs, only the most switch-prone consumers continue to switch. Being more switch-prone, these consumers may be less profitable for the firm in the long run. In the view of the current move in several countries towards deregulation of utility markets, the sequential-move segment pricing competition between an incumbent and an entrant, and the corresponding entry deterrence game with consumer switching costs would be interesting to analyze.

Generalizing the demand functions is another direction to follow. In our model consumers take a passive role and are not even fully rational. A more realistic model should also include consumer search. Competitor-based price commitment policies, such as low-price guarantees, are the types of long-term contracts which have not been analyzed in this setting (for a study on the endogenous choice of long-term and short-term contracts by firms in a two-period model with consumer switching see Fudenberg & Tirole (1999)). This chapter is also related to the literature on consumer loyalty and its determinants (for example, Chen & Rosenthal (1996), Shaffer & Zhang (2000), and Arbatskaya (2000)). If the firms can undertake certain investments to increase or decrease switching costs, one could find the optimal level of such investments. In general, the behavior-based price discrimination and consumer switching, although important, is still a relatively under-researched area.

NOTES

1. The U.K. market for natural gas was opened up to competition in 1996, and the deregulation of the U.K. electricity market followed in 1998. A few states, including Georgia, Ohio and Pennsylvania, led the way by actively promoting competition in U.S. natural gas and electric power industries in 1999.

2. According to a 1999 study by *Cambridge Energy Research Associates*, switchers are saving an average of 5–15% off their utility bills in U.S. areas open to consumer choice.

3. Among the reasons for a slow penetration of new utility providers into the markets, the industry analysts and researchers mention the complexity of the pricing schemes which make direct comparisons difficult, lack of information, and misleading information provided by former monopolies (arguing, for example, that the switch may result in a temporal disconnection of the service. On top of these information-related reasons, other factors are cited, such as the hassle of switching suppliers, fear of service interruption, the value of the established relationship with the company, uncertain prospects of new entrants, and expectations of price cuts from the former monopoly. See Parmar et al. (2000) for an investigation into the consumer switching in the U.K. residential gas markets.

4. See Shaffer & Zhang (2000) for more examples. Other business strategies that increase customer loyalty include frequent flyer programs, trade-ins, and gift certificates.

5. With more and more transaction-level data becoming available, the firms can potentially trace a longer history of each customer's consumption. Compared to traditional retail structures, electronic commerce provides sellers with the environment that makes a finer segmentation of the market possible (based, among other factors, on the past consumer behavior). For more on market segmentation in electronic commerce, see Choi et al. (1997).

6. A hair stylist or a dentist would immediately know whether their customer uses the services of another firm.

7. An alternative explanation for why switching costs may lower firm profits is that consumers are reluctant to try a firm's product, being afraid to be locked in and exploited by the firm. The idea is similar to that discussed in the literature on the choice of product compatibility. Firms choose to make their product compatible with the rival's product to ensure consumers that competition will prevail in the market, and the firm will not be a monopoly charging high prices.

8. See, for example, Holmes (1989) and Corts (1998).

9. Several authors have recently studied infinite-horizon duopoly competition with consumer switching costs: Rosenthal (1982), Farrell & Shapiro (1988), Padilla (1995), Chen & Rosenthal (1996), and To (1996). Closer to our analysis are the two-period models of Klemperer (1987b), Caminal & Matutes (1990), Padilla (1992), Bester & Petrakis (1996), Fudenberg & Tirole (1999), and the multi-period model of Taylor (1999). Entry into the markets with consumer switching costs is discussed in Klemperer (1988, 1989), Beggs & Klemperer (1992), and Wang & Wen (1998).

10. Alternatively, the first-period and second-period consumers may be interpreted, for example, as a sequence of tenants who are located in the same apartment. The apartment may, by default, be assigned to a particular gas provider. If new tenants wanted to switch to an alternative service provider, they would have to incur switching costs associated with finding the new supplier and establishing the relationship with the supplier: new tenants often continue to receive discounts sent out by local stores to their customers, the former tenants.

11. The exogenous consumer switching costs include the costs of learning to use the new product, as well as any costs induced by the firms prior to the price competition: fees for switching the supplier (for example, the payment for breaching the contract

with the previous supplier and the fee for contracting with a new one), discounts, coupons and rebates obtained by consumers as a result of switching.

12. More formally, at prices $p_{1,2}$ and $p_{2,2}$, the locations of the marginal firm 1 and firm 2's consumers are $x = \frac{1}{2t}(p_{2,2} - p_{1,2} + t\theta + s_2)$ and $y = \frac{1}{2t}(p_{2,2} - p_{1,2} + t\theta - s_1)$. It follows that $x > y$ for non-negative switching costs.

13. Formally, solving for SPE under the assumption of one-sided switching yields $p_{it} = \frac{2}{3}c_i + \frac{1}{3}c_j + t\theta$ and firm 1's marginal consumers $\alpha_{1,1} = \frac{1}{6t}(c_2 - c_1 + 3t\theta)$ and $\alpha_{1,2} = \frac{1}{6t}(c_2 - c_1 + 3t\theta - s_1)$. Then, $\alpha_{1,1} > \alpha_{1,2}$ and we arrive at a contradiction.

14. One has to be careful in stating and interpreting results on price discrimination in this setting. Since the product is differentiated, charging the same price by the firm to its own and its rival's consumers is, in fact, price discrimination against the rival's customers since they are located further away from the firm and have to incur higher traveling costs. In what follows we refer to mill prices set by the firms, not the total prices.

ACKNOWLEDGMENTS

An earlier version of this chapter was presented at the SEA meeting in November, 1999. The author is grateful to the seminar participants, especially to Mark R. Frascatore, for their helpful comments. I thank the editor for the suggestions on the chapter and Ning Liu, Anthony F. Radich and Michael R. Hammock for research assistance.

REFERENCES

Arbatskaya, M. (2000). Competition in Mature Markets with Consumer Loyalty, unpublished manuscript.

Bester, H., & Petrakis, E. (1996). Coupons and oligopolistic price discrimination. *International Journal of Industrial Organization*, *14*, 227–242.

Beggs, A., & Klemperer, P. (1992). Multi-period Competition with Switching Costs. *Econometrica*, *60*, 651–666.

Caminal, R., & Matutes, C. (1990). Endogenous switching costs in a duopoly model. *International Journal of Industrial Organization*, *8*, 353–373.

Chen, Y. (1997). Paying Customers to Switch. *Journal of Economics & Management Strategy*, *6*, 877–897.

Chen, Y., & Rosenthal, R. W. (1996). Dynamic duopoly with slowly changing customer loyalties. *International Journal of Industrial Organization*, *14*, 269–296.

Choi, S. Y., Stahl, D. O., & Whinston, A. B. (1997). Intermediation, Contracts, and Micropayments at http://cism.bus.utexas.edu/works/articles.html.

Corts, K. S. (1998). Third-degree price discrimination in oligopoly: all-out competition and strategic commitment. *Rand Journal of Economics*, *29*, 306–323.

Farrell, J., & Shapiro, C. (1988). Dynamic Competition with Switching Costs. *Rand Journal of Economics*, *19*, 123–137.

Fudenberg, D., & Tirole, J. (1999). Customer Poaching and Brand Switching, working paper.

Holmes, T. J. (1989). The Effects of Third-Degree Price Discrimination in Oligopoly. *American Economic Review*, *79*, 244–250.

Klemperer, P. (1987a). Markets with Consumer Switching Costs. *Quarterly Journal of Economics, 102*, 375–394.

Klemperer, P. (1987b). The Competitiveness of Markets with Switching Costs. *Rand Journal of Economics, 18*, 138–150.

Klemperer, P. (1988). Welfare Effects of Entry Into Markets with Switching Costs. *Journal of Industrial Economics, 37*, 159–165.

Klemperer, P. (1989). Price Wars Caused by Switching Costs. *The Review of Economic Studies, 56*, 405–420.

Klemperer, P. (1995). Competition when Consumers have Switching Costs: An Overview with Applications to Industrial Organization, Macroeconomics, and International Trade. *The Review of Economic Studies, 62*, 515–539.

Padilla, A. J. (1992). Mixed pricing in oligopoly with consumer switching costs. *International Journal of Industrial Organization, 10*, 393–411.

Padilla, A. J. (1995). Revisiting Dynamic Duopoly with Consumer Switching Costs. *Journal of Economic Theory, 67*, 520–530.

Parmar, M., Waddams Price, C., & Waterson, M. (2000). Exercising Consumer Choice: Switching gas suppliers in the U.K. residential market. *Research paper 00/1*. Center for Management under Regulation, University of Warwick.

Rosenthal, R. W. (1982). A Dynamic Model of Duopoly with Customer Loyalties. *Journal of Economic Theory, 27*, 69–76.

Shaffer, G., & Zhang, J. Z. (2000). Pay to Switch or Pay to Stay: Preference-Based Price Discrimination in Markets with Switching Costs. *Journal of Economics and Management Strategy, 9*, 397–424.

Taylor, C. R. (1999). Supplier Surfing: Competition and Consumer Behavior in Subscription Markets, working paper.

To, T. (1996). Multi-period competition with switching costs: an overlapping generations formulation. *Journal of Industrial Economics, 44*, 81–87.

Wang, R., & Wen, Q. (1998). Strategic Invasion in Markets with Switching Costs. *Journal of Economics & Management Strategy, 7*, 521–549.

TRANSACTION-COST ECONOMICS AND THE ORGANIZATION OF AGRICULTURAL TRANSACTIONS

Scott E. Masten

ABSTRACT

This chapter discusses the logic, hypotheses, empirical methods, and principal findings of the transaction-cost approach to economic organization as a foundation for analyzing the organization of agricultural transactions. In contrast to textbook characterizations of agriculture as the quintessential spot market, agricultural transactions display a broad range of governance arrangements. The chapter traces the use of these arrangements to the location-specific nature of investments and, especially, to temporal specificities associated with the perishability of many agricultural products. Case studies of the governance of fruit, vegetable, and dairy processing; the emergence of multinational firms in the banana trade; the evolution of contractual relations between tuna harvesters and processors; and the governance of transactions between cattle feedlots, slaughtering, and beef fabrication operations illustrate the arguments.

I. INTRODUCTION

Why do manufacturers buy some components and make others? What influences the duration and structure of contractual agreements? When will a company distribute its product through an independent distribution network

Industrial Organization, Volume 9, pages 173–195.

ISBN: 0-7623-0687-4

rather than use commissioned employees? Why did franchising emerge as an organizational form? When does foreign direct investment dominate exporting or licensing?

It was in attempting to answer questions of this type that transaction-cost economics originated and evolved. Although the choice among alternative organizational arrangements seems on the surface to be a natural and straightforward extension of neoclassical methods – organizational form being simply another dimension of an economic actor's overall optimization problem – efforts to identify and enumerate factors that lead one transaction to be integrated and another left to market mediation revealed that conventional assumptions about human behavior and the economic environment were not conducive to analyzing such questions. What emerged from those efforts was not just another extension of neoclassical price theory but a whole new paradigm for analyzing the structures that govern economic activity.

At the heart of the transaction cost approach is the insight, due originally to Coase (1960), that all potential gains from trade would be realized but for transaction costs and legal impediments to exchange. Hence, in assessing the relative merits of alternative organizational arrangements the focus of attention becomes the nature and size of the barriers preventing transactors from securing those gains. The main substantive contribution of transaction-cost economics has been to relate the limitations and costs associated with organizational alternatives to the attributes of transactions in a discriminating fashion (Williamson, 1975, 1985), which is to say, in a way that permits hypotheses about organizational form to be formulated and tested. Indeed, the success over the past fifteen or so years in developing and testing the theory has made transaction-cost economics arguably the dominant methodology for analyzing problems of economic organization.

Though the logic and predictions of transaction-cost economics have been shown to apply to a broad range of institutions and an increasing number of industries,[1] there has been little systematic analysis of agricultural transactions in transaction-cost terms. This void is not for lack of interesting organizational questions to explore. Even a cursory scan of the agricultural landscape reveals considerable diversity and variation in the way agricultural transactions are governed. Why do some grain elevator operators contract with railroads? What accounts for the growing use of long-term contracts between feed lots and meat processors? Why does McDonald's subcontract its food processing operations in the U.S. and why has it integrated those operations in the former Soviet Union? Why did highly integrated multinational firms dominate the banana industry for most of its history?

What I will do here is lay out the basic foundations of transaction-cost reasoning and discuss the principal hypotheses, empirical methodologies, and findings of the literature to date. In Section II, I elaborate on the logic of focusing on transaction costs and discuss the relation between legal rules and three particular organizational alternatives: simple market exchange, contracting, and internal organization. In Section III, I review the main positive predictions of the transaction-cost literature. Section IV then describes the methods employed and the problems encountered attempting to test transaction-cost propositions and offers a brief overview of empirical results and recent advances in estimating models of economic organization.

One of the salient lessons of attempts to apply transaction-cost arguments has been the need for detailed knowledge of the industry being studied. Although the conditions of bounded rationality and opportunism that generate transactional frictions may be universal, the manifestations of those frictions and the factors that influence their incidence vary from industry to industry. Even within the agricultural sector, there is considerable diversity among commodities, making a comprehensive analysis of the organization of agricultural transactions impractical. Nevertheless, in Section V, I will identify some distinctive aspects of agricultural commodities that are likely to influence the organization of their production and distribution and will illustrate these concerns with a small number of examples drawn from existing studics and some casual observations.

II. THE PROBLEM OF ECONOMIC ORGANIZATION

Problems of coordinating vertical exchange arise wherever two or more parties transact across a technologically separable interface. Links in the vertical chain of production include research and development; raw material production; various stages of intermediate processing or assembly; packaging; transportation; storage; and marketing and distribution. At each juncture, transactors face a decision of how to govern their relationship. Their joint interest is to undertake activities that enhance total surplus. But each transaction also involves a potential source of conflict: Given the gains to be derived from trade, each party will wish to arrange the terms of trade in such a way as to appropriate for himself as large a share of those gains as possible. Inevitably, efforts to influence their distribution dissipate some of the gains potentially obtainable.

Organizations and institutions are, from the transaction-cost perspective, the means through which transactors attempt to regulate this behavior. The broad goal is to adopt governance arrangements that promote efficient adaptations while economizing on the costs of reaching agreements and resolving disputes. Thus, processes as well as outcomes matter (Williamson, 1988).

A. *An Illustration: Efficient Breach*[2]

The role of process can be usefully illuminated by considering the problem of efficient breach of contract. At the time parties enter a contract for future exchange, there is usually some uncertainty about the opportunity costs to each of performing the stipulated transaction. It is thus possible that at the date at which the exchange was scheduled to take place, one or the other of the parties will regret having entered the agreement and will wish to breach the contract. Whether or not it is efficient for the transaction to proceed, however, depends on the realized value of performance to each party. Thus, if the value of the seller's assets in an alternative application were to exceed their value supplying the intended buyer, nonperformance of the contract would in fact be efficient. In some cases, however, one or the other party may wish to breach even though efficiency dictates performance.

Economists typically view this as an incentive problem, that of defining appropriately scaled damages to induce efficient performance.[3] A legal rule requiring specific performance of each contract, it is argued, would prevent breaches even when nonperformance is efficient and thus cause parties to trade too often relative to the optimum. Requiring the breaching party to pay expectation damages or 'lost profits', in contrast, would induce him to breach only when doing so would leave him better off even after compensating the other party and would therefore lead to breach only when efficient.

A problem with this formulation of the issue surfaces, however, as soon as one recognizes that buyers and sellers are free to negotiate mutually advantageous adjustments – regardless of the legal rule governing breach – whenever nonperformance is the desired result (see Macneil, 1982). Thus, under a specific performance rule, if exchange with a third party is the efficient thing to do, that result can be achieved by, for instance, the seller negotiating a release from his contractual obligation to the original buyer. If the new trade is in fact efficient, then the seller could induce the buyer to release him from his obligation by offering the buyer some share of the additional surplus generated by the new opportunity – regardless of the legal rule in place.

So stated, it becomes apparent that an assumption underlying standard economic models of efficient breach is that negotiated adjustments are costly; otherwise no legal rule would have an advantage. That bargaining around specific performance requirements would be costly seems a reasonable assumption. That the costs generated in such negotiations necessarily exceed the costs of determining fault and assessing damages under a damage rule, however, is less transparent.

The central message is this: The substantively rational action for the parties to take is the same regardless of whether specific performance or damages is the legal rule. What varies with the rule adopted is the process or procedures, and the tactics or strategies, through which the transactors can seek to realize mutually beneficial adaptations. Hence, differences in the process and associated costs of achieving desired adjustments are critical to determining the relative merits of alternative legal rules.

B. Legal Regimes and Organizational Forms

The logic that makes process concerns central to analyzing the efficiency of alternative legal rules also applies to the analysis of organizational forms. The substantively rational course of action is the same regardless of organizational form. Given the same technology, information, and actors, anything that could be accomplished within the firm could, in principle, also be accomplished in a market transaction, and vice versa. What varies as transactions are relocated from one organizational form to another are the duties, procedures, and sanctions available to the transactors and, hence, the tactics they can employ and the processes through which adaptations are realized. Since it is the legal system that regulates those duties and delimits and enforces sanctions, distinctions among organizational forms find their ultimate basis in the law.[4]

Simple market exchange

The legal system affords independent firms or individuals transacting without the protection of a formal contract great autonomy and flexibility in the periods both leading up to and following the actual transaction (Williamson, 1991). In simple market transactions, parties are generally free to bargain or not bargain as they please. Moreover, once a transaction is consummated, the parties have relatively few ongoing obligations and may, for example, use or dispose of the items procured in whatever manner they choose.[5] The relative latitude afforded transactors in simple exchanges provides them both the ability and, as importantly, the incentive to adapt their behavior to events and information as they arise.

That latitude, however, also provides transactors a tactic to extract rents. Parties to a simple exchange may haggle, stall, or walk away from the deal altogether if they are dissatisfied with the terms currently tendered; courts will rarely intervene to force the parties to come to terms. Similarly, in the absence of concrete agreements by the parties themselves, courts typically refrain from regulating the post-exchange behavior of the transactors, even if that behavior imposes costs on other parties to the transaction. Thus, although flexible, market transactions are relatively exposed to the potential for hold ups and free riding.

Contracting

Where the hazards of simple market exchange are severe, parties may seek to use contracting or integration to protect their relationship. By contracting, transactors can attempt to limit free riding by defining the range of acceptable behavior. Contracts also serve to attenuate hold up problems by stipulating terms of trade ex ante, thereby reducing the prospect of costly repetitive bargaining. In support of these objectives, courts assess damages or employ other remedies to enforce contracts, restrict the ability of one party to extort concessions from the other by threatening not to perform, and perhaps most importantly, provide an explicit and detailed set of rules and procedures for resolving disputes.

Contracting, however, is also subject to a variety of well-known hazards and limitations. First, contracts that cover the full range of possible contingencies are prohibitively expensive to write. Contractual agreements may be incomplete "because of the inability to identify uncertain future conditions or because of inability to characterize complex adaptations even when the contingencies themselves can be identified in advance" (Goetz & Scott, 1981: 2). Second, contract *enforcement* is imperfect and costly. Even where the intentions of the parties are clear, courts may be unable to determine whether those intentions have been fulfilled. The inability to define the precise obligations of each party in response to changing events in ways that can be enforced in a low-cost fashion means that contracts will, on the one hand, tend to be inflexible and, on the other, leave considerable opportunity to cheat on the agreement or to redistribute contractual surpluses by forcing an explicit renegotiation of the terms of trade (see Goetz & Scott, 1983; Goldberg, 1985; Williamson, 1985).

Internal organization

Internal organization has traditionally been viewed as offering greater control and superior access to information relative to market exchange. Again, the source of these advantages rests in the legal relationships of the transactors to each other and to the assets required for production and exchange. By allocating residual rights of control over the use and disposition of assets, ownership restricts the ability of nonowners to withhold assets from production and thus limits hold-up opportunities.[6] In addition, differences in the legal defaults, sanctions, and procedures governing transactions between employers and employees (vis-à-vis independent contractors) support greater discretion and control in response to change and superior access to information within the firm compared to independent commercial transactions (see Masten, 1988b; Williamson, 1991). Thus, for instance, whereas transactions between independent contractors require mutual agreement before adaptations can be effected, courts generally defer to managerial judgement in internal transactions, giving managerial decisions a presumptive validity that enhances flexibility and control within the firm relative to market exchange.

The benefits of integration are limited, however, by the bureaucratic inefficiencies that inevitably plague large organizations. Whereas market exchange establishes 'high-powered incentives' to economize on costs, maintain assets, and acquire and use information productively, internal organization substitutes weaker indirect incentives dependent on managerial oversight.[7] As successive transactions are brought within the firm, additional demands are placed on managers' scarce time and attention, increasing the probability of mistakes and the costs of organizing (Coase, 1937; Williamson, 1975).

III. TRANSACTION-COST DIFFERENTIALS

Once the features that distinguish organizational forms from one another have been identified, the remaining task is to match organizations and transactions or, more precisely, to "align transactions (which differ in their attributes) with governance structures (the costs and competencies of which differ) in a discriminating (mainly, transaction cost economizing) way" (Williamson, 1988: 73).

Much of the positive contribution of transaction-cost reasoning over the last 15 years has been in identifying the critical dimensions that influence the relative efficiency of various organizational forms.

Asset specificity

When assets are not specialized to any particular transactor, exchange is regulated by the ability of the parties to turn to alternative partners should one seek to gain at the expense of the other. Often, however, realization of cost economies or design benefits requires investment in relationship-specific assets such as special dies or the expertise to produce or use them. Once the die is cast, only imperfect market alternatives exist and both the buyer and seller are locked, in effect, into a bilateral monopoly relationship – within the bounds set by those imperfect alternatives.

This 'Fundamental Transformation' from a situation of ex ante competition to small numbers bargaining is the primary force motivating the adoption of specialized governance structures (Williamson, 1985). The more difficult it is to find substitute performance or to redeploy investments to alternative uses, the larger the appropriable quasi-rents, and the greater the risk that gains from trade will be dissipated in contention over their distribution. Accordingly, transactors are likely to be reluctant to make such investments without some form of protection.

Asset specificity can take at least five forms. The first three – physical, human, and site specificity – are well-known and need little explanation. In each case, capital is designed or located for a particular use or user and has a discretely lower value in its next best use. A fourth type, dedicated assets, refers to investments made for a particular customer that, though not specific to that customer, are made at such a level that their release on the market would depress the market value of the assets (Williamson, 1985). Thus, despite the general nature of such investments, an opportunistic buyer (seller) could seek to extract a more favorable price by threatening to discontinue trading, thereby forcing the seller (buyer) into the market. The larger and more durable the investments in relationship-specific assets of any of these four types, the greater the incentive for transactors to attempt to influence the terms of trade through bargaining or other rent-seeking activities once the investments are in place.

Finally, the threat of hold ups may also arise, or be exacerbated, where the timing of performance is important. In such settings, delay may become an effective strategy for eliciting price concessions – even if none of the assets required for performance are specific under the preceding definitions – because of the difficulty of arranging substitute performance on short notice (see Masten et al., 1991; Pirrong, 1993). 'Temporal specificity' may arise because a product's value is inherently time dependent, like newspapers; because of the serial nature of production, as in construction projects; or because the product is perishable, as is the case, of course, with agricultural commodities.

Asset specificity poses the greatest hazards in simple market transactions because of the relative autonomy transactors possess in such relationships. Refusals to deal by independent transactors (absent contractual protections) are subject to neither sanctions nor limits. Consequently, the time and resources dissipated attempting to secure a favorable distribution of the quasi-rents associated with relationship-specific investments is, other things the same, likely to be greater in simple market transactions than in contractual or employment relationships.

Externalities

Economists have long recognized that externalities present potential organizational problems, the core element being that the party bearing the cost (reaping the benefits) of some action does not appropriate the full benefits (bear the full costs) of that action. Since Coase tackled the issue in 'The Problem of Social Cost' (1960), economists have also known that efficient solutions to externality problems turn on transaction-cost considerations. Only considerably more recently, however, have the existence and role of externalities in vertical transactions been recognized and studied.

Externalities among vertically related transactors are most likely to arise where an upstream firm, often a manufacturer, serves many downstream firms, usually retailers or distributors. The most common examples involve situations where efforts by one dealer to promote the manufacturer's product or to enhance the value of the brand name also benefit other dealers. The incentive of each dealer to free ride on the others' expenditures tends to result in a suboptimal level of presale services and in the debasement of the reputation of branded products. Externality problems may also arise, however, in the context of manufacturer services that can be profitably employed with competitors' products (Marvel, 1982) and for nonpatentable innovations (Masten & Snyder, 1993). Contracting and integration provide alternative means for regulating such behavior. The more important the external effects of a particular activity, the more likely transactors will choose to regulate that activity through one of these alternatives.

Complexity/uncertainty

The relative merits of contracting and integration as alternative solutions to asset specificity and externality problems depend also on the nature of the transaction. Among the factors affecting that comparison is the degree of complexity and uncertainty associated with the exchange. The more complex the transaction, for instance, the harder it will be to describe fully and accurately the responsibilities of each party in a contract, and the more difficult it will be for a court to assess whether those obligations have been fulfilled. Failure to account adequately for uncertainty, meanwhile, may leave gains from adaptation unrealized or, if one party is greatly disadvantaged by ensuing developments, cause that party to attempt to escape performance, raising the risk of costly litigation.

Complexity and uncertainty also increase costs of internal organization. More complex transactions are more difficult to manage, while the need to keep track of and respond to changing circumstances places greater demand on management's limited attention.[8] Nevertheless, the differential effect of complexity and uncertainty on the costs of contracting and internal organization are likely to favor the latter: Whereas contracting demands prior anticipation of potential problems, internal organization facilitates adaptation to changing circumstances as they unfold (Williamson, 1975: 25).

Similarity

According to Coase, "[t]he costs of organizing and the losses through mistakes will increase with an increase in the spatial distribution of the transactions organized [and] in the dissimilarity of the transactions . . .". (1937: 395). In other words, internal organization costs are likely to be higher for transactions that are differentiated by either their location or characteristics from other activities in which the firm is engaged. Hence, the costs of supervising and managing employees will be greater where managers are unfamiliar with the production process or geographically removed from the operations.

Similarity is unlikely to generate counteracting effects on the costs of market exchange for two reasons. First, parties engaged in bargaining across a market interface care primarily about final outcomes and not the manner in which goods or services are produced. Provided that the attributes of items exchanged can be adequately verified, high-powered incentives relieve the parties of the need for detailed knowledge about the production operations of their trading partners.[9] Second, to the extent that market exchange relies more heavily than internal organization on third-party dispute resolution, the knowledge and experience of judges and juries rather than of the transactors themselves will have a greater impact on the course and cost of litigation. Even if familiarity of the transactors with each others' techniques increases the probability of detecting opportunistic efforts to evade contractual obligations, it may do nothing to improve the prospects of successful enforcement.[10] Hence, similarity serves primarily to lower the costs of internal organization and, other things constant, to increase the likelihood of vertical integration.

International Transactions

Coordination of vertical transactions across national boundaries involves the same concerns and tradeoffs as transactions within a single country, but a number of features of international transactions serve to make those tradeoffs more acute. On the one hand, geographical distances and language and cultural differences tend to raise monitoring costs and thus make integration relatively less desirable. Stated conversely, market-mediated exchange offers greater (high-powered) incentives for independent agents to acquire and use knowledge about local conditions in foreign markets. On the other hand, greater political uncertainty and underdeveloped or even nonexistent legal protections tend to increase appropriability hazards in market transactions across national borders. In assessing the organization of international transactions, one would thus want to consider, in addition to the factors listed above, the proximity and cultural similarity of the foreign market and the sophistication and stability of foreign political and legal institutions.[11]

IV. EMPIRICAL METHODOLOGIES AND OVERVIEW OF RESULTS

The literature is replete with case studies and anecdotes illustrating transaction-cost propositions.[12] These studies have been valuable, above all, for the crucial detail they offer. A limitation of case studies, however, is their vulnerability to charges of ex post rationalization. Attempts to compare the costs of organizing under alternative arrangements confront a basic selection problem: Transaction costs cannot be directly observed for organizational forms not chosen. Thus, even if the costs associated with internal organization could be observed for transactions organized within the firm, the costs that would have occurred had the same transaction been governed by, say, a long-term contract would not, by virtue of the integration decision itself, be observable. As a result, claims that a particular organizational arrangement minimizes transaction costs are easy to make and impossible to refute.[13]

To address this criticism, researchers have sought to apply more formal empirical tests. The empirical model implied by the theory can be represented quite generally as

$$\begin{aligned} G^* &= G^1, \text{ if } C^1 < C^2; \text{ and} \\ &= G^2, \text{ if } C^1 \geq C^2 \end{aligned} \tag{1}$$

where G^1 and G^2 represent alternative organizational arrangements, G^* the alternative chosen, and C^1 and C^2 the costs of organizing under the corresponding alternatives.

As noted previously, one of the main contributions of transaction-cost economics has been to relate the incidence of various transaction costs to attributes of the transaction. For empirical purposes, hypotheses regarding the effects of the various attributes on the costs of organizing can be represented as

$$C^1 = aX + e, \text{ and} \tag{2}$$

$$C^2 = bX + u, \tag{3}$$

where X is a set of exogenous attributes thought to influence transaction costs, a and b are coefficient vectors, and e and u are random variables.

In principle, propositions relating transaction costs and attributes could be tested directly by estimating equations (2) and (3). The difficulty of observing and measuring transaction costs, however, hampers such direct tests. Many hazards of exchange, such as inflexibility in response to changing circumstances or the need to litigate performance, are either implicit or latent to the transaction. And others, such as the increased demands placed on a manager's attention by bringing an additional transaction under his purview, or the losses due to withholding of information by either employees or subcontractors, while manifest, are difficult to quantify.[14]

Because of these measurement problems, empirical tests of transaction-cost propositions have typically been based on a reduced form of the model. Specifically, the choice of organizational form is estimated as a qualitative choice model (such as probit or logit) in which the probability that G^1 will be chosen equals $Pr(C^1 < C^2) = Pr(e\text{-}u < (b\text{-}a)X)$. Hypotheses are then based on the predicted sign of the coefficients b-a.

While this greatly reduces the information burden of testing transaction-cost models, researchers still face a number of practical hurdles. Implementing the theory requires detailed information on both organizational forms and the attributes of the product, assets, and environment for each transaction. Objective measures for factors like specificity and uncertainty are hard to devise; and even though variables representing organizational form pose fewer measurement problems, data on organizational form are rarely available in convenient form.

Measurement and Testing

To surmount these problems, studies of organizational form have tended to restrict attention to a single industry or firm and to base hypothesis tests on ordinal rankings of components with respect to the independent variables. The earliest and best known of the formal tests of transaction-cost hypotheses examined procurement and integration decisions in manufacturing settings. The first systematic empirical evidence came from Monteverde & Teece's (1982a) study of automobile parts procurement. Their finding that the amount of engineering effort involved in development of a component positively correlated with the probability that that component would be procured internally is consistent with the proposition that relationship-specific knowhow generated in the process of developing new products creates quasi-rents that favor integration. In a related study (Monteverde & Teece, 1982b), they also showed that automobile manufacturers were more likely to own the tooling used by suppliers the more specialized and expensive that tooling. Shortly thereafter I found similar evidence of the role of specificity in a study of aerospace procurement using a measure of the degree to which components used in production were adaptable for use by other firms (Masten, 1984). My results also supported the prediction that the probability that a component would be procured internally increased with the complexity of the component being procured.

Anderson & Schmittlein's (1984) article on the decision of whether to integrate the sales function in the electronic components industry demonstrated that the same forces that drive backward integration in manufacturing influence forward integration decisions as well. In particular, their data revealed that electronic components manufacturers were significantly more likely to integrate marketing of their products, i.e. use direct, employee salespeople rather than independent sales representatives, as human-asset specificity and uncertainty increased. In a similar but more recent study, George John & Barton Weitz (1988) surveyed a sample of industrial goods manufacturers about their distribution practices and found again that the percent of sales made through direct rather than independent channels (wholesalers, distributors, or retailers) increased with the level of firm-specific skills required of the salesperson and the level of uncertainty associated with the transaction.

Although each of these studies focused on organization decisions in a particular firm or industry, they provide cumulative support for the central transaction-cost hypotheses that asset specificity and complexity/uncertainty are important determinants of the decision to integrate. The effects of externalities on organizational choice, in contrast, have received less systematic analysis. Nevertheless, systematic empirical evidence is beginning to emerge. Gatignon & Anderson (1988), for instance, find that advertising has a positive influence on both the decision to integrate and the degree of control parent firms retain over subsidiaries, lending support to the hypothesis that there is greater need for integration where brand name capital creates free rider problems.[15]

Methodologically, all of these studies are similar, using observed attributes of transactions to predict observed organizational forms. Reduced-form tests have a number of drawbacks, however. In particular, such indirect tests of the hypotheses are unable to distinguish between the influence of variations in internal and market organization costs on the decision to integrate and hence whether the underlying hypotheses regarding the incidence of these costs are supported.[16] In addition, such estimations leave open the question of the magnitude of transaction costs and hence of the importance of organizational form to overall efficiency.

Using a new data set and techniques borrowed from the empirical labor economics literature, James Meehan, Edward Snyder and I (Masten et al.,

1991) attempted to address these concerns by estimating the structural model underlying the choice of organizational form. Specifically, we first estimated a standard transaction-cost model of the choice between internal and market procurement for a sample of components and tasks involved in the construction of a ship. We then went a step further, however, and included a measure of the costs of internal organization for those transactions organized within the firm. Using censored regression techniques, we were able to isolate the effects of attributes of transactions on the costs of both internal and market organization and also to provide dollar estimates of those costs. With regard to the latter, our results indicated that overall organization costs represented about 14% of total costs for the components in our sample. But perhaps more importantly, we found that these costs varied substantially with the type of transaction and the organizational form adopted. For example, our estimates indicated that the mistaken integration of the typical subcontracted component in our sample would increase organization costs by approximately 70%, while subcontracting work that would have been more efficiently organized within the firm would, on average, generate market organization costs almost three time those that would have arisen had that work been managed internally.

The results of this study also provided evidence on the role of temporal specificity and similarity on integration decisions. As in other construction settings, timing of performance is crucial to efficient production; delays at one stage can reverberate throughout the entire project. Knowing this, an opportunistic subcontractor could seek to extract a larger share of the gains from trade by threatening to suspend performance at the last minute. The results indicate that such scheduling concerns significantly raised the hazards of subcontracting and increased the probability of integrating the corresponding activities. The evidence also supported the proposition that internal organization costs would be lower, and the probability of integration higher, the more similar a particular transaction to the firms primary area of expertise.[17]

Finally, evidence is increasing that transaction-cost considerations influence the *design* as well as the choice of governance arrangements. Paul Joskow (1987), for instance, has shown that the duration of contracts between coal mines and electric utilities is positively related to variables serving as proxies for physical asset specificity, site specificity, and dedicated assets. Moreover, contract duration appears to be sensitive to the costs as well as the benefits of contracting. My research with Keith Crocker on the duration of natural gas contracts offered evidence that uncertainty and distortions in contract incentives increased contracting hazards and induced gas producers and pipelines to adopt agreements of shorter duration (Crocker & Masten, 1988). A number of recent empirical studies have also analyzed the quantity and pricing structure of contractual agreements in transaction-cost terms (see Masten & Crocker 1985; Mulherin, 1986; Joskow, 1988a; Crocker & Masten, 1991; Leffler & Rucker, 1991).

V. VERTICAL COORDINATION IN AGRICULTURE

Although there has been little systematic statistical analysis of the organization of agricultural transaction from a transaction-cost perspective, it is possible to frame some broad generalizations regarding the features that are likely to be operative in governing agricultural production and exchange from case studies and casual observation. For example, the nature of agricultural products and production would lead us to expect physical and human asset specificities to play a less important role in agricultural transactions than in most other contexts. Even though equipment and facilities used in the cultivation, processing, and transportation of agricultural products may be highly specialized, their design is rarely specific to, and hence their use limited to, a particular customer or supplier. Similarly, the skills necessary for growing and processing agricultural products, however sophisticated and specialized, will seldom be relationship-specific and are thus unlikely to generate quasi-rents that would expose the transactors to the threat of hold ups.

Probably the most conspicuous attribute distinguishing agricultural goods from other commodities is their perishability. Virtually all unprocessed and many processed foodstuffs have limited shelflives. Much of the coordination task in agriculture is related either directly or indirectly to assuring that foodstuffs are produced, processed, and distributed in a timely fashion. As discussed earlier, the prospect of strategic delay is a major concern where timing of performance is critical. Variations in the inherent stability of agricultural products and the costs of preserving them are thus likely to be important determinants of organizational form in agriculture.

Because of the high weight-to-value ratio of many agricultural products, site (or location) specificity is also likely to be a potential concern in agricultural transactions. To the extent that processing reduces that ratio, one would expect processing facilities to be located in proximity to input sources. Inasmuch as the need for refrigeration or other special care increases transportation costs, perishability is likely to be an aggravating factor here as well.

Finally, attributes like uncertainty, complexity, and similarity should play a role in agricultural transactions analogous to that in other settings. Externality problems arising in the distribution of branded processed food products, for instance, are likely to parallel those of other branded consumer products. Similarly, variations in cost and demand conditions and product heterogeneity will tend to exacerbate contracting hazards, while geographical and process disparities will tend to impede internal administration.

A number of case studies illustrate these issues. The connection between perishability and hold-up problems has been discussed explicitly in Charles Knoeber's (1983) analysis of relations between growers and processors of fruits and vegetables. According to Knoeber, "at harvesttime, the grower has a very perishable crop and so default by the processor would mean either loss of the crop or a quick (costly) sale in the thin spot market (and likely some crop deterioration as well)" (1983: 338). At the same time, perishability also prevents processors from holding inventories of unprocessed inputs. As a result, "[d]efault by a grower will either interrupt the processing run or require quick (costly) replacement from a very thin spot market" (id.). The resulting exposure to hold ups explains the widespread use of forward contracts and the protective role of agricultural bargaining cooperatives in governing fruit and vegetable transactions.[18]

Similar institutional arrangements are also found in the dairy industry.[19] Raw milk is again an extremely perishable commodity. Accordingly, it is not surprising that sales of raw milk to milk processors are typically governed under long-term contracts. In addition, however, producer cooperatives often maintain some internal processing capacity. On purely technical grounds, this would appear to be inefficient. Individual cooperatives cannot achieve the economies of scale and specialization in processing that large regional processors possess. Given that contractual protections against hold ups are inherently imperfect, however, maintaining some processing capacity – in effect, partial forward integration – provides a measure of protection against processor opportunism.

Perishability also appears to have played an important role in the evolution of organizational arrangements in the U.S.-Caribbean banana trade.[20] The banana trade between the U.S. and Caribbean countries began in the mid-1800s. From the start, the extreme perishability of bananas posed acute logistical problems. Once the fruit was cut, it had to be delivered to port, loaded, transported (often great distances), and distributed promptly. But ships were slow and their arrival at a given port unpredictable. Growers faced uncertainty about the availability of buyers for their product, and shippers about the availability of bananas at a given port. A grower that chose to wait for another ship rather than accept a current offer risked loosing his entire crop. Shippers, on the other hand, had to weigh 'the potential returns from the fruit currently on offer against possible returns from proceeding to another port to obtain better fruit' (Read, 1983: 182) – or a better price. The attendant risks of hold ups in market transactions threatened to undermine the appeal of the banana trade. Highly integrated multinational firms came to dominate the industry as banana traders integrated both production and distribution to mitigate these problems.[21]

Perhaps the most interesting and detailed analysis of organizational responses to transactional frictions where perishability plays a significant role is Edward Gallick's (1996) study of the organizational arrangements governing relations between tuna harvesters and processors. The potential for hold ups in tuna transactions derives in part from the use of exclusive dealing contracts between harvesters and processors. Although the value of tuna varies with, among other things, its size, species, and freshness, the value of tuna *of a particular quality* did not vary *among tuna processors*. Despite this, each processor in a competitive auction would have had an incentive to inspect the tuna before bidding for it.[22] On the other hand, if a processor agreed to forego inspection and purchase tuna as 'run of the catch', tuna captains would have had an incentive to sort the tuna themselves beforehand and sell the best tuna to another processor at a higher price.

Exclusive dealing contracts reduced duplicative, and therefore socially wasteful, inspections and sorting by requiring captains of tuna vessels to deliver all tuna to the contracting processor throughout the contract's term (Gallick, 1996; also see Kenney & Klein, 1983: 505–509.) A side effect of the agreements, however, was to transform tuna boats and harvests into specialized assets, exposing captains to the threat of hold ups by processors. Because of limited refrigeration capacity, processors needed flexibility in off-loading tuna from vessels.[23] But processors could also use this flexibility to seek price concessions from captains. Failure to accede to such demands could cost the captains additional time in port and risk of spoilage of his cargo.

Prior to 1967, the small size and frequent stops in port made by 'bait boats' restrained processors' incentives to hold-up harvesters. The conversion to mechanized 'purse-seine' vessels in the late 1960s, however, resulted in much larger boats making less frequent deliveries, increasing the incentive for processors to hold up captains. In response, harvesters and processors made a number of organizational adjustments designed to increase the cost and reduce the benefits of processor hold ups, among which were: (i) the introduction of demurrage fees (or fines) on processors that failed to unload vessels within 10 days; (ii) joint ownership of tuna boats by captains and processors (where boats had previously been owned by captains independently), and (iii) processor guarantees on vessel mortgage loans (Gallick, 1996: 218). The need to leave captains the incentive to exercise care and judgement in the use of boats and choice of fishing grounds, however, weighed against full integration given the difficulty of monitoring performance over the 10,000 mile range and up to three month periods these ships were at sea (Gallick, 1996: 220).

Finally, the growth of long-term contracting between feedlots and beef processors and the prevalence of integrated slaughter and beef fabrication operations can also be understood in transaction-cost terms. Because of transportation costs, beef processing plants tend to be located near feedlots, which, in turn, tend to be located close to feed grain supplies.[24] Hence, assets are to some degree site specific. In addition, "[m]odern kill lines and fabricating operations are designed to handle a specific quantity flow, and per unit costs increase rapidly at other volume levels" (Purcell, 1990: 1216). To assure the cost savings of high-speed production lines, processors need a steady flow of inputs. Efficient management of cattle stocks, meanwhile, requires immediate processing of fed cattle once they reach optimal slaughter weight; holding cattle in feed lots beyond that point is purely a holding cost. The cost of storage and the importance of timely processing becomes even greater once cattle are slaughtered. Absent transaction costs, there is no reason that scheduling economies could not be realized by simple communication and agreement between independent operators. What makes this prospect unlikely, and makes formal protective mechanisms desirable, is the prospect of hold ups occasioned by the quasi-rents accruing to prompt performance.[25]

VI. CONCLUSIONS

Transaction cost economics is just emerging from its infancy. As new theoretical and empirical work emerges, the theory is being refined and extended. The nature and sources of asset specificity, for instance, are being fleshed out and expanded. Much of that development has resulted from applying basic transaction cost insights to new industries or novel organizational arrangements. Agricultural transactions provide a rich and largely unexplored area for application and refinement of transaction-cost theory. If the past is any indication, research on the organization of agricultural transactions will serve to advance transaction-cost theory as well inform agricultural economics about an important input into agricultural production and distribution.

The illustrations of the preceding sections are suggestive of how that analysis might proceed. The problem with such illustrations, however, is that, no matter how compelling, they are subject to the criticism of being ex post rationalizations – a complaint that can only be tempered through more formal statistical tests.[26] Variations in organizational forms and in the attributes of agricultural commodities would appear to present abundant opportunities for such analyses.

NOTES

1. For recent surveys of the empirical literature, see Shelanski & Klein (1995) and Crocker & Masten (1996).
2. This subsection draws on my discussion of these issues in Masten (1988a).
3. See, for instance, Shavell (1980).
4. See Masten (1988b), and Williamson (1991). The question of what distinguishes internal organization from market exchange is still a matter of considerable controversy among economists. For a sampling of opinions on the nature of the firm, see Williamson and Winter (1991). Also see Hart (1989).
5. The existence of product liability and implied warranty obligations are exceptions to this generalization.
6. Integration of physical assets may also influence incentives to create, employ, and maintain assets (Grossman & Hart, 1986; Hart, 1989; Hart & Moore, 1990).
7. On the relationship between integration and high-powered and low-powered incentives, see Williamson (1985).
8. Complex and uncertain transactions are also more costly for independent transactors to administer. When transactions are brought within the firm, however, the manager must oversee the new operations as well as pursue whatever activities he was previously engaged in. The limits to internal organization derive ultimately from increasing demands on management's finite time and attention as successive transactions are integrated.
9. Moreover, independent contractors have formal legal rights only over the results and not over the means by which the work of subcontractors is performed. See, for instance, Masten (1988b: 186). Notice that acquisition of expertise through hiring of a manager familiar with the relevant processes does not eliminate the added administrative burden incurred when unrelated activities are integrated within the firm; top management's ability to assess the performance of that manager – hence, the quality of the low-powered incentives they can bring to bear on him – will depend on their comprehension of the operations he oversees.
10. In other words, familiarity may help to make performance 'observable' to the contracting parties but is unlikely to affect whether or not it is 'verifiable' to a third party such as a court.
11. A more comprehensive overview of multinational firms from a transaction-cost perspective can be found in Hennart (1991).
12. Among the most interesting and detailed of these are Stuckey (1983), Joskow (1985), Goldberg & Erickson (1987), and Klein (1988).
13. Lest this criticism of case studies seem to harsh, let me point out that in many instances case studies can be far more informative than formal statistical evidence, particularly given the quality of data that are often available for more 'rigorous' empirical analyses. Indeed, statistical measures are incapable of adequately capturing the richness of real-world contracts, and exclusive reliance on statistical analyses risk faulty inferences. My own view is that a combination of case studies and formal empirical analysis is necessary to advance knowledge in this area.

14. Estimations of organization costs must also control for the selection problem mentioned above. See Masten, Meehan & Snyder (1991).

15. There is also a growing empirical literature on franchising in which the externalities associated with brand names plays a role. See, for instance, Norton (1988), and Lafontaine (1992).

16. For example, transaction-cost arguments suggest that assets that are located closely to one another tend to be site specific, which raises the potential for hold ups. An alternative hypothesis that yields the same prediction, however, is that monitoring costs are lower for operations that are geographically close-by. Reduced-form estimations cannot distinguish between these hypotheses.

17. Other evidence that attribute and geographic similarity favor integration can be found in Davidson & McFettridge (1985a, b), who looked at the choice of international technology transfer modes. John & Weitz (1988) also included a measure of territorial sparseness, measured as the percent of a sales persons time spent on travel, in their study of forward integration into distribution channels. Although they interpret this variable differently, the result that firms are less likely to integrate sales channels covering sparse territories is consistent with the greater difficulty of monitoring sales people that are on the road covering broad territories.

18. Knoeber cites figures for 1970 that 85% of vegetables for processing were grown under forward contracts, with an additional 10% grown by vertically integrated processors (1983: 337).

19. I am grateful to Keith Crocker for providing me with this example. Knoeber notes that bargaining cooperatives are also important in the market for 'milk to creameries' (1983: 339).

20. See Read (1983).

21. An alternative view is that integration solved an information asymmetry problem. To the extent that the condition of and care exercised in handling bananas could be determined only once the banana had fully ripened and not through inspection by the buyer at port, sellers had an incentive to mishandle and misrepresent the quality of bananas. Vertical integration can be seen as a means of reducing the incentives to cheat in these ways. See Read (1983), and Hennart (1991: 91–92).

22. In terms of auction theory, this would be a common value auction. A drawback of such auctions is that each party has private incentives to try to improve their information about the value of the item being auctioned, but such efforts are socially wasteful. See McAfee & Macmillan (1987), and Goldberg (1977).

23. In the mid-60s, the monthly average unloading time ranged from 3 to 33 days (Gallick, 1996: 212).

24. Charles Knoeber (1989) cites similar considerations as influencing the location of feed suppliers, growers, and integrators of broilers (chickens), the relationships among which are almost exclusively contractual.

25. The relatively greater reliance on contracts between feed lot and slaughter operators and the greater tendency toward integration of slaughter and fabricating operations is consistent with the greater specificity and similarity of assets and operations in the latter than the former.

26. Recent empirical research by Stuart Frank and Dennis Henderson (1992) on variations in vertical coordination across sectors of the food industry is a commendable start in that direction.

ACKNOWLEDGMENTS

This chapter has benefitted from discussions with Keith Crocker and Craig Pirrong and from the comments of Stuart Frank.

REFERENCES

Anderson, E., & Schmittlein, D. (1984). Integration of the Sales Force: An Empirical Examination. *Rand Journal of Economics. 15*, 385–395.

Coase, R. (1937). The Nature of the Firm. *Economica N.S.. 4*. In: O. E. Williamson & S. E. Masten (Eds). *The Economics of Transaction Costs* (pp. 386–405). Cheltenham, U.K.: Edward Elgar Publishing. (1999).

Coase, R. (1960). Problem of Social Cost. *The Journal of Law and Economics* 3: 1–44.

Crocker, K. J., & S. E. Masten. (1988). Mitigating Contractual Hazards: Unilateral Options and Contract Length. *Rand Journal of Economics, 19*, 327–343.

Crocker, K. J., & Masten, S. E. (1991. Pretia ex Machina? Prices and Process in Long-Term Contracts. *Journal of Law and Economics, 34*, 69–99.

Crocker, K. J., & Masten, S. E. (1996). Regulation and Administered Contracts Revisited: Lessons from Transaction-Cost Economics for Public Utility Regulation. *Journal of Regulatory Economics, 9*, 5–39.

Davidson, W. H., & McFetridge, D. (1985a. International Technology Transactions and the Theory of the Firm, Journal of Industrial Economics, *32*, 253–264.

Davidson, W. H., & McFetridge, D. (1985b). Key Characteristics in the Choice of International Technology Transfer Mode. *Journal of International Business Studies, 16*, 5–22.

Frank, S. D., & Henderson, D. R. (1992). Transaction Costs as Determinants of Vertical Coordination in the U.S. Food Industries. *American Journal of Agricultural Economics, 74*, 941–950.

Gallick, E. (1996). Exclusive Dealing and Vertical Integration: The Efficiency of Contracts in the Tuna Industry. In: S. E. Masten (Ed.), *Case Studies in Contracting and Organization*. New York: Oxford University Press.

Gatignon, H., & Anderson, E. (1988). The Multinational Corporations Degree of Control over Foreign Subsidiaries: An Empirical Test of a Transaction Cost Explanation. *Journal of Law, Economics, and Organization, 4*, 305–336.

Goetz, C., & Scott, R. (1981). Principles of Relational Contracts. *Virginia Law Review, 67*, 1089–1150.

Goetz, C., & R. Scott. (1983). The Mitigation Principle: Toward a General Theory of Contractual Obligation. *Virginia Law Review, 69*, 967–1024.

Goldberg, V. P. (1977). Competitive Bidding and the Production of Precontract Information. *Bell Journal of Economics, 8*, 250–261.

Goldberg, V. P. (1985). Price Adjustment in Long-Term Contracts. *Wisconsin Law Review, 1985*, 527–543.

Goldberg, V. P., & Erickson, J. (1987). Quantity and Price Adjustment in Long-Term Contracts: A Case Study of Petroleum Coke. *Journal of Law and Economics, 30*, 369–398.

Grossman, S., & Hart, O. (1986). The Costs and Benefits of Ownership: A Theory of Vertical and Lateral Integration. *Journal of Political Economy, 94*, 691–719.

Hart, O. (1989). An Economists Perspective on the Theory of the Firm. *Columbia Law Review, 89*, 1757–1774.

Hart, O., & Moore, J. (1990). Property Rights and the Nature of the Firm. *Journal of Political Economy, 98*, 1119–1158.

Hennart, J.-F. (1991). The Transaction Cost Theory of the Multinational Enterprise. In: C. N. Pitelis & R. Sugden (Ed.), *The Nature of the Transnational Firm* (pp. 81–116). Rutledge, New York.

John, G., & Weitz, B. (1988). Forward Integration into Distribution: An Empirical Test of Transaction Cost Analysis. *Journal of Law, Economics, and Organization, 4*, 337–355.

Joskow, P. (1985). Vertical Integration and Long-Term Contracts: The Case of Coal Burning Electric Generating Plants. *Journal of Law, Economics, and Organization, 1*, 33–80.

Joskow, P. (1987). Contract Duration and Relationship-Specific Investments: Empirical Evidence from Coal Markets. *American Economic Review, 77*, 168–185.

Joskow, P. (1988a). Price Adjustment in Long-Term Contracts: The Case of Coal. *Journal of Law and Economics, 31*, 47–83.

Joskow, P. (1988b). Asset Specificity and the Structure of Vertical Relationships: Empirical Evidence. *Journal of Law, Economics and Organization, 4*, 95–117.

Kenney, R., & Klein, B. (1983). The Economics of Block Booking. *Journal of Law and Economics, 26*, 497–540.

Klein, B. (1988). Vertical Integration as Organizational Ownership: The Fisher Body-General Motors Relationship Revisited. *Journal of Law, Economics, and Organization, 4*. In: O. E. Williamson & S. G. Winter (Eds). *The Nature of the Firm: Origins, Evolution, and Development* (pp. 199–213). New York: Oxford University Press (1991).

Knoeber, C. (1983). An Alternative Mechanism to Assure Contractual Reliability. *Journal of Legal Studies, 12*, 333–343.

Knoeber, C. (1989). A Real Game of Chicken: Contracts, Tournaments, and the Production of Broilers. *Journal of Law, Economics, and Organization, 5*, 271–292.

Lafontaine, F. (1992). Agency Theory and Franchising: Some Empirical Results. *Rand Journal of Economics, 23*, 263–283.

Leffler, K. B., & Rucker, R. R. (1991). Transaction Costs and the Efficient Organization of Production: A Study of Timber-Harvesting Contracts. *Journal of Political Economy, 99*, 1060–1087.

Macneil, I. (1982). Efficient Breach of Contract: Circles in the Sky. *Virginia Law Review, 68*, 947–969.

Marvel, H. (1982). Exclusive Dealing. *Journal of Law and Economics, 25*, 1–25.

Masten, S. E. (1984). The Organization of Production: Evidence from the Aerospace Industry. *Journal of Law and Economics, 27*, 403–417.

Masten, S. E. (1988a). Equity, Opportunism, and the Design of Contractual Relations. *Journal of Institutional and Theoretical Economics, 144*, 180–195.

Masten, S. E. (1988b). A Legal Basis for the Firm. *Journal of Law, Economics, and Organization, 4*. In: O. E. Williamson & S. G. Winter (Eds), *The Nature of the Firm: Origins, Evolution, and Development* (pp. 181–198). New York: Oxford University Press (1991).

Masten, S. E., & Crocker, K. J. (1985). Efficient Adaptation in Long-Term Contracts: Take-or-Pay Provisions for Natural Gas. *American Economic Review, 75*, 1083–1093.

Masten, S. E., Meehan, J. W., & Snyder, E. A. (1991). The Costs of Organization. *Journal of Law, Economics, and Organization, 7*, 1–25.

Masten, S. E., & Snyder, E. A. (1993). United States v. United Shoe Machinery Corporation: On the Merits. *Journal of Law and Economics, 36*, 33–70.

McAfee, R. P., & Macmillan, J. (1987). Auctions and Bidding. *Journal of Economic Literature, 25*, 699–738.

Monteverde, K., & Teece, D. J. (1982a). Supplier Switching Costs and Vertical Integration in the Automobile Industry. *Bell Journal of Economics*, *13*, 206–213.

Monteverde, K., & Teece, D. J. (1982b). Appropriable Rents and Quasi-Vertical Integration. *Journal of Law and Economics*, *25*, 321–328.

Mulherin, J. H. (1986). Complexity in Long-Term Contracts: An Analysis of Natural Gas Provisions. *Journal of Law, Economics, and Organization*, *2* 105–118.

Norton, S. W. (1988). An Empirical Look at Franchising as an Organizational Form. *Journal of Business*, *61*, 197–218.

Pirrong, S. C. (1993). Contracting Practices in Bulk Shipping Markets: A Transaction Cost Perspective. *Journal of Law and Economics*, *36*, 937–976.

Purcell, W. D. (1990). Economics of Consolidation in the Beef Sector: Research Challenges. *American Journal of Agricultural Economics*, *72*, 1210–1218.

Read, R. A. (1983). The Growth and Structure of Multinationals in the Banana Export Trade. In: M. C., Casson (Ed.), *The Growth of International Business* (pp. 180–213).

Shavell, S. (1980). Damage Measures for Breach of Contract. *Bell Journal of Economics*, *11*, 466–490.

Shelanski, H., & Klein, P. G. (1995). Empirical Research in Transaction Cost Economics: A Review and Assessment. *Journal of Law, Economics, and Organization*, *11*, 335–361.

Stuckey, J. (1983). *Vertical Integration and Joint Ventures in the Aluminum Industry.* Cambridge: Harvard University Press.

Williamson, O. E. (1975). *Markets and Hierarchies: Analysis and Antitrust Implications*,. New York: Free Press.

Williamson, O. E. (1985). *The Economic Institutions of Capitalism.* New York: Free Press.

Williamson, O. E. (1988). The Logic of Economic Organization. *Journal of Law, Economics, and Organization*, *4*. In: O. E. Williamson & S. G. Winter (Eds), *The Nature of the Firm: Origins, Evolution, and Development* (pp. 65–91). New York: Oxford University Press, 1991.

Williamson, O. E. (1991). Comparative Economic Organization: The Analysis of Discrete Structural Alternatives. *Administrative Science Quarterly*, *36*, 269–296.

Williamson, O. E., & Winter, S. G. (Eds) (1991). *The Nature of the Firm: Origins, Evolution, and Development.* New York: Oxford University Press.

A STRATEGIC SEARCH MODEL OF TECHNOLOGY ADOPTION AND POLICY

Heidrun C. Hoppe

ABSTRACT

This chapter analyzes the link between adaptive R&D and the timing of new technology adoption in a strategic search model with heterogeneous firms. It is shown that the subgame-perfect equilibrium is in stopping rules with a reservation property. The model is used to examine the effect of rivalry, and whether R&D and adoption subsidies can increase social welfare and generate strategic advantage in international technological competition. It is found that the answers depend critically upon the relative magnitude of first-mover and second-mover advantages in the timing of adoption.

I. INTRODUCTION

One of the key determinants of industrial performance are the incentives to adopt cost-reducing new technologies and introduce new products in the marketplace. There exists a large literature on adoption and diffusion of new technologies (see Reinganum, 1989, and Karshenas & Stoneman, 1995, for a comprehensive survey). However, very little formal work has been done to explore the process of adaptation of new technologies to individual firms' requirements. The purpose of this chapter is to develop a formal framework that

---

Industrial Organization, Volume 9, pages 197–214.

ISBN: 0-7623-0687-4

allows to study the link between adaptive R&D and the timing of new technology adoption. The analysis and insights of this model should provide some new answers to the questions of whether more competition may or may not force firms to improve and better exploit new technology, and how particular policy options may affect welfare and generate strategic advantage in international technological competition.

As Rosenberg (1976) notes, "it is an often-told tale in the history of inventions that they have to sit on shelves long after their initial conceptualization because of the absence of the appropriate mechanical skills, facilities, and design and engineering capacity required to translate them into a working reality" (p. 199). Imperfections in the new technology need to be overcome and bypassed by gradual inventive efforts. Modifications are required to meet the needs of the individual firms. Human skills have to be developed for an effective exploitation of the new technology. And often a series of complementary technological and organizational inventions has to be awaited before an innovation will appear attractive to potential adopters. Hence, the profitability of adopting a new technology may be increased by various development activities, however, these activities generally take time. Firms may therefore have an incentive to postpone the adoption of new technology until a later date. This incentive may be enhanced by late-mover advantages to adoption due to learning from the other firms' adoption experience (as in Hoppe, 2000), and offset by early-mover advantages because of strategic interaction in the product market (as in Fudenberg & Tirole, 1985). To find a firm's optimal adoption date these considerations have to be balanced.

The chapter presents a duopoly model in which firms may develop the profitability of adopting a new technology over time by active search for technological and adaptive information, taking potential first-mover and second-mover advantages into account. To accomplish this, I developed a strategic search model patterned after Chikte & Deshmukh (1993) and Lippman & Mamer (1993), but with important differences. In Lippman and Mamer's model, only one firm can adopt the new technology, i.e. the winner takes all. This chapter extends their model to allow for a follow-on adoption. Nevertheless, the first mover may be able to gain advantage for various reasons, e.g. better positioning in geographic and product characteristics space or preemptive investment in plant and equipment. However, second-mover advantages are likewise possible due to informational spillovers to adoption. In Chikte and Deshmukh's model of market entry and quality competition a second-mover advantage is not ruled out a priori. However, their analysis reveals that the competition always takes the form of a contest for being first.

In the model of the present chapter, firms employ reservation level strategies in the subgame-perfect equilibrium, i.e. they stop R&D and adopt the new technology if and only if the expected profitability of the new technology exceeds a certain level. At the equilibrium reservation level, the opportunity cost to delay – the risk of being preempted plus foregone earnings from the use of the new technology – are equal to the benefits of waiting – new information gained by active search or observing the adoption experience of other firms.

The analysis reveals that the firms' reaction functions in reservation levels slope upwards [downwards] when first-mover advantages are relatively more [less] important than second-mover advantages. This result may seem to contradict the findings by Gal-Or (1985) on the relationship between the slopes of reaction curves and the presence of first-mover and second-mover advantages. The explanation for the difference is that Gal-Or considers a two-stage game in which the order of moves is predetermined. In this chapter, the order of moves in the adoption game is made completely endogenous by adding a pre-play stage of search processes. The reaction curves considered in this chapter result from the decision in that pre-play stage.

The model is used to analyze the effects of rivalry on a firm's decision when to adopt a new technology. Contrary to Lippman & Mamer (1993), a firm's equilibrium threshold for the profitability of adopting first may be higher in a duopoly than a monopoly. This result is always obtained in the case of a strong second-mover advantage. The reason is that each firm's benefits from waiting for higher profitability are increased by the possibility that another firm may move first in the meantime. But even in markets with a strong first-mover advantage, a duopolist may choose to wait until it is more profitable to adopt the new technology, where a monopolist would have adopted. The reason is that the opportunity costs of waiting in terms of foregone profits are lower for a duopolist than a monopolist. This profit-dissipation effect is ruled out in Lippman and Mamer's model by the winner-take-all assumption.

The chapter also contributes to the surprisingly small literature on welfare issues and public policies with regard to adoption and diffusion of new technology.[1] The analysis reveals that R&D policies may influence the timing of adoption as well diffusion policies. In particular it is found that R&D and adoption subsidies differ in their impact on welfare, and should adequately depend on the relative magnitudes of early-mover and late-mover advantages in the timing of adoption. Furthermore, policy questions in the field of international technological competition are considered. In that context, R&D and adoption subsidies are shown to have opposing effects on the distribution of the leader's and follower's role.

This chapter bridges a gap between timing games and models of search. Search and game theory have been first combined by Reinganum (1982a) to analyze R&D competition. Since then, search processes have been more and more used to model development and information acquisition activities of competing firms (see, for instance, Reinganum, 1983, Mamer & McCardle, 1987, and Taylor, 1995). Nevertheless, to the best of my knowledge, only the following strategic search models allow for an analysis of timing questions: Chikte & Deshmukh (1993), Lippman & Mamer (1993), and Fershtman & Rubinstein (1997). However, all of them focus on first-mover advantages.

The chapter is related to the growing literature on real options or irreversible investment whose value stochastically evolves over time. However, in this literature the value of waiting is mostly analyzed in competitive settings without any externalities (see Dixit & Pindyck, 1994, for a survey).[2]

The model is set up in Section II. In Section III, the existence of closed-loop search equilibria is demonstrated. Comparative statics results and effects of rivalry are discussed in Section IV. Section V examines policy implications, and Section VI concludes. All proofs are placed in the appendix.

II. THE MODEL

There are two firms ($i = 1, 2$) contemplating to adopt a new technology. Each firm knows that, depending on the sequence of adoption, it may reap a potential first-mover or second-mover advantage of some sort. Apart from the adoption order, the expected value of adoption for a firm depends on how well the new technology fits in with the firm's working methods and production capabilities. This fit in turn is determined by the innovation's technical ambitiousness and the firm's competence in resource endowments and organizational capabilities (see the empirical study by Langowitz, 1988). Before adopting the new technology, each firm may therefore engage in adaptive R&D to enhance the fit and thus the profitability of adoption, e.g. by accumulating information from internal projects, technical consultants and academic research centers, exploring organizational innovations, training the labor force, and testing a prototype of the production process of a new good.[3] These activities generally involve time, costs and uncertainty. They are therefore best captured by sequential search processes (with recall). We assume that if a firm searches, it will obtain technological and adaptive information that is used to increase the (firm-specific) value of adoption over time.[4] The outcome of adaptive R&D is assumed to be private information.

Let $\Pi_1(n)$ and $\Pi_2(n)$ denote the value of the new technology for the first adopter (the leader) and second adopter (the follower), respectively, when n

firms have adopted the new technology. Assume further that $\Pi_2(2)=\alpha\Pi_1(2)$, where $\Pi_1(2)>0$, $0\leq\alpha\leq 1$. Hence, we allow for a potential first-mover advantage of some sort.[5] Adoption involves a fixed cost of K_i for firm i, where $0<K_i<\Pi_1(2)$. Adaptive R&D is assumed to result in a stochastic decrease of the adoption costs to $K_i-\theta$, or equivalently, a stochastic increase in the profitability of adoption by θ, where θ measures the value of the adaptive information, with $0\leq\theta\leq K_i$. Confining the impact of R&D to the firm-specific value of adoption, i.e. ignoring any effects on the strategic advantage α, is restrictive but makes the analysis tractable. Moreover, this assumption enables us to link the search model to the timing game of new technology adoption by Fudenberg & Tirole (1985) in which technological progress is represented by an exogenous and deterministic decrease in the cost of adoption over time (see the discussion in Section III). θ is a random variable representable as a draw from a continuous distribution F_i defined on $[0, K_i]$, with non-negative density f_i everywhere on its domain and finite mean. The draws are mutually independent. New information from adaptive R&D is assumed to arrive according to a Poisson process with rate $\lambda>0$. Let $r_i>0$ be firm i's discount rate. Hence search is costly due to the opportunity cost of foregone earnings from the use of the new technology.[6] To rule out corner solutions, assume that a firm initially prefers to engage in adaptive R&D, i.e. that $\lambda\int_0^{K_i}\theta dF_i(\theta)$ is sufficiently high.

Apart from the active search for information, each firm may also passively acquire relevant information by observing the other firm's adoption experience. We assume that adoption by one firm gives rise to an informational spillover. Let ε_i denote the value for firm i of the information disclosed by j's adoption. For simplicity, assume, $\varepsilon_i=K_i-x_i$, where x_i is the increment to profitability developed at a given point in time. Hence, we consider the polar case, in which firm i learns the best way to adapt and adopt the new technology by observing j's experience such that further search activities become completely unattractive. The adoption problem reduces thereby to a two-stage problem, in which the order of moves is endogenously determined in the continuous-time game of adaptive R&D. This specification is extremely stylized, but it is sufficient to analyze the impact of differences in the adoption order on the firms' incentives to engage in adaptive R&D. We could specify other forms of potential second-mover and first-mover advantages, e.g. by introducing uncertainty, information lags, or allowing for a continuation of search activities by the follower and a temporary monopoly position for the leader. However, these details are not crucial for my purpose. The main point is that adoption is viewed as occurring sequentially, and in response to the outcome of adaptive R&D.

III. SUBGAME-PERFECT NASH EQUILIBRIUM

A strategy for each firm is a sequence of state-dependent decisions: when new information arrives, each firm decides whether to adopt or to wait until it becomes more profitable to adopt, contingent upon the increment to the value of adoption developed to date and the rivals' previous adoption decisions. I shall focus on subgame-perfect Nash equilibria in which the order of moves is not predetermined (subgame-perfect closed-loop equilibria). In the subgames in which one firm (the leader) has already adopted, the other firm's (the follower's) decision problem is solved by choosing a best response. Taking the follower's best response into account, we can solve for the equilibrium behavior in the subgames *prior* to the first adoption. For this, we need to find one firm's best response to the rival firm's strategy, given no previous adoption.

It is a well-known result from search theory for the case of an individual firm that a firm's optimal policy is to invest in adaptive R&D until x exceeds a reservation value ξ, such that

$$V^W(\xi) - V^A(\xi) = 0 \tag{1}$$

where $V^A(x)$ is the firm's expected payoff from adopting the new technology, given that the increment to profitability developed already is x, and $V^W(x)$ is the expected discounted payoff from an additional round of information acquisition activities (see for example McMillan & Rothschild, 1994). In the following it will be shown that each firm will employ such a reservation level strategy in the subgame-perfect equilibrium of the two-firm game. Lemma 1 will demonstrate that firm i's best response to a reservation level strategy of firm j, $\varphi_i(\xi_j)$, possesses the reservation level property, i.e. $\varphi_i(\xi_j) = \xi_i$, and is a continuous function. Lemma 2 will characterize the best response function further. Proposition 1 will then establish the existence of a subgame-perfect closed-loop equilibrium in reservation level strategies. Note that there cannot be any equilibria in which firms do not employ reservation-level strategies. This follows from Lemma 1 and the assumption that search is initially attractive for each firm.

Let $V_i^A(x_i, \xi_j)$ be firm i's expected return from adopting as a function of the highest x_i at hand and the rival's reservation level ξ_j, given that no firm has adopted previously. Hence,

$$V_i^A(x_i, \xi_j) = \Pi_1(2) - K_i + x_i, \tag{2}$$

Let $V_i^W(x_i, \xi_j)$ denote firm i's expected return from additional search activities, given no previous adoption. By waiting, firm i receives new information before firm j's adoption with probability $\lambda/[\lambda+\lambda(1-F_j(\xi_j))]$, but is preempted by firm j with probability $[\lambda(1-F_j(\xi_j))]/[\lambda+\lambda(1-F_j(\xi_j))]$. We know that in the simplest case of our model the follower's best response to the leader's adoption is immediate adoption. Hence, we have

$$V_i^W(x_i, \xi_j) = \frac{\lambda}{\lambda+\lambda[1-F_j(\xi_j)]+r_i}\left[\Pi_1(2) - K_i + \int_{x_i}^{K_i} \theta dF_i(\theta) + x_i F_i(x_i)\right] + \frac{\lambda[1-F_j(\xi_j)]}{\lambda+\lambda[1-F_j(\xi_j)]+r_i}\Pi_2(2) \tag{3}$$

From Reinganum (1982a) we know that firm i's problem in the subgames prior to the first adoption is isomorphic to (1). Firm i's best response to ξ_j, $\varphi_i(\xi_j)$, is therefore an optimal stopping rule of the form: stop R&D whenever $x_i \geq \xi_i$, where ξ_i solves

$$V_i^W(\xi_i, \xi_j) - V_i^A(\xi_i, \xi_j) = 0; \tag{4}$$

otherwise continue. $\varphi_i(\xi_j)$ is obtained by substituting (2) and (3) into (4). To simplify notation, we define the value of the next round of adaptive R&D to be $H(x_i) \equiv \int_{x_i}^{K_i} (\theta - x_i) dF_i(\theta)$, and $\Pi \equiv \Pi_1(2)$.

Lemma 1. *(i) Firm i's best response to firm j's optimal stopping rule with a reservation level* ξ_j, *is to adopt whenever* $x_i \geq \xi_i$, *where* ξ_i *is the unique value such that*

$$\lambda H(\xi_i) = r_i(\Pi - K_i + \xi_i) + \lambda[1 - F_j(\xi_j)][(1-\alpha)\Pi - K_i + \xi_i]; \tag{5}$$

otherwise continue R&D. (ii) Firm i's best response function $\varphi_i(\xi_j)\colon[0, K_j] \to [0, K_i]$ *is continuous in* ξ_j.

The LHS of equation (5) represents the value of the next piece of information, while the first term of the RHS is the opportunity cost in terms of foregone profits. The remaining term gives the impact of the rival firm's strategy. Depending on whether the potential first-mover advantage outweighs the potential second-mover advantage or vice versa, the existence of a rival increases either the marginal costs or benefits of waiting.

The following lemma relates the slopes of the reaction curves to the advantages in timing. This lemma may be of some independent interest, for it derives properties of reaction functions in reservation level strategies that are not restricted to the specific context analyzed in this chapter.

Lemma 2. *(i) For* $\alpha = 0$ *i.e. if there is a dominant potential first-mover advantage, each firm's reaction function slopes upwards. (ii) For* $\alpha = 1$, *i.e. if there is dominant potential second-mover advantage, each firm's reaction function slopes downwards.*

An upwards sloping reaction curve $\varphi_i(\xi_j)$ implies that an increase in the reservation level for the profitability of adoption by firm j induces firm i to undertake more development activity as well. The reason is that a higher value of ξ_j reduces the risk for i of being preempted by j in the case of potential first-mover advantages. By contrast, when potential second-mover advantages matter, a higher reservation level by j induces i to engage in less adaptive R&D. This is due to the smaller chance of benefiting from informational spillovers.

Proposition 1. *There exists a subgame-perfect Nash equilibrium in reservation level strategies* (ξ_i^*, ξ_j^*) *in* $[0, K_j] \times [0, K_i]$.

In this game, the equilibrium order of adoption is induced by chance due to the stochastic nature of the information acquisition process. Firms that are initially identical, will have an equal chance of being first or second in the adoption timing. This result resembles the finding of Fudenberg & Tirole (1985) for a timing game of new technology adoption in which adoption costs decrease deterministically over time. They show that the only subgame-perfect equilibrium is in mixed strategies, given that preemption gains are relatively large. With identical firms, each firm will be the leader with probability 1/2, and with probability 1/2 the roles of the firms are reversed – as in the present chapter. In fact, the stochastic nature of the search process replaces the need for mixed strategies in the present chapter.[7]

Another difference between the two models seems worth pointing out. In the model by Fudenberg and Tirole, the first adoption occurs immediately when the potential first-mover advantage outweighs any potential second-mover advantage. In the model of this chapter, however, there is a delay of the first adoption – even in the presence of a dominant potential first-mover advantage. The reason is that the stochastic nature of R&D generates an option value of waiting for new information for each firm. The risk of being preempted is therefore diminished. As a consequence, ex ante and ex post returns are equalized in Fudenberg and Tirole's adoption game, while there is an ex post first-mover advantage in the preemption equilibrium of the present chapter.[8]

IV. COMPARATIVE STATICS

As in the search model of Lippman & Mamer (1993), there may exist multiple equilibria. However, it can be shown that uniqueness is obtainable by assuming that the difference between the potential first-mover and second-mover advantage is not too high. I shall prove and state the following results for a unique (stable) equilibrium with $F_i = F_j$, $K_i = K_j$, and $r_i = r_j$.

Proposition 2. *An increase in the potential first-mover advantage, i.e. a decrease in* α*, leads to a new equilibrium in which both firms engage in less R&D. (ii) An increase in the potential second-mover advantage, i.e. an increase in* α*, leads to a new equilibrium in which both firms engage in more R&D.*

At a stable equilibrium, a higher advantage of moving first in the adoption stage induces firms to choose lower reservation levels in the R&D stage, while an increase in the potential second-mover advantages introduces the reverse incentives. Reinganum (1982b) obtains the opposite results in a patent race model. In her model, R&D efforts determine the hazard rate of the arrival of innovative information, and information is assumed to arrive only once. Conversely, in the model of the present chapter, the rate at which information arrives, λ, is assumed to be fixed, but firms may choose to engage in further information acquisition in order to increase the payoff from adoption. We may conclude that the net impact of the different forces stressed in each model upon the firms' incentives to engage in R&D is ambiguous.

Proposition 3. *(i) A decrease in* r_i *leads to a new equilibrium in which firm i undertakes more R&D, while firm j undertakes less [more] R&D if potential first-mover advantages are relatively less [more] important than potential second-mover advantages. (ii) A decrease in* K_i *leads to a new equilibrium in which firm i undertakes less R&D, while firm j undertakes more [less] R&D if potential first-mover advantages are relatively less [more] important than potential second-mover advantages.*

Proposition 3 shows that a firm's optimal reaction to a change in the rival's search and adoption costs is sensitive to the relative importance of first-mover and second-mover advantages. A decrease in r_i shifts up the reaction function of firm *i*, increasing both its own equilibrium reservation level and that of the firm *j*, provided the reaction functions are upwards-sloping. If the reaction functions are downwards sloping, firm *j*'s equilibrium reservation level is reduced. Conversely, a decrease in the cost of adoption K_i shifts firm *i*'s reaction function down. Thus we obtain exactly the opposite results.

The next proposition sheds further light on how firms' incentives to engage in R&D before adoption depend on the number of firms in the industry. Generally, one would expect that the following three effects matter. An additional competitor may lower the value of postponing adoption by increasing a firm's risk of being preempted (*the preemption effect*). On the other hand, it may enhance the value of waiting by increasing a firm's possibility of free-riding (*the information-spillover effect*) and by lowering a firm's opportunity costs of foregone profits due to increased competition in the product market (*the profit-dissipation effect*). The net effect of increased competition on adoption timing should hence be ambiguous. This ambiguity is reflected by the inconclusiveness of empirical research (e.g. a positive relation between market concentration and the speed of adoption is found by Hannan & McDowell, 1984, a negative relation by Levin et al., 1987, and no statistically significant one by Karshenas & Stoneman, 1993). In the theoretical literature, the issue is addressed by Lippman & Mamer (1993) who find that a firm's equilibrium reservation level for the value of adoption is decreasing in the number of potential adopters due to the preemption effect. In their model, only one firm is allowed to implement the new technology. This winner-take-all assumption eliminates both, the information-spillover effect and the profit-dissipation effect. By contrast, in situations with relatively more important potential late-mover advantages, the preemption effect is outweighed by the informational-spillover effect. An increase in the number of potential adopters would then increase a firm's reservation level for adoption.

The remaining question is hence whether the profit-dissipation effect can possibly outweigh the preemption effect in situations with relatively more important potential first-mover advantages. To see that this is possible, consider the case in which a firm is faced either by one competitor or by none. It is reasonable to assume that a monopolist does not make less profit than two non-colluding duopolists, i.e. $\Pi_1(1) \geq (1+\alpha)\Pi$. Let $\Pi_1(1) = A\Pi$, where $A \geq 1+\alpha$. The next proposition shows that under these circumstances the profit-dissipation effect may be greater than the preemption effect if the potential first-mover advantage is not too high.

Proposition 4. *Suppose there is no informational spillover to adoption. Then there exists a unique value of* $\alpha \in [0, 1]$, *denoted by* $\hat{\alpha}$, *such that a monopolist engages in less R&D than a duopolist if* $\alpha > \hat{\alpha}$.

V. POLICY IMPLICATIONS

In the following, particular policy options aimed at tuning the speed of adoption shall be examined (for an excellent survey of the diffusion policy literature and actual policy initiatives see Stoneman & Diederen, 1994). It will be shown that the policies of subsidizing R&D costs and subsidizing adoption costs are not equivalent, and that their effects depend crucially on the relative importance of first-mover and second-mover advantages.

V.1. Corrective Policies

The purpose of this subsection is to examine policies to correct for the negative externalities (business-stealing) and positive externalities (informational spillovers) in the firms' adoption behavior. For this, we will compare the equilibrium reservation level with that chosen optimally by a social planner whose objective is to maximize the firms' expected contribution to social welfare. Consumer surplus is hence neglected in our analysis in order to find Pigouvian corrective policies (as in Dixit, 1988). The next proposition reveals that, depending on the relative importance of first-mover and second-mover advantages, R&D subsidies or adoption subsidies may be used to deal with the externalities.

Proposition 5. *(i) If $\alpha = 0$, i.e. in the case of strong potential first-mover advantages, corrective policy takes the form of an R&D subsidy. (ii) If $\alpha = 1$, i.e. in the case of strong potential second-mover advantages, corrective policy takes the form of a subsidy of the adoption costs.*

Both policies, R&D and adoption subsidies influence the equilibrium timing of adoption by altering the marginal value of continuing search for technological and adaptive information. Subsidizing R&D reduces the marginal costs of search. This tends to delay technology adoption which is socially beneficial in the case of strong first-mover advantages. On the other hand, subsidizing adoption costs reduces the marginal saving from waiting in terms of interest earned. This tends to speed up adoption, and can therefore be used to correct positive externalities in the case of strong second-mover advantages.

V.2. Strategic Policies

The comparative static properties of the equilibrium reservation levels (Proposition 3) suggest that the firm-specific cost of R&D and adoption may be manipulated for strategic advantage. This issue may be particularly important in the context of international technological competition. In what follows, I consider a simple government policy, designed to put the domestic firm in the leader position when there are relatively more important first-mover advantages, and in the follower position otherwise. The purpose is to compare such a rent-seeking policy with the corrective policies derived above.

Assume that one firm is located in the home country and the other in the foreign country. The point of interest concerns the effects of R&D and adoption subsidies by one country on the distribution of the leader's and follower's role. The other country might, of course, responds to such a policy. For our purposes, we will ignore the strategic interaction between governments. The following proposition states that, in situations with relatively more important potential first-mover advantages, subsidizing the home firm's R&D will make the *foreign* firm more likely to be the first-mover, while subsidizing the home firm's adoption costs will make the *home* firm more likely to be the first-mover. The opposite result is obtained in the case of relatively more important potential second-mover advantages.

Proposition 6. *Consider a unique (stable) equilibrium with $F_i = F_j$, $K_i = K_j$, and $r_i = r_j$. A decrease in r_i [in K_i] makes firm i less [more] likely to reap potential first-mover advantages, and more [less] likely to reap potential second-mover advantages.*

The intuition behind this result is straightforward. The lower a firm's search costs, the higher its expected value of waiting for an additional piece of information. Hence, an R&D subsidy credibly reduces the home firm's threat of preempting its rival. On the other hand, a decrease in the adoption costs lowers the marginal saving from waiting, and thus credibly reduces the home firm's incentive to wait. What is perhaps surprising in these results is that the strategic policies tend to counteract the corrective policies.[9] In markets with strong potential first-mover advantages, an adoption subsidy would establish a strategic advantage in international competition. But the corrective policy for that country turns out to be an R&D subsidy. In markets with strong potential second-mover advantages, the opposite holds.

Apart from the policy implications, the comparative statics result of Proposition 6 contributes to the large (mainly empirical) literature focusing on how differences in firm-specific factors such as firm size, R&D expenditures, or firm ownership may determine the identity of the leader and follower firm (see Karshenas & Stoneman, 1995 for a survey). The result of Proposition 6 suggests that firms facing lower per-period R&D costs, for example, due to increasing returns to scale in information acquisition activities, would be expected to move later in the sequence of adoption, while firms with lower cost of adoption, for example, because of scale effects in marketing and distribution, would be expected to move earlier.

VI. CONCLUSIONS

This chapter has extended the strategic search models by Chikte & Deshmukh (1993) and Lippman & Mamer (1993) to allow for second-mover advantages to adoption. It has been demonstrated that the subgame-perfect equilibrium is in reservation level strategies. The slopes of the reaction functions in reservation levels have been shown to be determined by the relative importance of potential first-mover and second-mover advantages to adoption.

The model has been used to examine whether an increase in the number of potential adopters induces firms to expand or reduce adaptive R&D activities, and how R&D and adoption subsidies affect welfare and the determination of the identity of the first-mover. The analysis has revealed that the answers to these questions depend critically upon the relative magnitude of potential first-mover and second-mover advantages.

It seems worth further exploring the link between R&D and the timing of new technology adoption, as well as other options of technology policy.

NOTES

1. Exceptions are Stoneman & David (1986), Stoneman (1987), Jensen (1992), Riordan (1992), and Magnac & Verdier (1993).
2. An exception is Choi (1994) who considers a game-theoretic model of irreversible technology choice in which technologies stochastically evolve over time, but his focus is on network externalities.
3. That R&D activities play an important role in the process of adoption has been previously stressed by Cohen & Levinthal (1989).
4. One could alternatively model R&D as sequential search resulting in a resolution of uncertainty regarding the profitability of the new technology over time (see Jensen, 1982, for the one-firm context, and Bhattacharya et al. (1986) for a discussion of strategic interaction). Due to the underlying structural similarities between the two classes of search frameworks (see DeGroot, 1970, Section 13.9, Theorems 1 and 2), the results derived in this paper should also hold for the uncertainty resolution approach.
5. Examples are the Stackelberg equilibrium of a game of quantity competition or the Stackelberg equilibrium in a game of spatial competition (see Anderson & Engers, 1997).
6. Explicit search costs are neglected in this paper. Nevertheless, the main principles derived below should hold when such costs are included in the analysis.
7. See Proposition 6 for an analysis of how asymmetries between the firms define (stochastically) a particular order of adoption.
8. An ex post first-mover advantage in the preemption game of the Fudenberg & Tirole (1985) is also obtained by Hendricks (1992) by allowing for uncertainty about the innovative capabilities of the rival firm.
9. A similar observation is made by Dixit (1988) for a completely different model of R&D competition. See also Beath et al. (1989) for an examination of strategic R&D policies in international competition.

ACKNOWLEDGMENTS

This chapter was completed while I was visiting the University of Southampton, whose hospitality and financial support are gratefully acknowledged.

I would like to thank Ulrich Doraszelski, Richard Jensen, Ulrich Lehmann-Grube, Stephan Panther, and seminar participants the Universities of Munich and Copenhagen and the Royal Economic Society Conference in Warwick for useful comments.

REFERENCES

Anderson, S. P., & Engers, M. (1997). *Preemptive entry in differentiated markets*. Working paper, University of Virginia.

Beath, J., Katsoulacos, Y., & Ulph D. (1989). Strategic R&D policy. *Economic Journal, 99*, 74–83.

Bhattacharya, S., Chatterjee, K., & Samuelson, L. (1986). Sequential research and the adoption of innovations. *Oxford Economic Papers, 38*, Suppl, 219–243.

Chikte, S. D., & Deshmukh, S. D. (1993). *To be the first or to be the best: New product quality and timing in R&D competition*. Discussion Paper No. 1056, Kellogg Graduate School of Management, Northwestern University.

Choi, J. P. (1994). Irreversible choice of uncertain technologies with network externalities. *RAND Journal of Economics, 25*, 382–401.

Cohen, W. M., & Levinthal, D. A. (1989). Innovation and learning: The two faces of R&D. *Economic Journal, 99*, 569–596.

DeGroot, M. H. (1970). *Optimal Statistical Decisions*. New York: McGraw-Hill.

Dixit, A. K. (1988). A general model of R&D competition and policy. *RAND Journal of Economics, 19*, 317–326.

Dixit, A. K., & Pindyck, R. S. (1994). *Investment under Uncertainty*. Princeton, New Jersey: Princeton University Press.

Fershtman, C., & Rubinstein, A. (1997). A simple model of equilibrium in search procedures. *Journal of Economic Theory, 72*, 432–441.

Fudenberg, D., & Tirole, J. (1995). Preemption and rent equalization in the adoption of new technology. *Review of Economic Studies, 52*, 383–401.

Gal-Or, E. (1985). First mover and second mover advantages. *International Economic Review, 26*, 649–653.

Hannan, T. H., & McDowell, J. M. (1984). The determinants of technology adoption: The case of the banking firm. *RAND Journal of Economics, 15*, 328–335.

Hendricks, K. (1992). Reputations in the adoption of a new technology. *International Journal of Industrial Organization, 10*, 663–677.

Hoppe, H. C. (2000) Second-mover advantages in the strategic timing of new technology adoption under uncertainty. *International Journal of Industrial Organization, 18*, 315–338.

Jensen, R. (1982) Adoption and diffusion of an innovation of uncertain profitability. *Journal of Economic Theory, 27*, 182–193.

Jensen, R. (1992). Innovation adoption and welfare under uncertainty. *Journal of Industrial Economics, 15*, 173–180.

Karshenas, M., & Stoneman, P. (1993). Rank, stock, order, and epidemic effects in the diffusion of new process technologies: An empirical model. *RAND Journal of Economics, 24*, 503–528.

Karshenas, M., & Stoneman, P. (1995). Technological diffusion. In: P. Stoneman (Ed.). *Handbook of the Economics of Innovation and Technological Change*. Oxford: Blackwell.

Langowitz, N. S. (1988). An exploration of production problems in the initial commercial manufacture of products. *Research Policy, 17*, 43–54.
Levin, S. G., Levin, S. L., & Meisel, J. B. (1987). A dynamic analysis of adoption of a new technology: The case of optical scanners. *Review of Economics and Statistics, 69*, 12–17.
Lippman, S. A., & Mamer, J. W. (1993). Preemptive innovation. *Journal of Economic Theory, 61*, 104–119.
Magnac, T., & Verdier, T. (1993). Welfare aspects of technological adoption with learning. *Journal of Public Economics, 52*, 31–48.
Mamer, J. W., & McCardle, K. F. (1987). Uncertainty, competition, and the adoption of new technology. *Management Science, 33*, 161–177.
McMillan, J., & Rothschild, M. (1994). Search. In: R. J. Aumann & S. Hart (Eds), *Handbook of Game Theory*. Amsterdam: North-Holland.
Reinganum, J. F. (1982a). Strategic search theory. *International Economic Review, 23*, 1–17.
Reinganum, J. F. (1982b). A dynamic game of R&D: Patent protection and competitive behavior. *Econometrica, 50*, 671–688.
Reinganum, J. F. (1983). Nash equilibrium search for the best alternative. *Journal of Economic Theory, 30*, 139–152.
Reinganum, J. F. (1989). The timing of innovation: Research, development and diffusion. In: R. Schmalensee & R. D. Willig (Eds), *Handbook of Industrial Organization*. Amsterdam: North-Holland.
Riordan, M. H. (1992). Regulation and preemptive technology adoption. *RAND Journal of Economics, 23*, 334–349.
Rosenberg, N. (1976). *Perspectives on Technology*. Cambridge: Cambridge University Press.
Stoneman, P. (1987). *The Economic Analysis of Technology Policy*. Oxford: Clarendon Press.
Stoneman, P., & David, P. A. (1986). Adoption subsidies versus information provision as instruments of technology policy. *Economic Journal, 96*, 142–151.
Stoneman, P., & Diederen, P. (1994). Technology diffusion and public policy. *Economic Journal, 104*, 918–930.
Taylor, C. R. (1995). Digging for golden carrots: an analysis of research tournaments. *American Economic Review, 85*, 872–890.

APPENDIX

Proof of Lemma 1. The proof of this lemma is analogous to that of Lemma 1 in Chikte & Deshmukh (1993). Substituting (2) and (3) into (4) and rearranging terms yields equation (5). It is straightforward to show that the LHS of (5) is zero for $x_i = K_i$, and decreasing in x_i with slope $-\lambda[1 - F_i(x_i)]$. The RHS is smaller than the LHS for $x_i = 0$ (since corner solutions have been ruled out), non-negative for $x_i = K_i$, and increasing with slope $\lambda[1 - F_j(\xi_j)] + r_i$. Thus, by the continuity of $V_i^W(x_i, \xi_j) - V_i^A(x_i, \xi_j)$ on $[0, K_j] \times [0, K_i]$, there exists a unique value ξ_i such that $V_i^W(x_i, \xi_j) - V_i^A(x_i, \xi_j) \gtreqless 0$ as $x_i \lesseqgtr \xi_i$. Furthermore, continuity of $\varphi_i(\xi_j)$ on $[0, K_j]$ is ensured. ■

Proof of Lemma 2. By Lemma 1 and the implicit function theorem, firm i's reaction function $\varphi_i(\xi_j)$ is continuously differentiable with

$$\frac{\partial\varphi_i(\xi_j)}{\partial\xi_j} = -\frac{\partial(V_i^W - V_i^A)/\partial\xi_j}{\partial(V_i^W - V_i^A)/\partial\xi_i}. \tag{6}$$

For $\alpha = 0$,

$$\frac{\partial\varphi_i(\xi_j)}{\partial\xi_j} = \frac{\lambda F_j'(\xi_j)[\Pi - (K_i - \xi_i)]}{r_i + \lambda[1 - F_i(\xi_i)] + \lambda[1 - F_j(\xi_j)]} > 0 \tag{7}$$

For $\alpha = 1$,

$$\frac{\partial\varphi_i(\xi_j)}{\partial\xi_j} = -\frac{\lambda F_j'(\xi_j)(K_i - \xi_i)}{r_i + \lambda[1 - F_i(\xi_i)] + \lambda[1 - F_j(\xi_j)]} < 0 \tag{8}$$

The proof is analogous for firm j. ■

Proof of Proposition 1. Lemma 1 and application of standard fixed point arguments establish the existence of a Nash equilibrium. ■

Proof of Proposition 2. The comparative statics result is first derived for the reaction functions. At a stable equilibrium, the result can straightforwardly be extended to the equilibrium reservation levels.

By the implicit function theorem,

$$\frac{\partial\varphi_i(\xi_j)}{\partial\alpha} = \frac{\lambda[1 - F_j(\xi_j)]\Pi}{r_i + \lambda[1 - F_i(\xi_i)] + \lambda[1 - F_j(\xi_j)]} > 0 \tag{9}$$

The results for j are analogous. ■

Proof of Proposition 3. The comparative statics results are first derived for the reaction functions. At a stable equilibrium, the results can straightforwardly be extended to the equilibrium reservation levels.

By the implicit function theorem,

$$\text{(i)}\quad \frac{\partial\varphi_i(\xi_j)}{\partial r_i} = -\frac{\Pi - K_i + x_i}{r_i + \lambda[1 - F_i(\xi_i)] + \lambda[1 - F_j(\xi_j)]} < 0 \tag{10}$$

$$\text{(ii)}\quad \frac{\partial\varphi_i(\xi_j)}{\partial K_i} = \frac{r_i + \lambda[1 - F_j(\xi_j)]}{r_i + \lambda[1 - F_i(\xi_i)] + \lambda[1 - F_j(\xi_j)]} > 0 \tag{11}$$

and similarly for firm j. ■

Proof of Proposition 4. Consider a unique (stable) equilibrium with $F_i = F_j$, $K_i = K_j$, and $r_i = r_j$, and assume that there are no informational spillovers. If firm i is the only potential adopter it chooses a reservation level ξ_i^M, which solves

$$\lambda_i H(x_i) = r_i(A\Pi - K_i + x_i). \tag{12}$$

Comparing equations (12) and (5) yields, that $\xi_i^M < \xi_i^*$, if

$$r_i[A\Pi - K_i + x_i] > r_i(\Pi - K_i + x_i) + \lambda[1 - F_j(\xi_j^*)][(1 - \alpha)\Pi]$$

or

$$1 - \alpha < \frac{r_i(A - 1)}{\lambda[1 - F_j(\xi_j^*)]} \equiv 1 - \hat{\alpha}. \tag{13}$$

Since $A > 1$, $r_i > 0$, and $\lambda < 1$, it follows that $\hat{\alpha} < 1$. ■

Proof of Proposition 5. Consider a unique (stable) equilibrium with $F_i = F_j$, $K_i = K_j$, and $r_i = r_j$.

(i) Suppose $\alpha = 0$. Firm i's optimal reservation level ξ_i^* is the solution of

$$\lambda H(x_i) = r_i(\Pi - K_i + x_i) + \lambda[1 - F_j(\xi_j^*)](\Pi - K_i + x_i). \tag{14}$$

Given two independent R&D processes, the time until new information arrives is exponential with parameter 2λ. The social planner's optimal reservation level ξ_S is therefore the solution of

$$2\lambda H(x_i) = r_i(\Pi - K_i + x_i)$$

or

$$\lambda H(x_i) = r_i(\Pi - K_i + x_i)/2 \tag{15}$$

Since the RHS of (15) is strictly smaller than the RHS of (14) for all x_i, $\xi_i^* < \xi_S$. By Lemma 2 and Proposition 3, the result follows immediately.

(ii) Suppose $\alpha = 1$. Firm i's optimal reservation level ξ_i^* is the solution of

$$\lambda H(x_i) = r_i(\Pi - K_i + x_i) - \lambda[1 - F_j(\xi_j)](K_i - x_i) \tag{16}$$

The social planner's optimal reservation level ξ_S is the solution of

$$2\lambda H(x_i) = r_i(\Pi - K_i + x_i) + r_i\Pi \tag{17}$$

or

$$\lambda H(x_i) = r_i(2\Pi - K_i + x_i)/2 \tag{18}$$

Since the RHS of (18) is strictly greater than the RHS of (16) for all x_i, $\xi_i^* > \xi_S$. By Lemma 2 and Proposition 3, the result follows immediately. ■

Proof of Proposition 6. Firm i's adoption date is an exponentially distributed random variable with mean $E[T_{\xi_i^*}] = 1/[\lambda(1 - F_i(\xi_i^*))]$. Starting from a unique (stable) equilibrium with $F_i = F_j$, $K_i = K_j$, and $r_i = r_j$. we know from Proposition 3 that a decrease in r_i results in an increase of both firms' equilibrium reservation levels in the case of upwards sloping reaction curves. But the movement to the new equilibrium is along firm j's reaction curve. Thus $\xi_i^* > \xi_j^*$ in the new equilibrium. Likewise, if the reaction functions are downwards sloping, we have $\xi_i^* > \xi_j^*$ in the new equilibrium. Hence, in the new equilibrium, $E[T_{\xi_i^*}] = 1/[\lambda(1 - F_i(\xi^*_i))] > 1/[\lambda(1 - F_j(\xi_j^*))] = E[T_{\xi_j}^*]$. The proof for a decrease in K_i is similar. ■

APPLYING THE RULE OF REASON TO MAXIMUM RESALE PRICE FIXING: *ALBRECHT* OVERRULED

Roger D. Blair, James M. Fesmire and Richard Romano

ABSTRACT

The imposition and social welfare consequences of maximum resale price fixing are analyzed when there is successive monopoly and a non-contractible service is provided downstream. An elasticity condition that weighs maximum resale price fixing's opposing effects on quantity, through the effects on final price and service provision, determines whether it is adopted. Wholesale price is increased along with the practice of resale price fixing, and the net effects on quantity and social welfare are ambiguous. This points to an output test and a rule of reason in the antitrust evaluation of this practice, consistent with the Supreme Court's overruling of Albrecht.

Few Supreme Court decisions have proved more anticompetitive than *Albrecht*.

Herbert Hovenkamp
Federal Antitrust Policy 422 (1994)

I. INTRODUCTION

Some 30 years ago, in Albrecht vs. Herald Company,[1] the Supreme Court characterized vertically imposed *maximum* resale prices as illegal per se under

Advances in Applied Microeconomics, Volume 9, pages 215–230.

ISBN: 0-7623-0687-4

Section 1 of the Sherman Act. In most instances, maximum resale price fixing occurs under conditions of successive monopoly.[2] For example, maximum resale prices have been imposed upon the distributors of newspapers, gasoline, prefabricated farm buildings, branded meats, soft drinks, and automobiles.[3] In each instance, there was an element of successive monopoly. In his classic article, Spengler (1950) showed that the vertical integration of successive monopolists enhanced social welfare as it resulted in lower prices and larger quantities. Since maximum resale price restraints are merely a contractual form of vertical integration, one would expect similar welfare effects. As a consequence, the *Albrecht* rule of per se illegality seemed obviously inconsistent with promoting social welfare, which ought to be the purpose of the antitrust laws. Criticism of the *Albrecht* rule began with the dissenting opinions of Justices Harlan[4] and Stewart[5] in the *Albrecht* case itself. This criticism has continued unabated for nearly 30 years.[6] At long last, the Supreme Court reconsidered the *Albrecht* rule in State Oil Company v. Khan.[7] In a unanimous decision, the Court candidly admitted that its thinking had changed in the 30 years since *Albrecht* had been decided: ". . . we find it difficult to maintain that vertically-imposed maximum prices could harm consumers or competition to the extent necessary to justify their per se invalidation."[8] It overruled *Albrecht* and, therefore, vertically imposed maximum resale prices are no longer illegal per se. But the Court did not find such restraints to be legal per se and, therefore, they remain subject to the rule of reason. This, we believe, is the right result.

As it turns out, the *Albrecht* Court's competitive concerns, which we review in the next section, were not entirely unjustified and, therefore, the conventional criticism of *Albrecht* is not completely warranted. This criticism relies on a special case in which the downstream firm provides no independent service. In the absence of downstream service, the application of Spengler's analysis to maximum resale pricing is correct. But when downstream service is present, the analysis changes as well as the implications for social welfare.

With independent downstream service, the upstream firm's positive price-cost margin distorts the downstream firm's choice of both price and service. The downstream firm still has an incentive to exploit its market power by raising the resale price above its marginal cost. In addition, the distributor will select a service level below the level that maximizes joint profits. Under some conditions, the upstream monopolist would want to impose a maximum resale price even though it leads to lower service provision. This might lead the upstream monopolist to increase the wholesale price, because the downstream firm cannot respond by raising the resale price, further reducing downstream service provision. We show that the latter can lead market output and social welfare to decline under the upstream monopolist's optimal choices. A sound antitrust policy must take into account these effects. A simple output test to guide a rule of reason in evaluating maximum resale price fixing is implied.

We consider the reversal of the *Albrecht* rule in this light. We conclude that one can unequivocally endorse jettisoning the *Albrecht* rule of per se illegality, and that the Supreme Court's decision to retain a rule of reason was correct. Allowing the pendulum to swing to per se legality as many economists seem to advocate would have been a mistake.

II. THE COURT'S COMPETITIVE CONCERNS IN *ALBRECHT*

The *Albrecht* scenario is fairly well known. The Herald Company published the only morning newspaper in St. Louis and to that extent had monopoly power. It distributed these newspapers through a system of exclusive distributorships thereby creating a successive monopoly situation in each exclusive territory. Albrecht was one of the distributors and, within his assigned territory, he enjoyed a monopoly of home delivery of the newspaper. When Albrecht refused to limit his prices to those advertised by the publisher, he was terminated. He sued under Section 1 of the Sherman Act. In its decision, the Supreme Court condemned maximum resale price fixing:

> . . . schemes to fix maximum prices, by substituting the perhaps erroneous judgment of a seller for the forces of the competitive market, may severely intrude upon the ability of buyers to compete and survive in that market.[9]

Without specifying what competitive market forces it thought were operative in Albrecht's exclusive territory, the Court went on to explain:

> Maximum prices may be fixed too low for the dealer to furnish services essential to the value which goods have for the consumer or to furnish services and conveniences which consumers desire and for which they are willing to pay. Maximum price fixing may channel distribution through a few large or specifically advantaged dealers who otherwise would be subject to significant non-price competition. Moreover, if the actual price charged under a maximum price scheme is nearly always the fixed maximum price . . . the scheme tends to acquire all the attributes of an arrangement fixing minimum prices.[10]

Our focus is on the services that might not be provided as a result of maximum resale pricing and the corresponding welfare effects. This is precisely the issue that Justice (then Judge) Breyer examined in his *Caribe* decision.[11] It is easy to dismiss these concerns. For example, Judge Posner felt that such a concern "implies that manufacturers misconceive their self-interest and is hardly an adequate basis for a per se rule."[12] Hence, Posner's interpretation was that any loss of service that lowers social welfare would be the result of a managerial mistake. We show, surprisingly, that this need not hold.

III. THE INCENTIVE FOR MAXIMUM RESALE PRICE RESTRAINTS

The Model

An upstream monopolist produces good q at a constant marginal cost of c. It sells this good to an exclusive distributor who resells it to ultimate consumers. For expositional convenience, we assume that the distributor's costs of performing the distribution function are zero. The distributor, however, can incur service costs at its discretion. Downstream service will shift the retail demand curve. We assume that the producer and the distributor both know the characteristics of retail demand including the influence of service by the distributor. Letting p denote price and x represent the units of service, we write the retail demand as

$$q = q\ (p, x).$$

Using subscripts to denote partial derivatives, we assume that

$$q_p < 0 \text{ and } q_x > 0.$$

We also assume that both firms know the cost function for downstream service, which is C(x) where $C'(x) > 0$.

We focus our analysis on the case of successive monopoly with downstream service provision under the assumptions that service cannot be contracted and that the upstream firm charges uniform wholesale prices. In this context, we are concerned with the manufacturer's incentive to impose maximum resale price restraints on its downstream distributor and the effects of such restraints should they be imposed. We are motivated by the fact that these restraints have been observed in a variety of industries – newspapers, automobiles, gasoline, and so on – where downstream service is clearly important to one extent or another. Our concern is not with explaining the relative advantages and disadvantages of contracts that rely exclusively on uniform prices. We do, however, wish to analyze these contracts in the most stark setting. In principle, vertical integration and two-part pricing contracts are superior to the simpler contracts that we study, but those contracts are not subject to the *Albrecht* rule.[13]

The Rationale for Price Restraints

As a benchmark, we start with the case where the upstream firm imposes no restraints on the distributor. The distributor will be free to select the price (p) and the service level (x) that maximize its profits (π^D) taking the producer's wholesale price (w) as given. In other words, the distributor solves

$$\max_{p,\,x} \pi^D = (p - w)q(p, x) - C(x). \tag{1}$$

The first-order conditions are[14]

$$q + (p - w)q_p = 0 \tag{2}$$

and

$$(p - w)q_x - C' = 0. \tag{3}$$

We define $\hat{x} = \hat{x}(p, w)$ as the x value that satisfies (3) and $p^* = p^*(w)$ as the value of p that jointly satisfies (2) and (3).

The upstream firm's problem is to choose w to maximize its profits (π^U) given the profit maximizing choices of p and x by the distributor.

This problem can be written as

$$\max_{w} \pi^U = (w - c)q(p^*(w), \hat{x}(p^*(w), w)). \tag{4}$$

For the upstream producer, the first-order condition is

$$q + (w - c)[(q_p + q_x\hat{x}_p)p^{*\prime} + q_x\hat{x}_w] = 0. \tag{5}$$

Let $w°$ denote the wholesale price that satisfies (5) and define $p° \equiv p^*(w°)$. In other words, $p°$ is the distributor's optimal resale price given a wholesale price of $w°$.

This solution is suboptimal for the upstream monopolist. Since w must exceed c for the upstream monopolist to be profitable, there are vertical externalities in the distributor's choices of p and x, which determine the output and the joint profits that the distributor only partially appropriates. If the manufacturer were to impose maximum resale prices in response to the distributor's mark up, the distributor will choose only the level of service. The distributor will select x according to (3), so that $x = \hat{x}(p, w)$ still. The upstream monopolist now solves

$$\max_{w,\,p} \pi^U = (w - c)q(p, \hat{x}(p, w)). \tag{6}$$

Let w^1 and p^1 denote the wholesale price and final price, respectively, that solve (6). The upstream monopolist will benefit from maximum resale prices if $p^1 < p^*(w^1)$. Proposition 1 presents the condition for the optimality of this business practice.

For Proposition 1, we must define three elasticities. First, let $\varepsilon_p \equiv -q_p p/q$ denote the usual price elasticity of (consumer) demand. Next, let $\varepsilon_x \equiv q_x x/q$

denote the service elasticity of demand. Finally, let $\eta \equiv \hat{x}_p p/x$ denote the price elasticity of 'induced service'.

Proposition 1. *The upstream monopolist will impose maximum resale price restraints on its distributors if and only if* $E \equiv -\varepsilon_p + \varepsilon_x \eta < 0$, *where the elasticities are evaluated at the optimal values of p, x, and w absent the practice of maximum resale price fixing.*

Proof of Proposition 1. *It is shown here that the upstream monopolist's local incentive is to force down the final price from* p° *whenever* $E < 0$.[15] *Differentiate* $\pi^U = (w - c)q(p, \hat{x}(p,w))$ *to obtain*

$$d\pi^U = \frac{(w - c)}{p} q(-\varepsilon_p + \varepsilon_x \eta)dp + (q + (w - c)q_x \hat{x}_w)dw. \tag{7}$$

Since this will be evaluated at (p°, w°), *we can substitute from (5) into (7) yielding*

$$d\pi^U(p^\circ, w^\circ) = \frac{(w - c)}{p} q\, Edp - \frac{(w - c)}{p} q\, E\, p^{*\prime}\, dw. \tag{8}$$

It is straightforward to show, using (2) and (3), that

$$p^{*\prime} = \frac{\pi^D_{xx}}{|H|}(-\varepsilon_p + \varepsilon_x \eta)q/p, \tag{9}$$

where H is the Hessian matrix associated with (1). Finally, substituting (9) into (8), one obtains

$$d\pi^U(p^\circ, w^\circ) = \frac{(w - c)q}{p}\left[E\, dp - \frac{\pi^D_{xx}}{|H|} E^2 \frac{q}{p} dw\right]. \tag{10}$$

Using the second-order condition that $\frac{-\pi^D_{xx}}{|H|} > 0$, *it follows from (10) that* π^U *can be increased by decreasing p and/or increasing w whenever* $E < 0$. *Hence, upstream profits can be increased by forcing down p from* p° *when* $E < 0$, *which conforms to maximum resale price fixing (see footnote 16). The incentive to increase w is discussed below.*

Proposition 1 provides a simple, intuitive condition that isolates the effects of maximum resale price restraints. Two vertical externalities are associated with the unconstrained distributor's choice of price and service. Since the distributor captures only a fraction of the joint profits, it will select a price that is too high and a level of service that is too low to maximize joint profits. By dictating the final price, the manufacturer can control the pricing externality. But using maximum resale price restraints to narrow the distributor's price-cost margin has an impact on the services provided. Thus, there are costs and benefits of this business practice. When the manufacturer restrains the resale price, there is an effect on the quantity demanded, which is measured by the usual price elasticity of demand (ε_p). At the same time, the restraint on the resale price induces a change in the level of service provided, which is captured by the price elasticity of induced service η. The induced change in the level of services influences the quantity demanded, which is measured by the service elasticity of demand (ε_x). The product of ε_x and η measures the effect on the service externality of restraining the resale price. Evaluating the elasticities at the unconstrained optimum, maximum resale pricing increases the monopolist's upstream profits when $\varepsilon_p > \varepsilon_x \eta$. In this case, the pricing externality outweighs the service externality. The alternative case is discussed below.

The conventional criticisms of *Albrecht* assume implicitly that there is no service externality or that service is irrelevant.[16] In this event, $\varepsilon_x = 0$ and the elasticity condition $E < 0$ will always hold. Absent service at the distributor stage, maximum resale price restraints always lead to an increase in output. When we generalize the model to include service by the distributor, however, the effect on output is ambiguous. We turn to that issue now.

The Effects on Wholesale Price and Output

The upstream monopolist's profits are given by $\pi^U(p, w) = (w - c)q(p, \hat{x}(p, w))$. Note that the retail price (p) enters the manufacturer's profit function only through its two effects on demand. Consequently, the upstream firm has an incentive to push the retail price down only if that will expand output, as determined by the elasticity condition. But the net result may not be an increase in output if the manufacturer also increases the wholesale price (w), since that would have the opposite effect. In fact, the (local) incentive will always be to increase the wholesale price. We examine this incentive here and show that output can rise or fall in response to maximum resale pricing. This is relevant to the normative evaluation of vertical price restraints. Those who believe the upstream monopolist's incentive in choosing whether to impose maximum resale price fixing complies with the social incentive seem to have ignored the accompanying incentive to increase wholesale price.

Proposition 2. *There is a local incentive to increase the wholesale price as part of the resale pricing strategy. That is,* $\partial\pi^U(p^\circ, w^\circ)/\partial w > 0$ *if* $E \neq 0$.

Proof of Proposition 2. *This can be seen by inspection of (10) since the coefficient on dw equals* $\partial\pi^U/\partial w$ *and is positive using the second-order condition* $-\pi^D_{xx}/\Lambda H > 0$.

Intuitively, a resale price restraint limits the distributor's flexibility in responding to increases in the wholesale price and thereby makes such increases profitable. Although output declines as a consequence of an increased wholesale price, it declines less than when the distributor is free to set final price.[17] Consequently, the manufacturer can capture more rents with a higher wholesale price. The upstream firm's optimal local departure from (p°, w°) has countervailing effects on output. When $E < 0$, the incentive is to reduce final price as that increases output and profits, but also to increase the wholesale price which decreases output. The incentive to increase the wholesale price is only a local one, but the logic suggests an ambiguous effect on output, which our next proposition confirms.

Proposition 3. *The imposition of maximum resale prices can lead to either an increase or a decrease in final output.*

The proof is in the appendix.

IV. THE WELFARE EFFECTS

The main result of this section is that welfare can increase or decrease when a manufacturer imposes maximum resale prices on its distributors. Two effects on social welfare can be conveniently isolated. One effect hinges on whether the price restraint raises output, which is socially desirable since the marginal consumer's valuation exceeds marginal cost. The other effect weighs the impact of changes in service provision on all consumers' valuations, marginal and inframarginal, against the associated change in service cost. We show that welfare can decline as a result of maximum resale price fixing without appeal to the second effect.

Measure social welfare by the sum of consumer and producer surplus. Since $x = \hat{x}(p, w)$ in our comparison, we can write social welfare, $W(p, w)$, as follows:

$$W(p,w) = \int_{p}^{\bar{p}(\hat{x}(p,\,w))} q(z, \hat{x}(p, w))dz + (p - c)q(p, \hat{x}(p, w)) - C(\hat{x}(p, w)), \quad (11)$$

where $\bar{p}$ is the price that just chokes off demand. Now, consider the change in welfare resulting from small changes in p and w. Using Leibnitz's Theorem, this is closely approximated by

$$\Delta W \approx (\hat{x}_p\, \Delta p + \hat{x}_w\, \Delta w) \int_p^{\bar{p}} q_x\, dz + (p - c)[q_p\, \Delta p + q_x(\hat{x}_p\, \Delta p + \hat{x}_w\, \Delta w)]$$
$$- C' \cdot (\hat{x}_p\, \Delta p + \hat{x}_w\, \Delta w) = (p - c)[(q_p + q_x\hat{x}_p)\, \Delta p + q_x\hat{x}_w\, \Delta w]$$
$$+ \left(\int_p^{\bar{p}} q_x\, dz - C' \right)[\hat{x}_p\, \Delta p + \hat{x}_w\, \Delta w]. \qquad (12)$$

Note that the first bracketed term on the right-hand side of (12) is Δq and the second bracketed term is Δx. The marginal change in welfare equals the change in output weighted by the gap between marginal valuation and marginal cost, plus the change in service weighted by the difference between the marginal value of service and its marginal cost. This intuition extends to the overall change in welfare resulting from maximum resale pricing since it is given by

$$W(p^1, w^1) - W(p^\circ, w^\circ) = \int_{p^\circ}^{p^1} \int_{w^\circ}^{w^1} dW.$$

A difficulty in determining the sign of the change in welfare is that the sign of the service effect captured by the second term on the right-hand side of (12) depends on the functional form of demand. We next state a lemma that is useful in this regard. It is clearer to state the result in terms of inverse demand, so let $\hat{p}(q, x)$ denote the inverse demand function.

Lemma 1.

(a) If $\hat{p}_{qx} = 0$, then the downstream firm chooses the socially optimal level of promotion *for the output it produces* in the absence of maximum resale pricing.

(b) This implies $\int_{p^\circ}^{\bar{p}} q_x(z, x^\circ)dz - C'(x^\circ) = 0$ absent maximum resale prices.

The economic content of Lemma 1 is contained in (a), which says that if increases in service result in vertically parallel shifts in demand, then a firm that chooses price and service makes the socially optimal choice of service.[18] Under the demand condition in (a), the downstream firm captures the entire increment to social surplus from increased service by increasing its price of the good. Part (b) of the lemma then follows from the downstream firm's first-order condition for x. Lemma 1 facilitates the proof of Proposition 4, which we now state.

Proposition 4. *The change in welfare resulting from maximum resale prices can be positive or negative.*

Proof of Proposition 4. *Assume that: (i) (p°, w°) and (p^1, w^1) are arbitrarily close; and (ii) $\hat{p}_{qx} = 0$. Then (i) implies that the change in welfare from maximum resale pricing is closely approximated by (12), and Lemma 1 implies that*

$$sign\ \Delta W = sign\left[q_p + q_x\hat{x}_w + \frac{\Delta w}{\Delta p}\right]^\circ \Delta p = sign\ \Delta q.$$

The result then follows from Proposition 3.

The ambiguity in the welfare effects of maximum resale price restraints flows from the ambiguity in the effect on output. Absent services by the distributor, the welfare effect is positive because output will rise. But this is not necessarily the case when the distributor provides services that are valued by the consumer. As part of the practice of resale price fixing, it is optimal for the upstream monopolist to increase wholesale price as we have seen. This squeeze only has redistributive effects in the absence of downstream provision of services. With downstream services, however, this squeeze lowers further downstream services, and can lower output and welfare.

If we depart from the presumption that service causes vertically parallel shifts in demand, then the net welfare effect of changes in service will not vanish even for small changes. This effect can also be positive or negative and, therefore, it adds another layer of ambiguity to the evaluation of maximum resale price restraints.[19]

If $E > 0$, the upstream monopolist would impose a minimum resale price as the service externality dominates the pricing externality. Propositions 2–4 continue to apply, so, again, welfare can decline.[20] Given our concern for *Albrecht* and *Khan*, we henceforth continue to discuss the case of maximum resale price fixing.

V. IMPLICATIONS FOR ANTITRUST POLICY

In a simple world with no discretionary distributor services, the *Albrecht* rule is unambiguously inconsistent with promoting social welfare and it is a good thing that *Khan* overruled it. It is clear that the old *Albrecht* per se rule was unjustified. In a large class of cases, maximum resale price restraints are welfare enhancing. Whenever there are no discretionary services provided by the distributor or in instances where the potential externality in the provision of distributor services can be resolved contractually, the conventional wisdom prevails. There is only a pricing externality. Removing the resale price distortion through a maximum resale price restraint leads to greater welfare. In this situation, the practice should not be illegal at all much less per se illegal. But if maximum resale price restraints are not going to be per se illegal, they must be judged under the rule of reason or be per se legal.

The results above imply that it would be dangerous to prescribe per se legality, which leaves us with the rule of reason.

Our welfare analysis provides fairly *practical* guidelines for applying a rule of reason.[21] Although social welfare can rise or fall even with a positive output effect as a general matter, welfare will increase with an increase in output when all demanders value services equally, i.e. in cases where Lemma 1 holds. This can be used to create a rebuttable presumption that vertically imposed maximum resale prices have positive welfare effects provided that output increases. In other words, in the face of proof that output has expanded, the burden of proof shifts to the plaintiff to demonstrate that welfare has declined. The reverse would apply if output is demonstrated to have declined. Antitrust violation is then presumed, unless the defendant can effectively demonstrate otherwise.

Hence, this chapter advocates a simple output test as the backbone of antitrust scrutiny of maximum resale price fixing. Unless sufficiently contradictory effects on inframarginal consumers can be demonstrated, output increases signal legality while output decreases signal illegality. At the same time, our analysis supports the *Khan* decision to move only to a rule of reason in rejecting the *Albrecht* rule.

VI. CONCLUDING REMARKS

In *Albrecht*, the Supreme Court identified several competitive concerns regarding maximum resale price restraints. One of these concerned the provision of services by the distributor that consumers valued. In order to reconsider *Albrecht's* per se prohibition of maximum resale price restraints, we examined a market structure characterized by successive monopoly with noncontractible downstream services. In that context, two vertical externalities arise: The distributor will choose a price too high and a level of service too low to maximize joint profits. Maximum resale pricing serves to resolve the pricing externality, but it worsens the service externality by decreasing the distributor's margin. Explicitly recognizing this tradeoff, we derived the condition under which the upstream monopolist will find it optimal to impose maximum resale price restraints. We found that such restraints will be employed when the pricing externality dominates the service externality.

In analyzing the welfare effects of maximum resale prices, we found that the welfare result depends on the net result of two effects. One effect depends on the change in output that results. Since price exceeds marginal cost, output increases are welfare enhancing while decreases in output diminish welfare. Although the manufacturer's incentive is to change the final price in such a way that output rises, there is also an incentive to increase the wholesale price to extract more profit. The effect of this is to reduce output. We have shown that on balance output can go up or down when maximum resale pricing is practiced. The second effect on welfare weighs the change in inframarginal valuations of output resulting from downstream service adjustments against the change in service cost. This effect can vanish in which case the overall welfare change is determined solely by the output effect. Consequently, welfare can rise or fall, and a rule of reason approach that employs an output test is indicated.

NOTES

1. The *Albrecht* decision can be found at 390 U.S. 145 (1968).

2. In their assessment of maximum resale pricing, Scherer & Ross (1990) at p. 558, conclude that "[t]he most plausible reason why a manufacturer would wish to prescribe price ceilings binding its resellers is to avoid repeated marginalization by vertically pyramided monopolists."

3. Cases involving these products are examined in Blair & Esquibel (1996).

4. Justice Harlan explained the economic logic of maximum resale prices and why the majority's lack of economic reasoning led to an unfortunate result. 390 U.S. at 156.

5. Justice Stewart concluded that "[t]he Court in this case does more, I think, than simply depart from the rule of reason. The Court today stands the Sherman Act on its head." 390 U.S. at 170.

6. For a sample of the criticism, see Areeda and Turner (1978, ¶734e); Blair & Kaserman (1981); Easterbrook (1981) at note 20; Blair & Fesmire (1986); and Ross, (1993) at 255.

7. 118 S. Ct. 275 (1997). In Khan & Associates vs. State Oil Company, 93 F. 3d 1358 (1996), Judge Posner grudgingly conceded that "... despite all its infirmities, its increasingly wobbly, moth-eaten foundations ..., Albrecht has not been expressly overruled ..." Consequently, he applied the *Albrecht* rule although he clearly disapproved of its impact.

8. *Id.* at 282.

9. 390 U.S. at 152.

10. 390 U.S. at 152–153.

11. Caribe BMW, Inc. vs. Bayerische Motoren Werke Aktiengesellschaft, 19 F. 3d 745, 753 (1994): At least in theory, if customers would have preferred a higher price and

consequently better quality or greater service, the agreement forced Caribe to provide less of what they wanted; the agreement thereby might have led to lower Caribe profits. And, at least in theory, if the agreement helped other, larger BMW dealers, Caribe is the firm that would have suffered.

12. Posner (1976) at 158. See also, Blair & Lang (1991).

13. For more general contracts in successive monopoly with downstream services that lead to similar incentives, see Romano (1994) and Blair & Lewis (1994).

14. It is assumed throughout that all functions are twice continuously differentiable and that the second-order conditions are satisfied.

15. A proof that this incentive carries over globally is available upon request.

16. Alternatively, one might assume that the potential moral hazard with respect to the distributor's provision of services can be resolved contractually. But this would be unrealistic in many cases due to monitoring costs.

17. Formally, $[\partial q(p, x(p, w))/\partial w]^{\circ} < [dq(p^{*}(w), x(p^{*}(w), w))/dw]^{\circ}$, the superscript indicating where arguments are evaluated.

18. This result can be traced to Swan (1970). See, also, Spence (1975) for proof in a closely related setting.

19. See Spence (1975).

20. The proofs of Propositions 2–4 continue to apply when $E > 0$.

21. Whether a rule of reason approach to deciding the legality of a business practice with ambiguous competitive effects is operational has been questioned by, among others, Judge Posner. See Posner (1977). Our interpretation is that it is useful for economic analysis to develop tests for such cases.

ACKNOWLEDGMENTS

The authors are Huber Hurst Professor, University of Florida; Professor of Economics, University of Tampa; and Gerald Gunter Professor of Economics, University of Florida, respectively. We thank our respective institutions for financial support. John Lopatka provided some very useful suggestions, but he is not responsible for what follows.

REFERENCES

Areeda, P., & Turner, D. (1978). *Antitrust Law, 3*.

Blair, R. D., & Esquibel, A. K. (1996). Maximum Resale Price Restraints in Franchising, *Antitrust Law Journal, 65*, 157–180.

Blair, R. D., & Fesmire, J. M. (1986). Maximum Price Fixing and the Goals of Antitrust. *Syracuse Law Review, 37*, 43–77.

Blair, R. D., & Kaserman, D. L. (1981). The Albrecht Rule and Consumer Welfare: An Economic Analysis, *University of Florida Law Review, 33*, 461–484.

Blair, R. D., & Lang, G. L. (1991). Albrecht after ARCO: Maximum Resale Price Fixing Moves Toward the Rule of Reason. *Vanderbilt Law Review, 44*, 1007–1039.

Blair, B., & Lewis, T. R. (1994). Optimal Retail Contracts Under Asymmetric Information and Moral Hazard. *RAND Journal of Economics, 25*, 284–296.

Easterbrook, F. (1981). Maximum Price Fixing. *University of Chicago Law Review, 48*, 886–910.

Posner, R. A. (1976). *Antitrust Law: An Economic Perspective.*

Posner, R. A. (1977). The Rule of Reason and the Economic Approach: Reflections on the Sylvania Decision. *University of Chicago Law Review, 45*, 1–20.

Romano, R. E. (1994). Double Moral Hazard and Resale Price Maintenance, *RAND Journal of Economics, 25*, 455–466.

Ross, S. F. (1993). *Principles of Antitrust Law.*

Scherer, F. M., & Ross, D. (1990). *Industrial Market Structure and Economic Performance* (3rd ed.).

Spence, M. (1975). Monopoly, Quality and Regulation. *Bell Journal of Economics, 6*, 417–429.

Spengler, J. J. (1950). Vertical Integration and Antitrust Policy. *Journal of Political Economy, 58*, 347–352.

Swan, P. (1970). Durability of Consumption Goods. *American Economic Review, 60*, 884–894.

APPENDIX

Proof of Proposition 3. We prove the more general result that output can rise or fall for both the cases of E positive and negative. Assume for the moment that (p°, w°) is arbitrarily close to (p^1, w^1). Let $\Delta p \equiv p^1 - p^\circ$, $\Delta w \equiv w^1 - w^\circ$, and $\Delta q \equiv q(p^1, \hat{x}(p^1, w^1)) - q(p^\circ, \hat{x}(p^\circ, w^\circ))$. Then

$$\text{sign } \Delta q \equiv \text{sign}\left[q_p + q_x \hat{x}_p + q_x \hat{x}_w \frac{\Delta w}{\Delta p} \right]^\circ \Delta p. \tag{A1}$$

Now we find expressions for Δp and Δw. For notational convenience, the U superscript on upstream profits is dropped in this proof; i.e. let $\pi(p, w) = (w - c)q(p, \hat{x}(p, w))$.

Doing Taylor expansions on π_p and π_w about (p°, w°) one obtains

$$\pi_p^1 - \pi_p^\circ \approx \pi_{pp}^\circ \Delta p + \pi_{pw}^\circ \Delta w \tag{A2}$$

and

$$\pi_w^1 - \pi_w^\circ \approx \pi_{ww}^\circ \Delta w + \pi_{wp}^\circ \Delta p. \tag{A3}$$

Substitute $\pi_p^\circ = \pi_w^1 = 0$ into these and solve for Δp and Δw yielding

$$\Delta p \approx -\frac{(\pi_{ww}\pi_p - \pi_{pw}\pi_w)^\circ}{|H|^\circ} \tag{A4}$$

and

$$\Delta w \approx -\frac{(\pi_{pp}\pi_w - \pi_{pw}\pi_p)^\circ}{|H|^\circ} \tag{A5}$$

Before substituting (A4) and (A5) into (A1), it is easier to consider first what is sufficient for (p°, w°) and (p^1, w^1) to be arbitrarily close. By 'arbitrarily close', we mean that Δp and Δw are arbitrarily small in the sense that they cannot be bounded away from zero. For this it is sufficient that

(i) π_p° and π_w° are both arbitrarily small;

and

(ii) π° is locally 'strongly concave', meaning $|H|^\circ$ can be bounded away from zero (on the positive side), and π_{pp}° and π_{ww}° can be bounded away from zero (on the negative side).

That (i) and (ii) are sufficient can be confirmed by inspection of (A4) and (A5).

To see the implications of (i)–(ii), from (10) we have

$$\pi_p^\circ = (w^\circ - c)q^\circ E^\circ / p^\circ, \tag{A6}$$

and

$$\pi_w^\circ = -(w^\circ - c)(q^\circ)^2 E^2 \pi_{xx}^D / [(p^\circ)^2 |H|^\circ] \tag{A7}$$

Then (i)–(ii) require that E° is arbitrarily small. This also implies that π_w° is of lower order than π_p°, so that terms with π_w° in (A4) and (A5) can be ignored.

It follows that

$$\frac{\Delta w}{\Delta p} \approx -\frac{\pi_{pw}^\circ}{\pi_{ww}^\circ} \tag{A8}$$

and, using local concavity of π,

$$\text{sign } \Delta p = \text{sign } \pi_p^\circ = \text{sign } E^\circ. \tag{A9}$$

Using these results and substituting into (A1), we have

$$\text{sign } \Delta q = \text{sign} \left[\left(\frac{Eq}{p} - q_x \hat{x}_w \frac{\pi_{pw}}{\pi_{ww}} \right) E \right]^\circ$$

$$= \text{sign} \left[-\hat{x}_w E \frac{\pi_{pw}}{\pi_{ww}} \right]^\circ \text{ (since E is small)} \tag{A10}$$

$$= \text{sign}[\hat{x}_w \pi_{pw} E]^\circ \text{ (since } \pi \text{ is locally concave).}$$

Using (3), it is straightforward to show that $\hat{x}_w^\circ = [q_x/\pi_{xx}^D]^\circ$, which must be negative by (local) concavity of π^D. Finally, then

$$\text{sign } \Delta q = \text{sign}[-\pi_{pw}E]^{\circ}.$$

Then the proof is complete if π°_{pw} can be positive or negative for E° positive or negative while also preserving (i) and (ii) (and concavity requirements). Although tedious, this is straightforward to confirm. As is not surprising, there are sufficient degrees of freedom in the parameters to permit it. This completes the proof.

RIGID MONOPOLY PRICES

Mark Stegeman

ABSTRACT

A consumer in the real world typically must visit (e.g. by phone) a monopolist to observe its price, even though the consumer may have correct expectations about that price. This causes monopoly prices to be higher and stickier than is predicted by the textbook model. If visiting costs are small, and consumers do not observe the firm's costs, then prices conform to the textbook model when costs are high but are downwardly rigid when costs drop below a threshold. Price flexibility increases as the fraction of ignorant consumers, who observe their idiosyncratic valuations of the product only after visiting the firm, increases. In a repeated game, a 'ratchet' equilibrium, in which price increases are permanent and price decreases temporary within a certain band, supports equilibria for which price is rigid on the equilibrium path but which Pareto dominate the fluctuating one-shot equilibrium. The ratchet equilibrium has the advantages that price is a continuous function of past prices near the equilibrium path and the price coordination problem is solved in a natural way: the equilibrium price is the lowest price that can be sustained by such a ratchet. This equilibrium price exceeds but is close to the average prediction of the textbook model. Additional results are derived for the case in which costs are fixed.

INTRODUCTION

In most markets the buyer approaches the seller to initiate a transaction. The buyer may approach the seller through a personal visit, a phone call, an online

---

Industrial Organization, Volume 9, pages 231–263.

ISBN: 0-7623-0687-4

inquiry, or a letter, but in each case the buyer must expend some effort. Such visiting costs (or search costs) have important and largely unstudied effects upon the behavior of a monopolist. Unlike menu costs and many other transactions costs, the effects do not vanish as the costs approach zero.

Diamond (1971) pointed out the discontinuous effects of small visiting costs by showing that they dramatically increase the equilibrium price in an otherwise competitive market. Suppose that all firms are identical, with constant marginal costs. If the cost of visiting a firm and observing its price is sunk, then the lowest-priced firm can raise its price slightly above what consumers expect without losing any sales, and the market price ultimately rises to the monopoly price. Diamond's result led many researchers to construct models generating more reasonable prices in markets having many sellers.[1] In contrast with these attempts to find lower equilibrium prices and price dispersion, Salop & Stiglitz (1982) described an outcome even less attractive than the monopoly price. Suppose that every firm sets the price that is the monopoly price given the entire market demand curve. Anticipation of this price may cause some consumers not to visit firms at all, and the monopoly price given the demand curve of persons who actually visit may be yet higher. Then each firm has a Diamond-like incentive to raise price, and if each person desires to buy at most one unit of the good then it is easily shown that there is no equilibrium in which positive sales occur.

The distinctive feature of the Salop and Stiglitz model, relative to many of the sequels to Diamond's model, is that some consumers choose not to search at all and their disappearance from the market drives each firm's price so high that no one searches. To put this in another way, the typical in-store (i.e. post-visit) demand curve is not simply a fraction of the aggregate demand curve. Despite the absurd outcome, the model of Salop and Stiglitz is important because the marginal consumers (consumers on the threshold of searching) that drive the outcome obviously exist in real markets. This chapter describes the effects of marginal consumers in a model rich enough to generate positive sales equilibria. The conclusions differ qualitatively from Diamond's. Diamond's point is that visiting costs cause competitors to act like monopolists; this study shows that the conjunction of visiting costs and marginal consumers cause prices to rise even higher. Because these effects obtain regardless of the presence of competitors, we simplify matters by exploring the relationship between a single firm and its consumers.

A central feature of the present model is that a fraction of consumers, $\alpha \in [0, 1]$, is ignorant of their reservation values before visiting the firm. The remaining 'knowledgeable' consumers know their reservation values a priori. Let $\delta > 0$ denote the cost of visiting, and we assume throughout that $\delta \approx 0$.[2] At one extreme, if $\alpha \approx 1$ (almost all consumers are ignorant), then the equilibrium price is similar to that predicted by the standard textbook model of monopoly. ('Textbook' model refers to the price implied by the standard MR = MC rule, where the market demand curve is calculated assuming that all consumers correctly anticipate the firm's price.)[3] At the other extreme, if $\alpha \approx 0$, then the (one-shot) market disappears in the fashion of Salop and Stiglitz. Between these extremes, the market does not disappear, but the monopolist's behavior is quite different from the prediction of the textbook model.

By dividing consumers into those who choose to visit the firm and those who do not, small visiting costs introduce an important informational asymmetry: visitors automatically update their information about the firm, and others do not. This asymmetry produces a kink in the market demand curve, because price hikes drive visitors away more effectively than price cuts attract non-visitors. The existence of such a kink seems to have been first anticipated by Okun (1981) and Stiglitz (1984).

Combining these insights, one expects even small visiting costs to make monopoly prices both higher and stickier than is predicted by the textbook model. The formal model of this chapter confirms that such effects appear discontinuously as visiting costs rise from $\delta = 0$ to small $\delta > 0$. Aside from underestimating prices and overestimating price flexibility, the textbook model of monopoly incorrectly predicts the impact of consumer ignorance. The textbook model predicts that consumer ignorance of own reservation values increases prices, because more consumers experiment; but ignorance lowers prices in the present model, because it mitigates the holdup problem that drives prices above their textbook monopoly values.

Sections I and II of the chapter study the one-shot market, and Section III studies the possibility that reputation effects can lower prices in a repeated market. In the equilibria studied, reputation effects do reduce prices but only by making them perfectly rigid.

In Section I, the firm's constant marginal cost is known to all, and in Section II it is known only to the firm. A comparison of the results shows that consumer ignorance about the firm's costs partially substitutes for ignorance about own reservation values, for the purpose of improving price performance. Recall that $\alpha \in (0, 1)$ denotes the fraction of consumers who are ignorant of their own reservation values, a priori. Equilibrium prices decrease in α, as already indicated. Holding α fixed at any level: if consumers observe costs, then the equilibrium price always exceeds the textbook prediction, but if costs are random and unobserved by consumers, then equilibrium prices match textbook predictions when costs are high but are downwardly inflexible when costs are low. Even when costs are low, the equilibrium price is generally lower than it would be if consumers observed costs.

These results have interesting dynamic implications. If costs are stable, then prices are fixed at a level exceeding the textbook prediction. If costs fluctuate widely, then prices are sticky but upwardly flexible in periods of high costs. If costs fluctuate widely but consumers receive strong signals of these fluctuations (effectively implying that the cost distribution is tight period-by-period), then prices fluctuate widely but at levels higher than the textbook prediction.

Section III of the chapter is motivated by a welfare question that does not arise in Diamond's (1971) model. In Diamond's model search costs push prices toward monopoly levels, helping firms but hurting consumers. In Sections I and II visiting costs push prices beyond monopoly levels, hurting everybody. For example, given $\delta \approx 0$, it is not difficult to show that everyone would be better off if the firm could commit itself to the textbook price t(c). The question for Section III is: what mechanisms can restore textbook prices or support other Pareto-improving prices?

Since the underlying problem is non-visitors' inability to observe prices, one possibility is to transform non-visitors effectively into visitors by reimbursing their visiting costs. This solution may be unrealistic, however, because complete and uncontingent reimbursement (the only kind that works, since reimbursement contingent on purchase is equivalent to a price cut), is likely to attract repeat visitors and consumers who are not even contemplating purchase. In reality, complete and uncontingent reimbursement schemes seem to be rare, though not unknown.

A more plausible way to improve welfare is to improve non-visitors' price information. In some cases visitors may buy guaranteed price lists from an outside source, but casual observation suggests that non-visitors receive most price information from the firm itself, either by observing its past prices or through costly advertising. Past prices are an important source of information in almost all markets, but the importance of advertising varies, depending upon its cost, its coverage, the advantages of being able to change price frequently, and the mechanism that sustains its credibility. To simplify matters, the model in Section III considers only price reputation as a Pareto-improving mechanism. The model could represent a market in which advertising is prohibitively expensive relative to its benefits, or it could represent a starting point towards the analysis of a market in which both price reputation and advertising are important sources of price information.

The model is replicated through time by assuming that the firm lives forever and sells to consumers who enter and leave the model at rate α. Consumers are ignorant when they enter, become knowledgeable during their first visit to the firm, desire to consume the product in each period and cannot store it, and remain knowledgeable until they exit. We study only equilibria in which new consumers visit the firm immediately upon entry, implying that the fraction of ignorant consumers is α in every period. The firm's costs are statistically independent across periods. The equilibrium concept presents a slight difficulty, because the game has hardly any subgames and is infinite, but the sequential equilibrium concept can be applied in a natural way.

The conventional wisdom embodied in the folk theorem suggests that it should be possible to replicate textbook pricing in the dynamic model, but imperfect information makes this impossible. In any given period the firm would be tempted to raise its price, and consumers who do not observe costs cannot detect isolated deviations. Given this observation, it is not surprising (and it is shown) that it is impossible to support any stationary price rule $p_t = w(c_t)$ that is continuous, non-constant, non-decreasing, and noticeably different from the one-shot equilibrium. The textbook rule $p_t = t(c_t)$ falls into this class.

If the textbook price is unsustainable even in the dynamic game, then how do monopolists set prices? The simplest way to improve upon the one-shot equilibrium (given the result described in the previous paragraph) is to adopt an equilibrium in which the monopolist holds price rigid, at a level generally lower than prices in the one-shot equilibrium. Very high variance in costs can preclude such an equilibrium, particularly when consumers enter and exit the market rapidly, but for a very broad range of parameter values there exist rigid price equilibria that Pareto dominate the one-shot equilibrium. In these equilibria the firm's constant marginal cost may occasionally exceed its price.

Even the constant price equilibria introduce a serious problem. Within the range of sustainable prices, which price is focal? There is no obvious answer, both because costs are random and because the players' interests conflict. This chapter proposes a novel solution to the problem of selecting a focal price. It is of course impossible to argue that the equilibrium proposed here, which we will a 'ratchet' equilibrium, is the *correct* equilibrium, but it has attractive properties. The first is that the firm chooses its own equilibrium price. In other words, the firm continues to charge whatever price it sets in its first period, within an interval of sustainable prices. Downward deviations from this price are ignored, but if the firm deviates upward to a new higher price, then it charges the new higher price thereafter. The firm's strategy is thus continuous, except that if it sets a price that is unsustainably high, then it switches to the semi-sticky one-shot pricing formula. It turns out that the best price for the firm to pick in the first period is always the lowest sustainable price.

One of the attractive properties of the 'ratchet' equilibrium is that it solves the price coordination problem in a reasonable way: the firm sets its own price when it enters the market. Although real-world prices obviously do not stay fixed forever, this easy means of coordination seems to reflect much of what happens in reality. Ignorant consumers need not guess the focal price, because they want to visit the firm anyway. Knowledgeable consumers also need not guess the focal price; they have already observed it. The firm and consumers must of course coordinate on the notion that the firm's price strategy is an upward ratchet, but that notion seems more intuitively appealing than somehow coordinating on an arbitrary price p^* and supporting it with highly discontinuous trigger strategies.

A second attractive property of the ratchet equilibrium is that the strategies require very little computation on the part of the consumer. The only slightly complicated part of the equilibrium is the possibility of reversion to the one-shot equilibrium, which occurs outside the interval of prices that can be sustained by the ratchet.

The final attractive property of the ratchet equilibrium is that computations suggest that the implied equilibrium price is typically quite close to the *average* textbook price. In other words, although it is impossible to sustain the flexible textbook price, this equilibrium generates a price that is fixed close to the average of what the textbook predictions would be. This is not surprising, because the firm picks its own price subject to the sustainability constraint. To this extent, the present analysis supports the textbook pricing model, with the caveat that prices fluctuate less than the textbook analysis suggests.

I. MODEL I (ONE-PERIOD, CONSUMERS OBSERVE COSTS)

A. The Assumptions

The supply side of the market is a single firm, which produces a homogeneous good at constant unit cost $c \in [0,1)$. The demand side of the market comprises infinitesimal consumers with heterogeneous reservation values. Consumer x receives utility x from one unit of the good and zero utility from additional units. Consumers' reservation values x are distributed uniformly on the [0, 1] interval. Consumers are further distinguished by their knowledge of x. A fraction α $(0 < \alpha < 1)$ fraction of the consumers are *ignorant*; the rest are *knowledgeable*. A knowledgeable consumer knows her x value before visiting the firm; an ignorant consumer does not know her x value but assumes that

$x \sim U[0, 1]$. Whether a consumer is knowledgeable is statistically independent of her reservation value.

The cost of visiting the firm is $\delta > 0$. This may represent the cost of a physical or online visit to a store, a phone call, or a letter. If consumer x visits the firm, then she: (1) observes her own x value (if she did not know it already), (2) observes the firm's price p, and (3) is entitled to purchase the good at price p. Fact (1) implies that the good is a search good in the sense of Nelson (1974).[4] Fact (3) implies that no additional transactions costs are attached to the decision to purchase, but that is unimportant. One may imagine that any such costs have already been deducted from consumers' reservation values. The consumer's utility thus equals $\max(0, x - p) - \delta$ if she visits the firm, and zero otherwise.

The exogenous parameters of the model are $c \in [0, 1)$, $\alpha \in (0, 1)$, and $\delta > 0$. We assume throughout that $\delta \approx 0$, to simplify the equilibria and to underscore the point that small visiting costs support the phenomena described.

The firm and consumers play a game in which the firm chooses a price p, after which each consumer decides whether to visit the firm. The decisions are effectively simultaneous, since a consumer observes p only after deciding to visit. Given $c \in [0, 1)$ and $\alpha \in (0, 1)$, we assume that δ is small enough to ensure that every ignorant consumer wants to visit the firm (because the benefit of experimentation outweighs the cost). The decisions of knowledgeable consumers can be summarized by a threshold value x^*, such that knowledgeable consumers $x < x^*$ do not visit the firm, and knowledgeable consumers $x > x^*$ do visit the firm. The actions of consumers $x = x^*$ may be ignored, since they have measure zero. Given x^*, the firm's profits if it sets price p are:

$$\pi(p; x^*, c) = (p - c) \cdot [\alpha \cdot (1 - p) + (1 - \alpha) \cdot \min(1 - p, 1 - x^*)]. \quad (1)$$

The pair (p^*, x^*) is defined to be an equilibrium if (a) $p = p^*$ maximizes $\pi(p; x^*, c)$, (b) x^* is optimal for knowledgeable consumers given p^*, and (c) visiting is optimal for ignorant consumers given p^*. These conditions are reexpressed in (2a)–(2c).

Definition 1. *(Nash equilibrium; fixed costs).* A Nash equilibrium is a price p^* [0, 1] and a reservation threshold x^* [0, 1] such that:

$$p = p^* \text{ maximizes } \pi(p; x^*, c); \quad (2a)$$

$$x^* - p^* = \delta; \quad (2b)$$

$$\int_{p^*}^{1} (x - p^*)\, dx > \delta; \text{ equivalently,} \quad (2c)$$

$$(1 - p^*)^2 > 2\delta. \quad (2c')$$

B. The Textbook Prediction

Before deriving the unique Nash price, it is instructive to derive the price predicted by the standard (textbook) model. The first step is to calculate the *market demand curve* D(p), defined as the fraction of consumers who purchase the good at price p, if consumers expect the firm to charge price p. D(p) is the demand curve that must be observed empirically if consumers have rational point price expectations. D(p) can be calculated as follows. If $p = 1 - \sqrt{2\delta}$, then (1c′) shows that ignorant consumers are indifferent to visiting the firm; assume arbitrarily that they do not visit the firm.[5] Then ignorant consumers visit the firm if and only if $p < 1 - \sqrt{2\delta}$. If they do visit, their demand, as a fraction of all consumers, is $\alpha \cdot (1 - p)$. A knowledgeable consumer x visits the firm only if $p + \delta < x$; the demand from knowledgeable consumers is thus $(1 - \alpha) \cdot (1 - p - \delta)$ if $p + \delta \leq 1$, and otherwise zero. If δ is small, then $1 - \sqrt{2\delta} \leq 1 - \delta$. From these facts it is easy to derive (3).

$$\begin{aligned} D(p) &= \beta - p && \text{if } p < 1 - \sqrt{2\delta} \\ &= \beta - p - \alpha \cdot (1 - p) && \text{if } 1 - \sqrt{2\delta} \leq p \leq 1 - \delta \\ &= 0 && \text{if } 1 - \delta \leq p. \\ &\text{where } \beta \equiv 1 - (1 - \alpha)\delta. \end{aligned} \tag{3}$$

If visiting costs δ are sufficiently small, then only the $\beta - p$ segment of the demand curve is relevant. The monopoly price, calculated using the textbook rule MR = MC, is

$$t(c) = (\beta + c)/2 \tag{4}$$

If the asymmetry in price information did not exist (consumers could observe the firm's price *without* visiting the firm), then $p = t(c)$ would indeed be the optimal rule for the firm. In the real world, however, as in this model, consumers typically do not have reliable observations of a firm's current price until they visit the firm. Then the appropriate equilibrium is the Nash equilibrium of section A. The Nash price is generally higher, sometimes much higher, than the textbook prediction t(c).

C. The unique Nash equilibrium

Proposition 1. If δ is sufficiently small, then the unique Nash equilibrium is $(p^*, x^*) = (n(c), n(c) + \delta)$, where

$$n(c) \equiv [1 + \alpha c - (1 - \alpha)\delta]/(1 + \alpha) = (\beta + \alpha c)/(1 + \alpha) \tag{5}$$

Proof. Suppose that (p^*, x^*) is an equilibrium. From (2a), $p = p^*$ must maximize $\pi(p; x^*, c)$. The function π has a concave kink at $p = x^*$ and is strictly concave on either side of the kink. Given (2b), it must be that $\partial\pi(p; p^* + \delta, c)/\partial p|_{p=p^*} = 0$, implying $p = n(c)$, and $(n(c), n(c) + \delta)$ is the only possible equilibrium. If δ is small enough to satisfy (2c) at $p^* = n(c)$, then $(n(c), n(c) + \delta)$ is in fact an equilibrium. ||

The Nash equilibrium has two interpretations. If the description of the game and the values of the exogenous parameters (c, α, δ) are common knowledge, then the players could deduce the equilibrium. Alternatively, if the game repeats for many periods, during which the values of the exogenous parameters remain substantially unchanged, then the players might learn the equilibrium.

If $\delta \approx 0$, as assumed throughout, then the Nash price exceeds the textbook price: $n(c) - t(c) \approx (1 - \alpha)(1 - c)/[2(1 + \alpha)] > 0$ (from (4) and (5)).[6] If most consumers are ignorant ($\alpha \approx 1$) then the discrepancy is small ($n(c) \approx t(c)$); if most consumers are knowledgeable ($\alpha \approx 0$) then the market almost disappears ($n(c) \approx 1$). The textbook prediction thus requires that consumers are unsure of their preferences for the product.

Regardless of the mix of ignorant and knowledgeable consumers, $n(c)$ does not approach $t(c)$ as visiting costs δ approach zero.[7] Arbitrarily small visiting costs thus have large price effects. Figure 1 illustrates $t(c)$ and $n(c)$ for $\alpha = 0.5$ and $\delta = 0.001$. $n(c)$ approaches $t(c)$ as c increases, but that is in part an artifact of the uniform distribution of reservation values. The $n(c)$ graph terminates at $c \approx 0.87$ because $p^* = n(c)$ violates (2c) for $c > 0.87$, causing the equilibrium to disappear (à la Salop and Stiglitz). Figure 2 shows the limits of $t(c)$ and $n(c)$ as $\delta \to 0$.

The discrepancy between textbook and Nash prices has empirical significance: given empirically observed demand and cost curves, the textbook prediction may substantially underestimate the actual (Nash) price. The demand and cost curves do not even contain enough information to calculate the Nash price. Conversely, the textbook estimate of the firm's costs, given the firm's chosen point of production on an observed demand curve, may substantially overestimate the firm's actual costs.

The textbook model is misleading also because it generates wrong comparative statics results for changes in α. The textbook model predicts that ignorance raises prices ($\partial t(c)/\partial\alpha > 0$), because increased experimentation shifts market demand $D(p)$ to the right. Actually, ignorance lowers prices ($\partial n(c)/\partial\alpha < 0$), because the temptation to exploit knowledgeable visitors declines as their numbers decline.

II. MODEL II (ONE-PERIOD, RANDOM COSTS)

The random cost model is identical to the fixed cost model, except that the firm's cost c is a random variable with probability distribution F(c) and support $C \subseteq \mathbb{R}_+$. F(c) is common knowledge, but only the firm observes c. The effect is that consumers perceive price to be uncertain. This seems realistic, because it seems unlikely that consumers who cannot reliably observe the firm's price could reliably observe its fluctuating costs. The timing of actions is as follows: (i) the firm observes c and chooses p; (ii) without observing c or p, each consumer decides whether to visit the firm; (iii) each consumer x who visits the firm purchases the good if $x > p$.

The pair $(p^*(\cdot), x^*)$, $p^*: C \rightarrow \mathbb{R}$, is an equilibrium if (a) $p = p^*(c)$ maximizes $\pi(p; x^*, c)$ for all $c \in C$, (b) x^* is optimal for knowledgeable consumers given $p^*(\cdot)$, and (c) visiting is optimal for ignorant consumers given $p^*(\cdot)$. Formally:

Definition 2 *(Nash equilibrium; random costs).* A Nash equilibrium is a price rule $p^*: C \rightarrow \mathbb{R}_+$ and a reservation threshold $x^* \in [0, 1]$ such that:

$$p = p^*(c) \text{ maximizes } \pi(p; x^*, c) \text{ for all } c \in C; \tag{6a}$$

$$\int_{p^*(c)}^{1} (x^* - p^*(c))\, dF(c) = \delta; \tag{6b}$$

$$\int_{0}^{1} \int_{p^*(c)}^{1} (x - p^*(c))\, dx\, dF(c) > \delta. \tag{6c}$$

For convenience assume that $F(c) = c$, implying that costs are distributed uniformly on the interval (0, 1). The standard textbook analysis predicts that the monopolist's price t(c) (from (4)) fluctuates uniformly on the interval $[\beta/2, (1+\beta)/2]$. Proposition 2 shows that the Nash equilibrium is quite different. Knowledgeable consumers having reservation values $x < \tau$ ((7)) never visit the firm, because the chance of finding a low price does not justify the cost of the visit. Knowledgeable consumers having reservation values $x > \tau$ always visit the firm. The resulting kink in the visitors' demand curve causes price to stick at $p = \tau$. If $\delta \approx 0$, then the probability of $p = \tau$ is approximately $(1 - \alpha)/(1 + \alpha)$. If $\alpha \approx 1$ (i.e. most consumers are ignorant), then the model is close to the textbook model and prices are almost completely flexible.

Proposition 2. Assume that $F(c) = c$. If δ is sufficiently small, then the unique Nash equilibrium is $(r(\cdot), \tau)$, where

$$\tau = (1 + \alpha 2\sqrt{\delta})/(1 + \alpha), \text{ and}$$

$$\begin{aligned} r(c) &= [1 + c\alpha - (1 - \alpha)\tau]/(2\alpha) && \text{if } c \le 2\sqrt{\delta} \\ &= \tau && \text{if } 2\sqrt{\delta} \le c \le 2\tau - 1. \\ &= (1 + c)/2 && \text{if } 2\tau - 1 < c. \end{aligned} \quad (7)$$

Proof. Suppose that $(p(\cdot), x^*)$ is an equilibrium. Given cost c, (6a) implies that the firm's expected profits equal $\pi(p(c); x^*, c)$. The function $\pi(\cdot; x^*, c)$ has a concave kink at $p = x^*$ and is strictly concave on either side of that kink. In the region $p < x^*$ $\partial\pi/\partial p = 1 + c\alpha - (1 - \alpha)x^* - 2p\alpha$. The right-hand expression is zero at $p = \hat{p}(c) \equiv [1 + c\alpha - (1 - \alpha)x^*]/(2\alpha)$. If $\hat{p}(c) < x^*$ (equivalently, $c < [(1 + \alpha)x^* - 1]/\alpha$), then it must be that $p(c) = \hat{p}(c)$. In the region $p > x^*$, $\partial\pi/\partial p = 1 + c - 2p$. The right-hand expression is zero at $p = \tilde{p}(c) \equiv (1 + c)/2$. If $\tilde{p}(c) > x^*$ (equivalently, $c > 2x^* - 1$), then it must be that $p(c) = \tilde{p}(c)$. If $\hat{p}(c) \ge x^*$ and $\tilde{p}(c) \le x^*$ then it must be that $p(c) = x^*$. Equation (6b) implies that the knowledgeable consumer $x = x^*$ is indifferent to visiting the firm. She does not benefit from the visit unless $c < c^* \equiv [(1 + \alpha)x^* - 1]/\alpha$, so her expected utility from the visit is $\int_0^{c^*} [x^* - \hat{p}(c)]dc - \delta = 0$. That equation implies $x^* = \tau$, which in turn implies $p(\cdot) = r(\cdot)$. Therefore, $(r(\cdot), \tau)$ is the only possible equilibrium. If δ is small enough to satisfy (6c) at $p(\cdot) = r(\cdot)$, then $(r(\cdot), \tau)$ is in fact an equilibrium. ||

Figure 1 illustrates r(c) for $\alpha = 0.5$ and $\delta = 0.001$. The textbook price prediction in the fluctuating cost model is t(c), just as in the fixed cost model (cf. fn. 3). Figure 1 indicates that t(c) is a good price predictor ($t(c) \approx r(c)$) when costs are high but a poor predictor when costs are low. t(c) is accurate at high c because consumers $x > \tau$ always visit, implying that the visitors' demand curve is close to the market demand curve at $p > \tau$. t(c) is inaccurate at low c because the consumers who would take advantage of low prices do not visit the firm. Figure 2 shows the limit of r(c) as $\tau \to 0$. As $\delta \to 0$, τ becomes a price floor. If it did not, then consumers having reservation values $x = \tau - \epsilon$, any small $\epsilon > 0$, would visit as $\delta \to 0$ in the hope of finding $p < \tau - \epsilon$, contradicting their equilibrium strategy. Prices are thus sticky when costs are low but not when costs are high. If consumers had small but continuously differentiated visiting costs bounded below by $\delta > 0$, then r(c) would not be horizontal at $p = \tau$ but would still be relatively flat in that region.

The price stickiness observed in the random cost model does not depend on menu costs, complicated strategic interactions, or a signalling role for prices. It arises from the asymmetric information induced by arbitrarily small visiting costs. Asymmetric information and small visiting costs clearly exist in real markets.

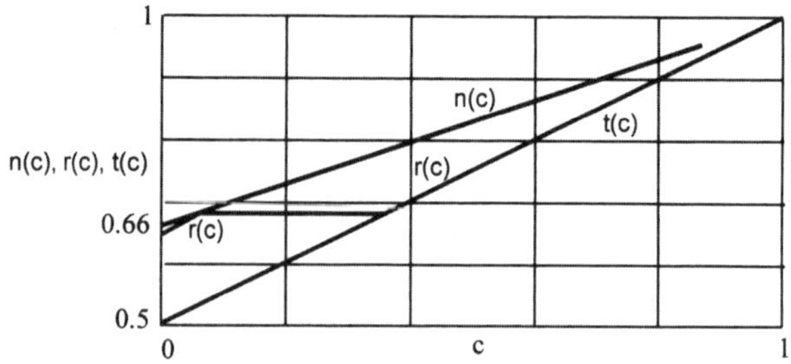

Fig. 1.

The random costs model has the surprising property that consumers and the firm are both better off if consumers do not observe the firm's costs. If consumers observe c_t (before visiting) then costs are effectively fixed period-by-period and the Nash equilibrium is $n(c_t)$. If consumers do not observe c_t the Nash equilibrium is $r(c_t)$. Although $r(2\sqrt{\delta}) > n(2\sqrt{\delta})$, it is obvious from Fig. 1 that $E[r(c_t)] < E[n(c_t)]$ for small δ, suggesting that consumers prefer $r(\cdot)$ to $n(\cdot)$. Since $t(\cdot)$ is closer to $r(\cdot)$ than to $n(\cdot)$, it also appears that the firm prefers $r(\cdot)$. In fact, it can be shown that if $\delta \approx 0$ then the firm and almost all consumers prefer the $r(\cdot)$ equilibrium, implying that almost everyone is better off if consumers do not observe the firm's costs.[8]

To summarize the one-period model, the textbook price predictions are generally inaccurate. If costs are fixed, then the textbook model underestimates prices; if costs are random and unobserved by consumers, then the textbook model is accurate when costs are high but fails to capture the downward price rigidity that emerges when costs are low, and which is most pronounced when most consumers are knowledgeable about their reservation values.

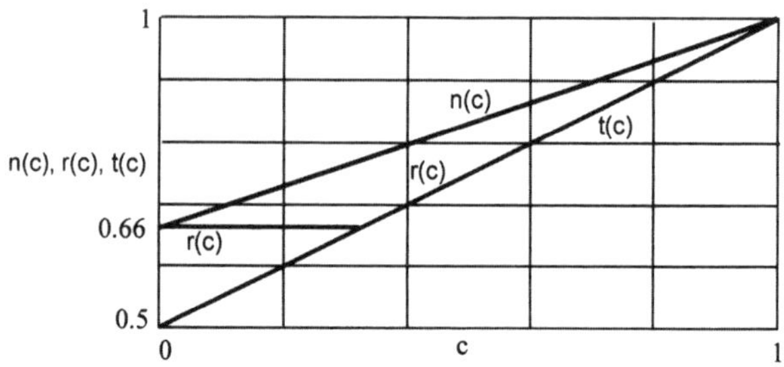

Fig. 2.

III. THE MANY PERIOD MODEL

The one-period Nash equilibria in Sections I and II are undesirable, because the firm and consumers would all benefit if prices were lower. One way to mitigate the information asymmetry that causes high prices is to allow the firm to send credible messages to non-visitors, but it is hard to send credible messages in a one-period model. This section extends the model to many periods.

The most elementary mechanism for sending messages is the firm's own price history. For example, a low p_t may represent a promise that $p_{t'}$ will be low in periods $t' > t$. One way to make that promise credible is to introduce menu costs that induce price inertia, but no such costs are introduced here. Another way is to extend the model to infinite periods, so that the firm's desire to encourage visits in periods $t'' > t'$ ensures a low $p_{t'}$. In other words, the firm keeps prices low to maintain a good reputation. This section shows that reputation effects support equilibria that Pareto-dominate the one-shot Nash equilibria. If costs are fixed over time then the textbook price is an equilibrium. If costs fluctuate then the reputation equilibria do not replicate the price flexibility implicit in the textbook prediction; indeed, the equilibria described below generate prices even more rigid than the Nash price. We shall see, however, that the rigid price equilibrium Pareto dominates the Nash price, because the rigid price supports a reputation that in turn supports a lower price.

A. Assumptions

To extend the one-period model to many periods ($t = 0, 1, \ldots + \infty$), assume that the firm enters the economy in period 0 and lives forever. In period 0 the consumer population has measure one, and consumers subsequently enter and exit the economy through a Poisson process. The consumers present in period 0 are called the original cohort. In each period $t > 1$ a new cohort, having measure $\hat{\alpha}$, enters the economy. Each consumer exits with probability $\hat{\alpha}$ in each period. Entrances thus equal exits, and the consumer population has measure one in every period. Every consumer is ignorant until she visits the firm, but she discovers her reservation value (i.e. becomes knowledgeable) during her first visit. Within each cohort reservation values are distributed uniformly on the [0, 1] interval. In each period each consumer purchases either one or zero units of the good. She cannot store the good.

We will consider only equilibria such that ignorant consumers visit the firm at the first opportunity. Then the fraction of ignorant consumers is $\hat{\alpha}$ in any period $t > 0$. If $\hat{\alpha} = \alpha$ then the consumer population in any period $t > 0$ is identical to the consumer population in the one-period model; to simplify notation, let $\alpha = \hat{\alpha}$. The one-period model is thus embedded in the many-period model. In the many-period model a consumer is distinguished by her cohort as well as by her reservation value and state of knowledge, but the constant probability of exit implies that a consumer's cohort has no independent effect on her behavior.

The firm's unit cost of production is c_t in period t. c_t is a random variable independently and identically distributed in every period, with probability distribution F(c) and support $C \subseteq [0, 1)$. (If costs are fixed then C is a singleton.) The timing in each period is: (i) new consumers enter the model (if $t > 0$); (ii) the firm observes c_t and chooses p_t; (iii) without observing c_t or p_t, each consumer decides whether to visit the firm; (iv) each consumer x who visits the firm purchases the good if $x > p_t$; (v) each consumer exits the model with probability α. The visiting cost is $\delta > 0$. A visit during period t allows consumer x to observe her own x (if she does not yet know it) and purchase the good at price p_t (i.e. if $x > p_t$). The discount rate for consumers and the firm is $\rho > 0$; the discount factor is $1/(1+\rho)$. The description of the model and the values of the exogenous parameters (α, δ, ρ, and F) are common knowledge.

The information structure is as follows. For $t > -1$, $p_{t-} = (p_t, p_{t-1} \ldots p_0, p_{-1})$ denotes the price history through period t. p_{-1} is a placeholder, and we adopt the convention $p_{-1} = 0$.[9] The firm recalls its past actions and hence knows p_{t-1-} in period t. A consumer who visits the firm in period t observes p_{t-} and remembers p_{t-} until she exits the model. The assumption that a consumer observes the entire price history of the firm during any visit simplifies equilibria by ensuring that a firm cannot rebuild its reputation by waiting for new consumers to enter the market. The possibility of rebuilding reputation is interesting but beyond the scope of this study. H_t, $t \geq -1$, denotes the set of possible price histories p_{t-}, and $I_t = \cup_{0 \leq j \leq t} H_j$ the possible realizations of a knowledgeable consumer's information in period t. The firm observes c_t in period t, but consumers never observe c_t. To simplify the exposition, assume that the firm forgets c_t in periods $t' > t$. (Since c_t is not observed by consumers and has no impact on the firm's profits in periods $t' > t$, there is no reason for the firm to remember c_t.) Assume also that the firm never observes the behavior of individual consumers, and consumers never observe the behavior of other consumers.[10]

The strategies defined below are pure.

Definition 3 *(summary of notation; strategies).* Notation: $\alpha \in (0, 1)$, probability of exit; $\delta > 0$, visiting cost; $\rho > 0$, discount rate; $c_t \in [0, 1)$, unit production cost in period t; $F(\cdot)$, distribution of c_t; p_t, price in period t; $p_{t-} = (p_t, p_{t-1} \ldots p_0, p_{-1})$; $p_{-1} = 0$; H_t, possible price histories p_{t-}; $I_t = \cup_{0 \leq j \leq t} H_j$.

A strategy for the firm is $A = \{a_t\}_0^\infty$, where $a_t: H_{t-1} \times \mathbb{R}_+ \rightarrow \mathbb{R}_+$; $a_t(p_{t-1-}, c_t)$ is the firm's action (price) in period t. A strategy for a consumer is a pair (B, Z), where $B = \{b_t\}_0^\infty$, $Z = \{z_t\}_0^\infty$, $b_t: I_t \times [0, 1] \rightarrow \{V, N\}$, and $z_t \in \{V, N\}$. 'V' means visit; 'N' means do not visit. $b_t(p_{t'-}, x)$ is the consumer's decision in period t if she last visited the firm in period t′ and has reservation value x.[11] z_t is the consumer's decision in period t if she has never visited the firm. If the consumer visits the firm in period t, then she buys the product if and only if $x > p_t$.

The payoffs have already been described implicitly and are now defined formally. Equations (9), (10), and (11) express each agent's expected discounted utility (or profits) in period t as a function of the agent's information in period t, the agent's strategy from period t forward, and the relevant actions of other agents (including nature).

Definition 4 *(payoffs).* $v_t: M \rightarrow \{V, N\}$ denotes consumers' visiting decisions in period t, where M is the set of consumers and $v_t(m)$ is consumer m's decision. Let $p_{t+} \equiv (p_t, p_{t+1}, \ldots)$, $c_{t+} \equiv (c_t, c_{t+1}, \ldots)$, and v_{t+} $(v_t, v_{t+1}, \ldots)$.

U_t denotes a knowledgeable consumer's expected discounted utility in period $t > 0$, as a function of her strategy B, her information $p_{t'}$, her reservation value x, and present and future prices p_{t+}.

$$U_t(B, p_{t'-}, x, p_{t+}) \equiv \sum_{j \in J} (1+\rho)^{t-j}[\max(0, x - p_j) - \delta], \tag{9}$$

where J denotes the index set of periods $j > t$ in which consumer n visits the firm, as implied by the arguments of U_t.

Y_t denotes an ignorant consumer's expected discounted utility in period $t > 0$, as a function of her strategy (B, Z), her reservation value x, and present and future prices p_{t+}.

$$\begin{aligned} Y_t(B, Z, x, p_{t+}) = {} & (1+\rho)^{t-j^*}[\max(0, x - p_{j^*}) - \delta] \\ & + (1+\rho)^{t-j^*-1} U_{j^*+1}(B, p_{j^*-}, x, p_{j^*+1+}), \end{aligned} \tag{10}$$

where j* is the first period $j \geq t$ such that $z_j = V$.

W_t denotes the firm's expected discounted profits in period $t \geq 0$, as a function of the firm's strategy A, the firm's information p_{t-1-} and c_t, present and future visiting decisions v_{t+}, and future costs c_{t+1+}.

$$W_t(A, p_{t-1-}, v_{t+}, c_{t+}) \equiv \sum_{j \geq t} (1+\rho)^{t-j}(p_j - c_j) \cdot s_j, \tag{11}$$

where s_j denotes the measure of consumers who purchase in period j, as implied by v_j and p_j, and p_j is the firm's price in period j, as implied by A, p_{j-1-}, and c_j.

B. Definition of equilibrium

Since the dynamic model replicates the one-period model in every period $t>0$, it has trivial equilibria that replicate the Section I and Section II equilibria period-by-period (given appropriate assumptions about $F(\cdot)$). The point of this section is to describe alternative equilibria in which the price history of the economy affects players' actions.

Before describing specific equilibria, however, it is necessary to define equilibrium. The criterion used here is similar to the sequential equilibrium of Kreps & Wilson (1982). In a sequential equilibrium each player n has a plausible set of beliefs concerning the history of the game up to her current information set i. The beliefs are derived from other players' equilibrium strategies or small perturbations of those strategies. Player n's strategy must be optimal at i given those beliefs, if all other players follow their equilibrium strategies from i forward. Unfortunately, the standard definition of sequential equilibrium does not apply to the present game, because the perturbations are defined only for finite games. It is possible to generalize the perturbations to infinite games, but that complicated exercise can be omitted here because the plausible beliefs are obvious.

Consider first the firm's beliefs. Referring to (11), the firm knows p_{t-1-} and c_t in period t, and the distribution of c_{t+1+}, but it does not know v_{t+}. The obvious conjecture is that consumers follow their equilibrium strategies in every period. Consider next consumers' beliefs. Referring to (9), a knowledgeable consumer knows x and $p_{t'-}$ for some $t'<t$, but it does not know $p_{t'+}$. The obvious conjecture is that the firm has followed its equilibrium strategy since period t'. An ignorant consumer would conjecture that the firm follows its equilibrium strategy in every period.

Given these beliefs, it is possible to define an equilibrium exactly. In Definition 5, $\hat{v}$ denotes the firm's beliefs, $\hat{p}$ denotes the consumers' beliefs, (12a) requires that A* be optimal for the firm given (B*, Z*), (12b) requires that B* be optimal for knowledgeable consumers given A*, and (12c) requires that Z* be optimal for ignorant consumers, given A* and B*.

Definition 5 *(equilibrium).* $\hat{v}_j\hat{p}(B, Z, p_{j-1-})$ denotes the value of v_j implied by (B, Z), given p_{j-1-}. For $j>t'$, $\hat{p}_j(A, p_{t'-}, c_{t'+1+})$ denotes the value of p_j implied by A, given $p_{t'-}$ and $c_{t'+1+}$.

Consider a set of strategies (A^*, B^*, Z^*), such that $z_{t*} = V$ for all t. (A^*, B^*, Z^*) is an equilibrium if:

$A = A^*$ maximizes $E[W_t(A, p_{t-1-}, \hat{v}_{t+}(B^*, Z^*, p_{t-1-}), c_{t+})]$ for all $p_{t-1-} \in H_{t-1}$, $c_t \in C$, and $t \geq 0$. The expectation is over c_{t+1+}. (12a)

$B = B^*$ maximizes $E[U_t(B, p_{t'-}, x, \hat{p}_{t+}(A, p_{t'-}, c_{t'+1+}))]$ for all $x \in [0, 1]$, $p_{t'-} \in I_t$, and $t \geq 1$. The expectation is over $c_{t'+1+}$. (12b)

$Z = Z^*$ maximizes $E[Y_t(B^*, Z, x, \hat{p}_{t+}(A, p_{-1}, c_{0+}))]$ for all $t \geq 0$. The expectation is over c_{0+} and x. (12c)

The equilibrium (A^*, B^*, Z^*) *supports* (the fixed) price p^* if $p_t = a_t^*(p_{t-1-}, c_t)$ for all t implies $p_{1+} = (p^*, p^*, \ldots)$, for any $c_{0+} \in C^\circ$.

C. *Trigger strategy equilibria in the constant cost model*

This section assumes that unit costs are fixed at $c_t = c$ for all t, hence $C = \{c\}$. Then the many-period model simply repeats the one-period model of Section I (in periods $t > 0$). Each equilibrium described below has the following general characteristics. There is a fixed equilibrium price $p^* \in (c, n(c)]$. On the equilibrium path $p_0 \leq p^*$ and $p_t = p^*$ for all $t > 0$. ($p_0 \leq p^*$ allows the firm to charge a lower price in period 0, to take advantage of the larger fraction of visitors who are ignorant and thus have lower average reservation values.) Every consumer visits the firm once to discover her reservation value x, and consumers having large x values continue to visit. A consumer who has never observed a price $p_t > p^*$ assumes that the firm will charge price p^* in all present and future periods (except perhaps period 0), and optimizes accordingly. A consumer who has observed a price $p_t > p^*$ assumes that the firm will charge the Nash price $n(c) > p^*$ ((4)) in all present and future periods, and optimizes accordingly. It is a convenient property of the model that n(c) is the one-shot Nash equilibrium after a deviation $p_t > p^*$, even though the fraction of consumers that has observed the deviation (and hence expect price n(c)) is less than one and changes over time.[12] The equilibrium price p^* is sustained by the threat of switching to $p_t = n(c)$.

The notation in (13) is useful for describing the interval of equilibrium prices.

$$\gamma \equiv 4\alpha/(1+\alpha)^2;$$
$$\theta \equiv (\sqrt{(1+\rho)(1-\gamma)} + \gamma - 1)/(\gamma + \rho). \qquad (13)$$

Note that $\gamma \in (0, 1)$ and $\theta \in (0, \frac{1}{2})$. (If $\gamma \approx 1$ then $\theta \approx 0$; if $\gamma \approx \rho \approx 0$ then $\theta \approx \frac{1}{2}$). If $\delta \approx 0$, then Proposition 3 (below) shows that any p^* satisfying (14) is an equilibrium price.

$$p^* \in [([1+c]/2 - \theta)/(1-\theta), n(c)]. \tag{14}$$

If $\delta \approx 0$, then (4) And $c<1$ imply

$$t(c) \in [(1+c-\theta)/(2-\theta), (1+c)/2],$$

which is a proper subset of (14). Therefore, the textbook price t(c) is an equilibrium price.

It may be surprising that the dynamic model supports the textbook price regardless of the discount rate ρ. The claim, stated precisely, is that for any given $\rho>0$, there exists $\delta^*>0$ such that t(c) is an equilibrium price for all $\delta<\delta^*$. The results can be summarized thus: if $\delta=0$, then the textbook model obtains; if $\delta>0$ but $\delta \approx 0$, then the textbook price does not obtain in a single period but can be supported in the dynamic model; if $\delta >> 0$, then the textbook price may be unsupportable even in the dynamic model.

The comparative statics for the equilibrium price interval (14) follow a simple pattern. As expected, anything that increases the penalty associated with switching to $p_t = n(c)$ expands the interval of equilibrium prices. Longer consumer lifespans (smaller α) expand the interval of equilibrium prices, because smaller α implies higher n(c). Lower discount rates (smaller ρ) also expand the interval, because future penalties receive more weight. Higher costs (larger c) raise the interval. Calculations in Section IV show that the interval is wide at reasonable parameter values.

Proposition 3. Assume that $C=\{c\}$ for some $c \in [0, 1)$. Consider $p^* \in \mathbb{R}$. If p^* satisfies (14) and δ is sufficiently small (given p^*), then there exists an equilibrium (A, B, Z) that supports price p^*. The strategies are $z_t = V$ for all $t>0$, $a_0(p_{-1}, c) = \min((1+c)/2, p^*)$, and $a_t = a^*$ and $b_t = b^*$ for all $t>0$,[13] where

$$a^*(h, c) = p^* \quad \text{if } \max(h) \le p^*$$

$$a^*(h, c) = n(c) \quad \text{if } \max(h) > p^*.$$

$$b^*(h, x) = V \quad \text{if } x - \delta > a^*(h, c).$$

$$b^*(h, x) = N \quad \text{if } x - \delta < a^*(h, c).$$

(h denotes a generic price history.)

The proofs of Propositions 3 through 5 are in the Appendix.

Within the interval of equilibrium prices, it is unclear which price is most likely. t(c) is the most profitable price for the firm in any single period (subject to the constraint that consumers have correct point price expectations) and is therefore the most profitable equilibrium for the firm in the dynamic model. t(c), however, is not the lowest price in the interval (14), and consumers prefer lower prices. The consumers and firm thus face a serious coordination problem, which has two manifestations: the original cohort and the firm must coordinate on an equilibrium price p^* before the firm sets p_0; the subsequent cohorts must deduce p^*, so that they know how to interpret their price observations.

D. Ratchet equilibria in the constant cost model

The strategies in Proposition 3 are trigger strategies because any upward price deviation causes a discontinuous switch to the one-shot Nash equilibrium. Trigger strategies are reasonable in this context because real-world consumers may react discontinuously to a perceived loss of reputation. Nevertheless, it is somewhat implausible that even a tiny deviation from the firm's equilibrium price causes a potentially huge price jump, especially given the difficulty of coordinating on the equilibrium price.

This section describes an equilibrium that does not rely on discontinuous strategies *and* solves the price coordination problem. In this equilibrium, the firm's price is a ratchet; price increases are permanent (up to the Nash price n(c)), and price cuts are temporary. Upward price deviations thus cause a continuous and gradual loss of reputation. The lowest price that can be sustained by such a ratchet is well-defined and denoted m(c). It is shown that $m(c) > t(c)$, which is not surprising, since one would expect the firm to deviate to the textbook price t(c) from any lower price. The firm chooses where to initialize the ratchet, so it is also unsurprising that the firm chooses to initialize the ratchet at $p = m(c)$, as close to the textbook price as possible. That solves the price coordination problem (in the sense that only one price is consistent with the general structure of the equilibrium). The equilibrium path is $p_t = m(c)$ for all $t > 0$. m(c) is defined as follows.

$$m(c) \equiv [1 + c + (1 + \alpha c)\rho]/[2 + (1 + \alpha)\rho]. \tag{15}$$

To see that $m(c) > t(c)$, observe that $m(1) = t(1)$ at $\delta = 0$, and $m'(c) < t'(c)$ for all $c \in [0, 1]$. Therefore $m(c) > t(c)$ for $c < 1$. Although $m(c) > t(c)$, Section IV shows that m(c) is close to t(c) for a wide range of reasonable parameter values. The ratchet equilibrium thus supports a price close to the textbook prediction. m(c) is closest to t(c) when consumer turnover is high (implying much consumer ignorance) and discount rates are low.

Proposition 4. Assume that $C = \{c\}$ for some $c \in [0, 1)$. If δ is sufficiently small, then there exists an equilibrium (A, B, Z) that supports m(c). The stategies are $z_t = V$ for all $t \geq 0$, $a_0 = \min((1 + c)/2, m(c))$,[14] where

$$a^*(h, c) = m(c) \quad \text{if } \max(h) \le m(c).$$

$$a^*(h, c) = \max(h) \quad \text{if } m(c) \le \max(h) \le n(c).$$

$$a^*(h, c) = n(c) \quad \text{if } n(c) \le \max(h).$$

$$b^*(h, x) = V \quad \text{if } x - \delta > a^*(h, c).$$

$$b^*(h, x) = N \quad \text{if } x - \delta \le a^*(h, c).$$

If m(c) is replaced by any value $\hat{m}(c) < m(c)$, then (A, B, Z) is not an equilibrium for δ sufficiently small.

The comparative statics results for m(c) are generally consistent with the earlier results. High costs (high c) and slow consumer turnover (low α) raise prices.[15] High discount rates (high ρ) also raise prices, because they reduce the value of a good reputation.

m(c) may be a realistic prediction of a monopolist's price if costs are fixed over time. The strategies supporting m(c) are simple, continuous, and intuitively plausible. Of course, it is not certain that the firm and consumers would coordinate on the ratchet equilibrium. Everyone in the economy would prefer the trigger strategy equilibrium $p^* = t(c)$, and consumers would prefer a trigger strategy equilibrium $p^* < t(c)$. Nevertheless, the difficulty of coordinating on a particular trigger strategy equilibrium is so severe that the ratchet equilibrium seems to be a more natural focal point.

E. Ratchet equilibria in the random costs model

If costs fluctuate from period to period, then it is still possible to construct a ratchet equilibrium that improves on the Nash equilibrium, but this equilibrium does not even come close to replicating textbook predictions. The textbook model predicts freely fluctuating prices, but the price in the ratchet equilibrium is fixed after period 0. A fixed price arises naturally because consumers cannot observe costs and so cannot distinguish justified from unjustified price increases.

Proposition 5 describes a ratchet equilibrium generally similar to the ratchet equilibrium of Proposition 4. Within a price interval $S \equiv [s_1, s_2]$ (defined below), price increases are permanent but price cuts are temporary. s_1 is the lowest price that can be sustained by the ratchet and exceeds the mean textbook prediction. After $p_0 \le s_1$ the firm initializes the ratchet at $p_1 = s_1$, because if the firm could earn higher profits at $p' \in S$, $p' > s_1$, then it could increase profits by switching from the path $p_t = s_1$ to the path $p_t = p'$, a contradiction.[16] The equilibrium path is $p_t = s_1$ for all $t > 0$, even when this requires pricing below cost. If the firm ever charges a price exceeding the upper bound of the ratchet ($p_t > s_2$), then the equilibrium switches to the period-by-period Nash equilibrium $p_t = r(c_t)$, where $r(c) \geq s_2$ for all $c \in (0, 1)$. The following description of the equilibrium price interval S can be used to calculate S for arbitrary parameter values. Let

$$\psi(p) \equiv [\alpha^3 + 3\alpha]/[6(1+\alpha)^3] - (1-p)(p-\tfrac{1}{2}) + \rho \cdot (1-p)$$
$$+ \delta \cdot [(1-\alpha)(p + p\rho - \rho - \tfrac{1}{2}) + 2];$$
$$q \equiv [2(1+\alpha)\rho + 3]/[2(1+\alpha)\rho + 4];$$
$$S \equiv [s_1, s_2] \equiv \{p : \psi(p) \leq 0,\ q \leq p \leq r(0)\}. \qquad (16)$$

The mean value of the textbook price $t(c_t)$, given $F(c_t) = c_t$, is 3/4. $q > 3/4$ implies $s_1 > 3/4$, but calculations in Section IV show that s_1 is close to 3/4 for a wide range of reasonable parameter values. If $s_1 = q$ (as is true in all of the examples), then (16) implies that s_1 must be close to 3/4 when turnover is high (α low) and discount rates are low. The ratchet equilibrium thus supports the textbook prediction 'on average'.

Proposition 5. Assume that $F(c_t) = c_t$. If δ is sufficiently small, then there exists an equilibrium (A, B, Z) that supports s_1. The stategies are $z_t = V$ for all $t \geq 0$, $a_0 = \min((1+c)/2, s_1)$, and $a_t = a^*$ and $b_t = b^*$, for all $t \geq 0$, where

$a^*(h, c) = s_1$	if $\max(h) \leq s_1$.
$a^*(h, c) = \max(h)$	if $s_1 \leq \max(h) \leq s_2$.
$a^*(h, c) = r(c)$	if $\max(h) > s_2$.
$b^*(h, x) = V$	if $\max(h) \leq s_2$ and $x - \delta > a^*(h, \cdot)$, or
	if $\max(h) > s_2$ and $x > \tau$.
$b^*(h, x) = N$	otherwise.

If s_1 is replaced by any $\hat{s}_1 < s_1$, then (A, B, Z) is not an equilibrium for δ sufficiently small.

The ratchet equilibrium Pareto-dominates the Nash equilibrium $r(c_t)$: the firm prefers $p_t = s_1$ because it declines to switch to the $p_t = r(c_t)$ regime, and consumers prefer $p_t = s_1$ because $s_1 \leq r(c_t)$ for all $c_t \in [0, 1]$. Pareto-dominance is unsurprising because the ratchet equilibrium supports reputation effects that allow the firm to set prices below one-shot Nash prices, but one does not usually expect fixed price equilibria to Pareto-dominate flexible price equilibria.

It is tempting to look for a fluctuating price equilibrium that dominates the Nash and fixed price ratchet equilibria. Unfortunately, any attempt to construct such an equilibrium faces a fundamental problem: consumers who do not observe costs cannot detect isolated deviations. It may still be possible to construct a fluctuating price equilibrium that improves upon the ratchet equilibrium, but the next paragraph shows that the price rule on the equilibrium path cannot take the most obvious form: it cannot be a stationary rule $p_t = w(c_t)$ for $t > 0$, where $w(\cdot)$ is continuous, non-decreasing, and $w(1) > w(0)$.

To demonstrate this claim, let $p_t = w(c_t)$ be such a rule. On the equilibrium path knowledgeable consumers follow a threshold rule $x^* > 0$, as before. Clearly $x^* > w(0)$. The proof of Proposition 2 shows that if $p_t < x^*$ is optimal for the firm in period $t > 0$, given arbitrary c_t, then $p_t \geq \hat{p}(0) = [1 - (1 - \alpha)x^*]/(2\alpha)$, because otherwise $p_t + \epsilon$ would be better than p_t. If $c_t = 0$, then $x^* > w(c_t)$, and the assumed optimality of $w(\cdot)$ implies that $w(0) \geq [1 - (1 - \alpha)x^*]/(2\alpha)$. Therefore $x^* > [1 - (1 - \alpha)x^*]/(2\alpha)$, implying $x^* > 1/(1 + \alpha)$. If $\delta \approx 0$ then $1/(1 + \alpha) \approx \tau$, implying that the consumers' visiting rule is approximately the visiting rule of the one-shot equilibrium of Proposition 2. Therefore the price rule $w(\cdot)$ must be approximately the price rule $r(\cdot)$, and behavior on the equilibrium path is very close to behavior in the Nash equilibrium. Since the ratchet equilibrium Pareto-dominates the Nash equilibrium, it must also Pareto-dominate the $w(\cdot)$ equilibrium.

These arguments show that the fixed price of the ratchet equilibrium is more than a modeling convenience. There may exist Pareto superior fluctuating price equilibria, but they are almost certainly discontinuous and much more complicated than the ratchet equilibrium.

IV. CALCULATED EQUILIBRIA

This section calculates examples of equilibrium prices for the dynamic model. Table 1 assumes that costs are fixed at $c_t = 0$ for all t. Table 2 assumes that costs fluctuate according to $c_t \sim U(0, 1)$. The other parameters of the model are α, ρ, and δ. All of the calculations are limits as $\delta \rightarrow 0$.

The parameters α and ρ are transformed into three more intuitive parameters: L, F, and R. Average lifespan (L) is the average number of years that a consumer remains in the model after her first purchase. Frequency (F) is the number of purchases that each consumer makes per year after her first purchase (as long as she continues to visit the firm). A consumer who purchases in every period makes $1 + LF$ average total purchases. The annual interest rate (R) applies to both consumers and firms. L, F, and R determine the original parameters α and ρ as follows.

$$\rho = (1 + R)^{1/F} - 1; \ \alpha = 1/(1 + LF).$$

Table 1. Monopoly Prices under Fixed Costs

	Consumers			
Average lifespan (L):	One year		Five years	
Purchases per year (F):	One	Twelve	One	Twelve
One period				
Textbook prediction (4)	0.50	0.50	0.50	0.50
Nash price (5)	0.67	0.93	0.86	0.98
Many periods				
Trigger strategies (14)	0.34–0.67	0.07–0.93	0.16–0.86	0.02–0.98
Ratchet equilibrium (15)	0.512	0.502	0.520	0.502

Common assumptions:
Cost of production, $c=0$
Visiting costs, $\delta \approx 0$
Aggregate demand, $D(p) \approx 1-p$
Annual interest rate, $R=0.10$

Explanation: The entries in the table are predictions of monopoly prices, depending on the characteristics of consumers (columns) and the model/equilibrium generating the predictions (rows). The numbers in parentheses refer to equations in the text. The entries in the third row correspond to the continuum of possible trigger strategy equilibria; each equilibrium generates only one price. All entries are limiting values as $\delta \to 0$.

Equilibrium prices are fairly insensitive to R at realistically low values of R. Tables 1 and 2 assume $R=0.10$, but all of the calculated values are unchanged to two decimal places if $R=0.05$. Prices are more sensitive to L and F. The columns of Tables 1 and 2 represent different assumptions about L and F.

Table 1 shows that the Nash price is much higher than the textbook prediction for plausible parameter values. The discrepancy increases as lifespan and frequency increase. The trigger strategy equilibria in the third row support a wide range of lower prices, centered approximately on the textbook price, but it is unclear which equilibrium is most likely (if any). The ratchet equilibrium supports a unique price that is close to the textbook prediction.

Table 2 shows that if costs fluctuate then the Nash price fluctuates less than the textbook model predicts and it becomes higher and stickier as lifespan and frequency increase. The ratchet equilibrium supports a fixed price that is lower than the Nash prices and close to the mean of the textbook prediction. The first column has no ratchet equilibrium because high consumer turnover reduces the importance of price reputation so much that a good reputation is not worth preserving when costs are high.

V. SUMMARY

The quite evident fact that real world consumers must exert themselves to observe the prices of real world firms creates information asymmetries that have profound implications for the theory of monopoly. If the monopolist's costs are fixed, then its prices are typically much higher than predicted by the textbook rule MR = MC, even if visiting costs are extremely small. If the model repeats indefinitely, then reputation effects can support lower prices. A trigger strategy equilibrium supports the textbook price, but it may be difficult to coordinate on the textbook price because consumers prefer lower prices. A ratchet equilibrium solves the price coordination problem and supports a price only slightly higher than the textbook price.

If the monopolist's costs fluctuate, then prices in the repeated one-shot equilibrium are higher and stickier than is predicted by the textbook rule, being sticky particularly in response to downward cost shocks. In the repeated model a ratchet equilibrium supports a price that is lower than the fluctuating Nash

Table 2. Monopoly Prices under Random Costs

	Consumers			
Average lifespan (L):	One year		Five years	
Purchases per year (F):	One	Twelve	One	Twelve
One period				
Textbook prediction (3)	0.50–1	0.50–1	0.50–1	0.50–1
Nash price r(c) (7)	0.67–1	0.93–1	0.86–1	0.98–1
(Prob[lowest value])	(0.33)	(0.86)	(0.71)	(0.97)
Many periods				
Ratchet equilibrium (16)	—	0.751	0.764	0.751

Common assumptions:
- Cost of production, $c \sim U[0, 1]$
- Visiting costs, $\delta \approx 0$
- Aggregate demand, $D(p) \approx 1 - p$
- Annual interest rate, $R = 0.10$

Explanation: The entries in the table are predictions of monopoly prices, depending on the characteristics of consumers (columns) and the model/equilibrium generating the predictions (rows). The integers in parentheses refer to equations in the text. The entries in the first two rows are the ranges in which the textbook and Nash prices fluctuate. The Nash prices tend to stick at the lowest price in the range; the probability attached to the lowest price is in parentheses. The ratchet equilibrium generates a fixed price, regardless of cost fluctuations. All entries are limiting values as $\delta \to 0$. The $L = F = 1$ case has no ratchet equilibrium.

prices, but the ratchet price is necessarily fixed. Nevertheless, the fixed price Pareto-dominates the Nash equilibrium, because it is lower than the fluctuating Nash prices. Calculations show that a broad range of reasonable parameter values support fixed price equilibria, even if costs fluctuate widely.

NOTES

1. This literature is surveyed in Stegeman (1994).

2. In a model of only one firm, it seems appropriate to call δ a 'visiting' cost rather than a 'search' cost.

3. Even in the random costs model of the present paper, the empirically observed demand curve is arbitrarily close to the curve that would be implied by rational point price expectations, as $\delta \rightarrow 0$.

4. If the good were an experience good, then it would be desirable to introduce some heterogeneity into the population of ignorant consumers (e.g. a signal of reservation value). Otherwise all ignorant consumers would buy at the same reservation price, creating an artificial discontinuity in the demand curve. There already exists on artificial discontinuity because all ignorant consumers make the same visiting decision (cf. note 5), but if visiting costs are small then the latter discontinuity is unimportant. Given such heterogeneity, results somewhat similar to the present results seem likely to obtain.

5. The behavior of indifferent ignorant consumers is unimportant because such consumers do not exist if visiting costs are small. In that case, any ignorant consumer strictly prefers to visit the firm to discover her reservation value.

6. The Nash price exceeds the textbook price even if $\delta >> 0$, but formulas (4) and (5) are inaccurate for large δ.

7. At $\delta = 0$ the set of Nash prices expands discontinuously to become the interval $[t(c), n(c)]$, and $t(c)$ is the only sequential equilibrium. The textbook model is thus perfectly correct at $\delta = 0$, regardless of the mix of ignorant and knowledgeable consumers.

8. Knowledgeable consumers having reservation values $x = \tau$ prefer the $n(c_t)$ equilibrium, because they obtain zero expected surplus in the $r(c_t)$ equilibrium but have a small chance of observing $p_t < x + \delta$ in the $n(c_t)$ equilibrium.

9. p_{-1} can be interpreted as the firm's price reputation when it enters the market. The convention $p_{-1} = 0$ endows the firm with a good reputation, which it can use however it wants. In the ratchet equilibrium of Proposition 4, $p_{-1} = 0$ allows the firm to initialize the ratchet wherever it wants.

10. Observing consumer behavior would make no difference on the equilibrium path, because observations would confirm what agents already assume in the sequential equilibrium described below, that consumers are following their equilibrium strategies. Observing consumer behavior would also make no difference on any path such that at most one consumer deviates from his equilibrium strategy in any given period, because one consumer is infinitesimal. Marschak & Selten (1978) suggest, in their definition of a paraperfect equilibrium, that behavior on other paths can be disregarded in the definition of equilibrium, because a player who contemplates a deviation in a non-cooperative game should not assume that other players will deviate simultaneously. In short, the sequential equilibrium criterion adopted below is essentially equivalent to a parasequential criterion if consumer behavior were observed.

11. A consumer who enters in a period $t > 0$ does not need a strategy for periods $t' < t$, but it simplifies the notation to assume that each consumer selects a strategy from the same strategy space.

12. The post-deviation price equals the one-shot Nash price because after a deviation every consumer acts as if he expects the one-shot Nash price. The consumers who do not observe the deviation are consumers who do not visit when they expect the price p, and they would still not visit if they expected the price $n(c) > p$.

13. This abuses notation slightly, because the domain of a_t does not equal the domain of a_{t+1}, implying $a_t \neq a_{t+1}$. Nevertheless $a_t = a^*$ and $a_{t+1} = a^*$ in the sense that a^* is defined over the union of the domains of a_t and a_{t+1} (e.g. $H_t \times R_+$), and a_t and a_{t+1} are restrictions of a^* to the appropriate subsets of $H_t \times R_+$. A similar remark applies to b_t and to the equilibrium strategies in Propositions 4 and 5.

14. $m(c) > (1 + c)/2$ for all $c \in [0, 1)$, but we express a_0 more generally to allow for the possibility $\hat{m}(c) < (1 + c)/2$, in the last statement of the proposition.

15. The one period equilibria from Sections I and II satisfy $\partial n(c)/\partial c > 0$, $\partial n(c)/\partial \alpha < 0$, $\partial p(c)/\partial c > 0$, and $\partial n(c)/\partial \alpha < 0$.

16. This argument is more general than the argument used in the fixed cost case. It is an inherent property of ratchet equilibria that the firm will tend to choose the lowest sustainable price.

ACKNOWLEDGMENTS

The results in this chapter appeared first in 'Monopoly Pricing with Visiting Costs, WP 89–5 (1989), University of North Carolina at Chapel Hill. I am grateful to James Friedman, Gary Biglaiser, Pat Conway, Charles Kahn, Helen Tauchen, and the participants of the UNC/Duke theory seminar for helpful comments on early versions of the chapter.

REFERENCES

Diamond, P. (1971). A Model of Price Adjustment. *Journal of Economic Theory, 3*, 156–168.

Kreps, D., & Wilson, R. (1982). Sequential Equilibria. *Econometrica, 50*, 863–894.

Marschak, T., & Selten, R. (1978). Restabilizing Responses, Inertia Supergames, and Oligopolistic Equilibria. *Quarterly Journal of Economics, 92*, 71–93.

Nelson, P. (1974). Advertising as Information. *Journal of Political Economy, 82*, 729–754.

Salop, S., & Stiglitz, J. (1982). The Theory of Sales: A Simple Model of Equilibrium Price Dispersion with Identical Agents. *American Economic Review, 72*, 1121–1130.

Stegeman, M. (1994). *Near-Walrasian Equilibria of a Model of Search and Advertising*, manuscript.

Stiglitz, J. (1984). Price Rigidities and Market Structure. *American Economic Review Papers and Proceedings, 74*, 350–355.

Okun, A. (1981). *Price and Quantities: A Macroeconomic Analysis*. Brookings Institution, Washington, D C.

APPENDIX

General Notes for all Proofs

It simplifies the proofs of Propositions 3 through 5 to observe at the outset that to establish (A) the optimality of a player's strategy s^* given any period and history of the game, it is sufficient to show (B) the action implied by s^* is optimal given any period and history of the game, if the player follows s^* in subsequent periods. The argument follows. Suppose that s^* satisfies (B) but strategy $\hat{s}$ is strictly preferred to s^* given period t and history x. Since there exists a constant upper bound on payoffs in any single period, there must exist some finite period T such that $\hat{s}$ would be strictly preferred to s^* even if the player switched back to s^* after period T. However, backward induction from period T shows that s^* must be optimal subject to the constraint that the player follows s^* after period T, a contradiction.

When verifying the optimality of the firm's actions, it shall be assumed that all consumers follow their equilibrium strategies in every period, since that is what the firm always believes to be true.

Proof of Proposition 3.

To simplify notation, let $a^*(h) \equiv a^*(h, c)$.

To show that b^* is optimal for knowledgeable consumers in a given period $t>0$: Suppose that consumer n last visited the firm in period $t'<t$. Let $p_{t''}$ represent her beliefs for period t'', for $t''>t'$; $p_{t''} \equiv \hat{p}_{t''}(A, p_{t'-}, c_{t'+1+}) = a^*(p_{t'-})$. Since consumer n does not expect a visit to improve her information, and no other agent observes her visit, she should maximize expected surplus in period t. Therefore she should visit iff $\max(0, x - p_t) - \delta > 0$, which holds iff $x - a^*(p_{t'-}) > \delta$, which holds iff $b^*(p_{t'-}, x) = V$. $b^*(p_{t'-}, x)$ is the action implied by b^*.

To show that $z^*(=V)$ is optimal for ignorant consumers in a given period $t \geq 0$: Let $p_{p''}$ represent an ignorant consumer n's beliefs for period t'', for $t'' \geq t$; $p_{t''} \equiv \hat{p}_{j''}(A, p_{-1}, c_{0+}) \leq p^*$. $p^* \leq n(c) < 1$ implies $E_x[\max(0, x - p_{p''}) - \delta] > 0$ for $t'' \geq t$, for sufficiently small δ, and it follows that $z_t = V$ is optimal.

To show that a^* is optimal for the firm in a given period $t>0$, if $p^* = n(c)$: $b^*(h, x) = V$ iff $x - \delta > n(c)$, implying that consumers' visiting decisions mimic their decisions in the one-shot Nash equilibrium. Proposition 1 shows that $p_t = n(c)$ is optimal for the firm. Since $a^*(p_{t-1-}) = n(c)$, a^* is optimal.

To show that a^* is optimal for the firm in a given period $t>0$, if $\max(p_{t-1-}) > p^*$: The firm's action (p_t) has no effect upon consumers' visiting decisions, because any consumer who observes p_t by visiting in a period $t' \geq t$ observes $p_{t'-}$, and $a^*(p_{t'-}) = n(c)$ regardless of p_t. Therefore the firm should choose p_t to maximize profits in period t. Consider a knowledgeable consumer x who last visited the firm in period $t'' < t - 1$. Since the consumer did not visit in period $t'' + 1$, B implies that $x - \delta \leq a^*(p_{t''-}) \leq n(c)$, which implies that the consumer never visits again. A knowledgeable consumer x who last visited the firm in period $t - 1$ visits the firm iff $x - \delta > a^*(p_{t-1-}) = n(c)$. Therefore any knowledgeable consumer x visits the firm in period t iff $x - \delta > n(c)$. Therefore all consumers' visiting decisions mimic their decisions in the one-shot Nash equilibrium, and Proposition 1 shows that $p_t = n(c)$ is optimal. Since $a^*(p_{t-1-}) = n(c)$, a^* is optimal.

To show that a^* is optimal for the firm in a given period $t > 0$, if $\max(p_{t-1-}) \leq p^* < n(c)$: For $p_t > p^*$, the choice of p_t has no effect on consumers' visiting decisions, because any consumer who visits the firm in a period $t' \geq t$ observes $p_{t'-}$, and $p_t > p^*$ implies $a^*(p_{t'-}) = n(c)$. Therefore, if the firm sets $p_t > p^*$, then it should set p_t to maximize profits in period t. $\text{Max}(p_{t-1-}) \leq p^*$ implies that any knowledgeable consumer x visits the firm iff $x - \delta > p^*$. Therefore the firm's profits in period t are $\pi(p_t; p^*, c)$ ((5)), which is strictly concave in p_t. In the region $p_t < p^* + \delta$ $\partial\pi(p_t; p^*, c)/\partial p_t = (1 - \alpha)(1 - p^* - \delta) + \alpha \cdot (1 - p_t) - \alpha \cdot (p_t - c)$. The right-hand expression is zero at $p_t = \hat{p}(p^*, c) = [1 + c\alpha - (1 - \alpha)(p^* + \delta)]/(2\alpha)$. Equation (4) implies $\hat{p}(p^*, c) = p^* + (n(c) - p^*) \cdot (1 + \alpha)/(2\alpha)$, and $n(c) > p^*$ implies $\hat{p}(p^*, c) > p^*$. $\hat{p}(p^*, c)$ is bounded away from p^* for small δ because $n(c)$ decreases in δ, implying $\hat{p}(p^*, c) > p^* + \delta$ for δ sufficiently small. Therefore profits are not maximized at $p_t < p^* + \delta$. In the region $p_t > p^* + \delta$ $\partial\pi(p_t; p^*, c)/\partial p_t = 1 + c - 2p_t$. The right-hand expression is zero at $p_t = (1 + c)/2$. If $(1 + c)/2 > p^* + \delta$, then profits are maximized at $p_t = (1 + c)/2$. If not, then profits must be maximized at $p_t = p^* + \delta$. Let $P(p^*, c)$ denote the p_t that maximizes $\pi(p_t; p^*, c)$, given $p_t > p^*$.

$$P(p^*, c) = \max(p^* + \delta, (1 + c)/2). \tag{18}$$

For $p_t \leq p^*$, the choice of p_t again has no effect on consumers' visiting decision in periods $t' > t$, because $a^*(p_{t'-1-})$ is unaffected by p_t. Therefore, if the firm sets $p_t \leq p^*$, then it should set p_t to maximize profits in period t. Since $\pi(p_t; p^*, c)$ is strictly concave in p_t, and since it has already been shown that $\pi(p_t; p^*, c)$ is maximized at $p_t = P(p^*, c) > p^*$ given $p_t > p^*$, it follows that $\pi(p_t; p^*, c)$ is maximized at $p_t = p^*$ given $p_t \leq p^*$. Therefore the firm should set either $p_t = p^*$ or $p_t = P(p^*, c)$.

If the firm sets $p_t = P(p^*, c)$, then $\max(p_{t-}) > p^*$, and A^* implies that $p_{t'} = n(c)$ for all $t' > t$. Call that strategy A'. Strategy A' yields profits $\pi(P(p^*, c); p^*, c)$ in period t and $\pi(n(c); n(c), c)$ in every period $t' > t$. Strategy A^* yields profits

$\pi(p^*;p^*, c)$ in every period. Adopting strategy A′ instead of strategy A* increases profits by $\phi(p^*, c)$ ((19)). To show that the firm prefers A* to A′, it is sufficient to show that (19) is satisfied.

$$\phi(p^*, c) = \pi(P(p^*, c);p^*, c) - \pi(p^*;p^*, c) + [\pi(n(c); n(c), c) - \pi(p^*;p^*, c)]/\rho < 0 \quad \text{for } \delta \text{ sufficiently small.} \tag{19}$$

Observe that

$$\begin{aligned}
\pi(p^*;p^*, c) &= (p^* - c)(1 - p^* - (1-\alpha)\delta) \\
\pi(P(p^*, c);p^*, c) &= (p^* + \delta - c)(1 - p^* - \delta) \qquad \text{if } c \le 2(p^* + \delta) - 1 \\
&= (1-c)^2/4 \qquad \text{if } c \ge 2(p^* + \delta) - 1 \\
\pi(n(c); n(c), c) &= [n(c) - c] \cdot [1 - n(c) - (1-\alpha)\delta] \\
&= [1 - (1-\alpha)\delta - c]^2 \alpha/(1+\alpha)^2.
\end{aligned} \tag{20}$$

Note that $\pi(P(p^*, c);p^*, c)$ is differentiable with respect to c at $c = 2(p^* + \delta) - 1$. Therefore ϕ is strictly convex in c.

For the moment suppose $\delta = 0$. In the region $c > 2p^* - 1$, $\phi(p^*, c)$ is a convex quadratic function of c. Since $\phi(p^*, 2p^* - 1) < 0$, the highest value of c that satisfies $\phi(p^*, c) = 0$ is the larger root of that quadratic function; $1 - 2(1-p^*)(1-\theta)$ is that larger root. In the region $c < 2p^* - 1$, $\phi(p^*, c)$ is again a convex quadratic function of c. The lowest value of c that satisfies $\phi(p^*, c) = 0$ is the smaller root of that function, which is $[(1+\alpha)p^* - 1]/\alpha$. Therefore $\phi(p^*, c) < 0$ if $[(1+\alpha)p^* - 1]/\alpha < c < 1 - 2(1-p^*)(1-\theta)$, but that is implied by (13) and $p^* < n(c)$. Since ϕ is continuous on δ, it must be that $\phi(p^*, c) < 0$ for δ sufficiently small. Therefore the firm prefers A* to A′, implying $p_t = p^*$ is optimal. Since $a^*(p_{t-1-}) = p^*$, a^* is optimal.

To show that a_0 is optimal for the firm in period 0: Since $p_{-1} \le p^* < n(c)$, the firm's profit maximization problem is similar to the firm's profit maximization problem in period $t > 0$ if $\max(p_{t-1-}) \le p^* < n(c)$. The only difference is that the firm's profits in period 0 are $\pi_0(p_0) \equiv (p_0 - c)(1 - p_0)$ instead of $\pi(p_0; p^*, c)$. $\pi_0(p_0) = \pi(p_0; p^*, c)$ for $p_0 \ge p^* + \delta$, but $\pi_0(p_0) = \pi(p_0; p^*, c) + (p_0 - c)(1-\alpha)(p^* + \delta - p_0)$ for $p_0 < p^* + \delta$. Since p* is optimal in the earlier problem, clearly optimal $p_0 < p^* + \delta$. Since p_0 is strictly concave and $\pi_0'(p_0) < \partial\pi(p_0; p^*, c)/\partial p_0$ for $p_0 \in [p^*, p^* + \delta)$ and δ small, clearly optimal $p_0 \le p^*$. Within that range p_0 has no effect on consumers' visiting decisions, so the firm should set p_0 to maximize $\pi(p_0)$ subject to $p_0 \le p^*$. Therefore optimal $p_0 = \min((1+c)/2, p^*)$, but that is just the rule a_0.

It has now been shown that (A, B, Z) is an equilibrium. (A, B, Z) obviously supports price p*.

Proof of Proposition 4.

To simplify notation, let $a^*(h) \equiv a^*(h, c)$.

The proof that b^* and z^* are optimal for consumers in any given period $t > 0$ is identical to paragraphs two and three of the proof of Proposition 3, except that '$m(c)$' replaces 'p^*'.

The proof that a^* is optimal for the firm in a given period $t > 0$, if $\max(p_{t-1-}) \geq n(c)$, is identical to paragraph five of the proof of Proposition 3.

To show that a^* is optimal for the firm in a given period $t > 0$, if $\max(p_{t-1-}) < n(c)$: Consider a knowledgeable consumer x who last visited the firm in period $t'' < t - 1$. Since the consumer did not visit in period $t'' + 1$, B implies that $x - \delta \leq a^*(p_{t''-}) \leq a^*(p_{t-1-})$ and implies that the consumer never visits again. A knowledgeable consumer x who last visited the firm in period $t - 1$ visits the firm iff $x - \delta > a^*(p_{t-1-})$. Therefore any knowledgeable consumer x visits the firm in period t iff $x - \delta > a^*(p_{t-1-})$. Therefore the firm's profits in period t are $\pi(p_t; a^*(p_{t-1-}), c)$ ((5)).

Suppose that the firm sets $p_t \geq n(c)$. Then the choice of p_t has no effect on consumers' visiting decisions in periods $t' > t$, because any consumer who visits the firm in a period $t' \geq t$ observes $p_{t'-}$, and $p_t \geq n(c)$ implies $a^*(p_{t'-}) = n(c)$. Therefore the firm should set p_t to maximize $\pi(p_t; a^*(p_{t-1-}), c)$. Paragraph six in the proof of Proposition 3 shows that $\pi(P(a^*(p_{t-1-}), c); a^*(p_{t-1-}), c)$ is an upper bound on $\pi(p_t; a^*(p_{t-1-}), c)$. $p_t \geq n(c)$ generates $p_{t'} = n(c)$ for all $t' > t$; alternatively, $p_t = a^*(p_{t-1-})$ would generate $p_{t'} = a^*(p_{t-1-})$ for all $t' > t$. Therefore the total discounted profits from $p_t \geq n(c)$ less the total discounted profits from $p_t = a^*(p_{t-1-})$ cannot exceed $\phi(a^*(p_{t-1-}), c)$ ((19)). $a^*(p_{t-1-})$ must satisfy (13), because $[(1 + c) - 2\theta]/[2 - 2\theta] < t(c) < m(c) \leq a^*(p_{t-1-}) \leq n(c)$, implying (by the proof of Proposition 3) that $\phi(a^*(p_{t-1-}), c) < 0$. Therefore $p_t \geq n(c)$ cannot be optimal.

Suppose that the firm sets $p_t \leq a^*(p_{t-1-})$. Then the choice of p_t has no effect on consumers' visiting decisions in periods $t' > t$, because any consumer who visits the firm in a period $t' \geq t$ observes $p_{t'-}$, and $p_t \leq a^*(p_{t-1-})$ implies $a^*(p_{t'-}) = a^*(p_{t-1-})$. Therefore the firm should set p_t to maximize $\pi(p_t; a^*(p_{t-1-}), c)$. Paragraph six in the proof of Proposition 3 shows that this function is strictly concave in p_t and reaches a maximum at $p_t = P(a^*(p_{t-1-}), c) > a^*(p_{t-1-})$, so $p_t = a^*(p_{t-1-})$ is optimal subject to the supposed constraint. $p_t < a^*(p_{t-1-})$ cannot be optimal.

Suppose that the firm sets $p_t \in [a^*(p_{t-1-}), n(c))$. Then $p_{t'} = p_t$ for all $t' > t$, and total discounted profits in period t are $\pi(p_t; a^*(p_{t-1-}), c) + \pi(p_t; p_t, c)/\rho$. This expression is strictly concave in p_t, and the derivative with respect to p_t, evaluated at $p_t = a^*(p_{t-1-})$, equals

$$1 + \alpha c - (1 + \alpha)\cdot a^*(p_{t-1-}) + [1 + c - 2a^*(p_{t-1-})]/\rho - (1 + \rho)(1 - \alpha)\delta/\rho$$
$$< 1 + \alpha c - (1 + \alpha)\cdot m(c) + [1 + c - 2m(c)]/\rho = 0.$$

Therefore $p_t = a^*(p_{t-1-})$ is optimal. If m(c) were replaced by some $\hat{m}(c) < m(c)$, then the derivative would be positive for $a^*(p_{t-1-}) = \hat{m}(c)$ and δ sufficiently small, implying that the optimal p_t would exceed $a^*(p_{t-1-})$.

To show that a_0 is optimal for the firm in period 0: Since $0 = p_{-1} < n(c)$, the firm's profit maximization problem is similar to the firm's profit maximization problem in period $t > 0$ if $0 = \max(p_{t-1-}) < n(c)$. The only difference is that the firm's profits in period 0 are $\pi_0(p_0) = (p_0 - c)(1 - p_0)$ instead of $\pi(p_0; m(c), c)$. The firm's profit $\pi_0(p_0) = \pi(p_0; m(c), c)$ for $p_0 \geq m(c) + \delta$, but $\pi_0(p_0) = \pi(p_0; m(c), c) + (p_0 - c)(1 - \alpha)(m(c) + \delta - p_0)$ for $p_0 < m(c) + \delta$. Since m(c) is optimal in the earlier problem, clearly optimal $p_0 < m(c) + \delta$. Since p_0 is strictly concave and $\pi_0'(p_0) < \partial\pi(p_0; m(c), c)/\partial p_0$ for $p_0 \in [m(c), m(c) + \delta)$ and δ small, clearly optimal $p_0 \leq m(c)$. Within that range p_0 has no effect on consumers' visiting decisions, so the firm should set p_0 to maximize $\pi_0(p_0)$ subject to $p_0 \leq m(c)$. Therefore optimal $p_0 = \min((1 + c)/2, m(c))$, but that is just the rule a_0.

It has now been shown that (A*, B*, Z*) is an equilibrium and would not be an equilibrium if m(c) were replaced by $\hat{m}(c) < m(c)$. ||

Proof of Proposition 5.

To show that b* is optimal for knowledgeable consumers in a given period $t > 0$: Let t′ denote the period in which the consumer last visited the firm. If $\max(p_{t'-}) \leq s_2$, then the proof is identical to paragraph two of the proof of Proposition 3. If $\max(p_{t'-}) > s_2$, then let $p_{t''}$ represent her beliefs for period t″, for $t'' \geq t$; $p_{t''} \equiv \hat{p}_{t''}(A, p_{t'-}, c_{t'+1+}) = r(c_{t''})$. Since consumer n does not expect a visit to change her probability distribution over $p_{t''}$, $t'' > t$, and no other agent observes her visit, she should maximize expected surplus in period t. Proposition 2 shows that she should visit iff $x > t$, which is the rule implied by b*.

The proof that z* is optimal for consumers in a given period $t > 0$ is identical to paragraph three of the proof of Proposition 3, except that 's_1' replaces 'p*'.

To show that a* is optimal for the firm in a given period $t > 0$, if $\max(p_{t-1-}) > s_2$: The firm's action (p_t) has no effect upon consumers' visiting decisions, because any consumer who observes p_t by visiting in a period $t' \geq t$ observes $p_{t'-}$, and $\max(p_{t'-}) > s_2$ regardless of p_t. Therefore the firm should choose p_t to maximize profits in period t. Consider a knowledgeable consumer x who last visited the firm in period $t'' < t - 1$. Since the consumer did not visit in period $t'' + 1$, B implies that the consumer never visits again. A knowledgeable consumer x who last visited the firm in period $t - 1$ visits the firm iff $x > t$. Therefore any knowledgeable consumer x visits the firm in period t iff $x > t$. Therefore all consumers' visiting decisions mimic their decisions in the one-shot Nash equilibrium, and Proposition 2 shows that $p_t = r(c_t)$ is optimal. Since $a^*(p_{t-1-}, c_t) = r(c_t)$, a^* is optimal.

To show that a^* is optimal for the firm in a given period $t > 0$, if $\max(p_{t-1-}) \le s_2$: To simplify notation, let $p = a^*(p_{t-1-}, c_t)$. Paragraph four of the proof of Proposition 4 shows that the firm's profits in period t equal $\pi(p_t; p, c_t)$.

Suppose that the firm sets $p_t > s_2$. Then the choice of p_t has no effect on consumers' visiting decisions in periods $t' > t$, because any consumer who visits the firm in a period $t' \ge t$ observes $p_{t'-}$, and $\max(p_{t'-}) \ge p_t > s_2$ implies $b^*(p_{t''}, x) = V$ iff $x > \tau$. Therefore the firm should set p_t to maximize $\pi(p_t; p, c_t)$. The 'gain' from action $p_t > s_2$, meaning profits earned in excess of profits earned from action $p_t = p$ (as required by b^*), is the sum of the gain in period t and the gain in subsequent periods. The maximum gain in period 1 obviously occurs if $c_t = 1$. In that case the optimal $p_t > s_2$ is $p_t = 1$, and the gain in period t equals $\pi(1; p, 1) - \pi(p; p, 1) = (1-p)^2 - (1-p)(1-\alpha)\delta$.

If the firm sets $p_t \ge s_2$, then A implies $p_{t'} = r(c_{t'})$ for all $t' > t$, and the profits in each period $t' > t$ equal the profits in the Nash equilibrium of Proposition 2. Those profits can be calculated as follows. Suppose that the firm followed the policy $p_t = t$ if $c_t \le 2t - 1$ and $p_t = (1 + c_t)/2$ if $c_t \ge 2t - 1$. Then expected profits in any single period would be

$$\int_0^{2\tau - 1} (t - c)(1 - \tau)\, dc + \int_{2\tau - 1}^{1} (1 - c)^2\, dc/4 = (1 - \tau)(2\tau - 1)/2 + (1 - \tau)^3 \cdot 2/3. \tag{21}$$

Expression (21) decreases in τ. An upper bound D for (21) is obtained by substituting $1/(1+\alpha)$ for τ.

$$D \equiv (\alpha^3 + 3\alpha)/[6(1+\alpha)^3].$$

The firm's actual policy $r(\cdot)$ generates slightly higher profits, because $r(c) < t$ for $c < 2\sqrt{\delta}$. The probability of that event is $2\sqrt{\delta}$, an upper bound on the extra sales is $t - r(0) = \sqrt{\delta}$, and an upper bound on the profit per sale is 1, so the additional profits do not exceed 2δ. Expected profits in the Proposition 2 equilibrium are therefore smaller than $D + 2\delta$.

If the firm sets $p_t = p$, then A implies $p_{t'} = p$ for all $t' > t$, and the expected profits in each period $t' > t$ equal $\pi(p; p, \frac{1}{2}) = (p - \frac{1}{2}) \cdot [1 - p - (1 - \alpha)\delta]$. Therefore the total gain from setting $p_t > s_2$ rather than $p_t = p$ is smaller than $\psi(p) \equiv (1-p)^2 - (1-p)(1-\alpha)\delta + (D + 2\delta - (p - \frac{1}{2}) \cdot (1 - p - (1-\alpha)\delta))/\rho$, and $p \in S$ ensures that $\psi(p) \le 0$. Therefore $p_t > s_2$ cannot be optimal. If $p \notin S$ and

$\psi(p) > 0$, however, then $p_t > s_2$ would yield higher expected profits than $p_t = p$ for $c_t = 1$ and δ sufficiently small.

The proof that $p_t < p$ cannot be optimal is identical to the sixth paragraph of the proof of Proposition 4.

Suppose that the firm sets $p_t \in [p, s_2]$. The gain from setting $p_t > p$ is obviously greatest if $c_t = 1$, so henceforth assume $c_t = 1$. Strategy A implies that $p_{t''} = p_t$ for all $t' > t$, and it follows that expected discounted profits in period t are $\pi(p_t; p, 1) + \pi(p_t; p_t, \frac{1}{2})/\rho$. This expression is strictly concave in p_t, and the derivative with respect to p_t, evaluated at $p_t = p$, equals

$$(1+\alpha)(1-p) + [3/2 - 2p]/\rho - (1+\rho)(1-\alpha)\delta/\rho < (1+\alpha)(1-q) + [3/2 - 2q]/\rho = 0,$$

because $p \geq q$ from (16). Therefore $p_t = p$, the action implied by b*, is optimal for the firm. If $p < q$, however, then the derivative would be positive for δ sufficiently small.

To show that a_0 is optimal for the firm in period 0: Since $0 = p_{-1} < s_2$, the firm's profit maximization problem is similar to the firm's profit maximization problem in period $t > 0$ if $0 = \max(p_{t-1-}) < s_2$. The only difference is that the firm's profits in period 0 are $\pi_0(p_0) = (p_0 - c_0)(1 - p_0)$ instead of $\pi(p_0; s_1, c_0)$. The firm's profit $\pi_0(p_0) = \pi(p_0; s_1, c_0)$ for $p_0 \geq s_1 + \delta$, but $\pi_0(p_0) = \pi(p_0; s_1, c_0) + (p_0 - c_0)(1 - \alpha)(s_1 + \delta - p_0)$ for $p_0 < s_1 + \delta$. Since s_1 is optimal in the earlier problem, clearly optimal $p_0 < s_1 + \delta$. Since π_0 is strictly concave and $\pi_0'(p_0) < \partial\pi(p_0; s_1, c_0)/\partial p_0$ for $p_0 \in [s_1, s_1 + \delta)$ and δ small, clearly optimal $p_0 < s_1$. Within that range p_0 has no effect on consumers' visiting decisions, so the firm should set p_0 to maximize $\pi_0(p_0)$ subject to $p_0 \leq s_1$. Therefore optimal $p_0 = \min((1 + c_0)/2, s_1)$, but that is just the rule a_0.

It has now been shown that (A, B, Z) is an equilibrium. If s_1 were replaced by $\hat{s}_1 < s_1$, then on the equilibrium path $p_t = a^*(p_{-1}, c_t) = \hat{s}_1$ for all $t > 0$, where either $\hat{s}_1 < q$ or $\psi(\hat{s}_1) > 0$, but in either case it has been shown that $p_t > \hat{s}_1$ would be optimal for $c_t = 1$ and δ sufficiently small. Therefore (A, B, Z) is not an equilibrium if s_1 is replaced by $\hat{s}_1 < s_1$. ||

MINIMUM QUALITY STANDARDS IN A HORIZONTALLY DIFFERENTIATED MARKET

John W. Maxwell

ABSTRACT

Labels that provide information about the environmental attributes seem to be designed to encourage firms to compete over environmental quality. However, it is sometimes possible to enhance the environmental attributes of products at the expense of other desirable product attributes. This chapter examines the effect of product labeling when environmental enhancements come at the expense of other desirable attributes. I find that hoped-for competition over product quality may not result. I go on to analyze the impact of a supplemental regulation of a minimum quality standard. I show that a minimum quality standard, if properly imposed, may benefit all consumers and improve social welfare.

INTRODUCTION

In recent years governments in Europe and North America have introduced product labeling schemes designed to highlight product differences based on environmental quality. Some labeling schemes have been crude; products deemed to meet an environmental standard are given a 'green' label. The trend, however, is toward a label that provides more informational content, such as the percentage of recycled paper, or fuel efficiency measures. The idea behind

Industrial Organization, Volume 9, pages 265–277.

ISBN: 0-7623-0687-4

these labeling schemes seems to be to promote competition based on environmental quality attributes. Firms are induced to compete over these attributes (in theory) because consumers desire environmental quality.

Many authors have studied models of product quality (see Bleath & Katsoulacos (1991) for an overview of models of product differentiation based on quality). Since most of these studies focus on overall product quality, measuring it in a single variable, they use models of vertical product differentiation. In these models all consumers agree that more quality is better, although they may differ in their willingness to pay for quality. If one considers the use of labels to promote competition based on environmental quality, however, it is less clear that vertical differentiation models are always appropriate. Environmental labels provide information about one (or just a few) product attribute(s). It is likely that these attributes can be enhanced at the expense of other desirable attributes. Fuel economy is a good example. The fuel economy of a car may be increased by reducing its weight, making it less safe. You may desire the extra fuel efficiency, but I may not.

I analyze the impact of product labeling schemes on a duopoly market in which consumer preferences over product quality are horizontal in nature. That is, each consumer differs according to his or her ideal level of environmental quality. I show that labeling schemes that inform consumers about environmental attributes need not generate competition over environmental quality, when firms compete on price. Rather than competing over environmental quality, firms desire maximal differentiation, thus relaxing price competition. The market will produce too much environmental quality at the top end and too little environmental quality at the bottom.

Since it is unrealistic to assume that a regulator would attempt to reduce the environmental quality of products, I focus on a supplemental policy of minimum quality standards. Early studies of minimum quality standards (see, e.g. Leland (1979), Shapiro (1983) and Besanko et al. (1988)) have produced less than favorable findings. These studies have found that minimum quality standards may harm some consumers. Favorite products, or firms may be barred from the market, thus reducing product choice and/or competition. Ronnen (1991) has challenged these results. He shows that all consumers can be made better off through a judicious choice of a minimum quality standard. These results suggest that minimum quality standards may be a useful companion regulation to product labeling. A key feature of Ronnen's model, however, is that preferences over product quality are vertically distributed across the consumer population. My purpose here is to examine the impact of imposing minimum quality standards when increasing product quality does not directly benefit all consumers. That is, preferences over environmental quality are horizontal in nature, because quality enhancement comes at the expense of other product attributes.

Despite the horizontal distribution of preferences, I find that appropriately set minimum standards can benefit all consumers. The minimum standard has the effect of rasing the environmental quality produced by both firms. I show that the high quality producer raises its quality by less than the increase in the minimum environmental standard. As a result, price competition intensifies. It is this competition that ensures that all consumers gain from the standard. I find that the imposition of a minimum standard harms both firms (due to the price competition). However, this negative implication is not so troubling when one recalls that it is the labeling regulation which permits firms to raise prices in the first place.

I find that an appropriately set minimum standard for environmental quality can increase social welfare. Raising the minimum standard brings the level of quality at the low end of the market closer to is social optimum, raising social welfare. The strategic effect of the standard (a further increase in the high quality firm's quality level) is negative, but is dominated by the direct (positive) effect. Thus a dual policy of product labeling and minimum standards are likely to have a desirable effect on social welfare. Product labeling will spur some environmental innovation, and minimum quality standards may be used to partially offset labeling's negative effects.

In the subsequent Section I develop the theoretical model and solve for the market equilibrium assuming that labeling is the only regulator policy. In Section II I analyze the effect of an imposition of a minimum standard on market prices, firms' profits, and social welfare. At the end of Section II I discuss how the results of the model carry over to a multi-firm setting in a straightforward manner. Section III concludes.

I. THE MODEL

In this Section I develop a simple model of horizontal product differentiation in order to capture the idea that enhancements in environmental quality may not be desirable for all consumers.[1] In the remainder of this chapter I will drop the term environmental, and refer simply to quality, where it does not cause confusion. The demand side of the market in this model consists of a continuum of consumers of mass 1 uniformly distributed over the unit interval and indexed by preference parameter θ. Let

$$v(\theta, p, q) = s - p - t(\theta - q)^2 \tag{1}$$

be the value (or indirect utility) that a consumer with preference parameter q receives from the purchase of 1 unit of a good with quality level q.[2] Consumers are assumed to be fully informed about q because of the labeling regulation. As is standard, I assume that each individual derives utility only from the first unit he buys. The final term in (1) denotes the disutility or loss in value that the consumer incurs if he is unable to purchase his most preferred product (θ). An increase in t implies a strengthening of consumer preferences for their most preferred brand.

On the supply side of the market I assume that there are two potential entrants to the market at the time product labeling is introduced. These two firms will differ only in the level of product quality they choose in equilibrium. In order to produce a product of quality level q a firm must incur a quality development cost $F(q)$ which is fixed with respect to the level of production. Furthermore, I assume that this development cost function obeys the following properties; $F(0)=0$, $F'(0)>0$, $F'(q)>0$, $F''(q)>0$ for all $q \in [0, 1]$. Marginal production costs are assume to be constant and independent of the level of quality.[3]

Competition between the two firms takes place in two stages. First, the firms simultaneously decide whether to enter the market and, if so, make their quality choice. In the second stage the two firms compete by simultaneously setting prices. I assume that entry and quality choices are observable by both firms prior to the price competition. As is standard with these types of models, the idea here is that prices are easier to change than product quality.

The game is solved from the last period in order to find the subgame perfect equilibrium. It is easy to verify that it is not an equilibrium for the two firms to produce the same quality product. The subsequent price competition would guarantee zero profits. I will denote the firm producing the higher quality product by h and its quality and price choices by q_h and p_h respectively. Similarly, the firm producing the lower quality product is denoted by l and the quality and price choices are q_l and p_l respectively.

I first derive the demand functions facing the two firms. Given the notation above the consumer who is indifferent between purchasing the high or low quality good is denoted by x^* where:

$$p_l + t(x^* - q_l)^2 = p_h + t(q_h - x^*)^2. \tag{2}$$

Firm l sells to all consumers to the left of x^* while firm h sells to all consumers to the right of x^*. Assuming an interior solution it is straight forward to show that, given their quality choices, firms l and h face the following demand functions:

$$D_l = q_l + \frac{q_h - q_l}{2} + \frac{p_h - p_l}{2t(q_h - q_l)}, \tag{3}$$

and

$$D_h = (1 - q_h) + \frac{q_h - q_l}{2} + \frac{p_l - p_h}{2t(q_h - q_l)}. \tag{4}$$

Given its quality choice firm i's price choice problem is:

$$\max_{p_i} (p_i - c)D_i(p_l, p_h);\ i = \{l, h\}. \tag{5}$$

Optimizing profits and solving for the Nash equilibrium prices yields

$$p_l(q_l, q_h) = c + t(q_h - q_l)\left(1 + \frac{q_l + q_h - 1}{3}\right),$$

and

$$p_h(q_l, q_h) = c + t(q_h - q_l)\left(1 + \frac{1 - q_h - q_l}{3}\right). \tag{6}$$

Substituting these reduced form prices into the expression for the location of the indifferent consumer yields:

$$x^* = \frac{2 + q_h + q_l}{6}. \tag{7}$$

Given the equilibrium prices, I can solve for each firm's second stage profit as a function of the quality choices. Because these profit functions are gross of quality choice costs, I denote the functions by $G^i(q_l, q_h)$, $i \in \{l, h\}$. Let

$$G^h(q_l, q_h) = \frac{1}{18}t(4 - q_l - q_h)^2(q_h - q_l), \text{ and}$$

$$G^l(q_l, q_h) = \frac{1}{18}t(2 + q_l + q_h)^2(q_h - q_l). \tag{8}$$

The duopoly profit functions for the two firms are

$$\pi^l(q_l, q_h) = G^l(q_l, q_h) - F(q_l), \text{ and } \pi^h(q_l, q_h) = G^h(q_l, q_h) - F(q_h). \tag{9}$$

Using (9), it is easy to verify that the low quality firm will always enter the market if there is no minimum quality standard. If the minimum quality standard is set high enough it could be a barrier to entry (potentially for both firms). I am interested, however, in determining whether rasing the minimum standard can benefit consumers, and raise social welfare in general. Since consumers are harmed by monopoly, I examine situations in which the minimum quality standard is not a barrier to entry. That is, I compare the duopoly equilibria under regulatory policies of product labeling, and labeling with minimum quality standards.

The following are easily verifiable properties of G^h and G^l, which will be used in the remainder of the article.

$$\frac{\partial G^l}{\partial q_l} < 0, \frac{\partial Gh}{\partial q_l} < 0, \frac{\partial G^l}{\partial q_h} < 0, \frac{\partial G^h}{\partial q_h} < 0, \tag{10}$$

$$\frac{\partial^2 G^l}{\partial q_l^2} < 0, \frac{\partial^2 G^h}{\partial q_h^2} < 0, \frac{\partial^2 G^h}{\partial q_h \partial q_l} > 0; \forall\, 0 \leq q_l < q_h \leq 1.$$

Note that the first four derivatives of (10), along with (7), imply that the price effects of quality changes dominate market share effects. Equation (7) implies that quality enhancements by the low quality firm will win it market share. However, the enhanced price competition will dominate this positive effect. Similarly, by increasing its product's environmental quality the high quality firm will lose market share, but the resulting relaxation in price competition makes such a move profitable (as long as the quality development costs are covered). The fifth and sixth derivatives in (10) illustrate increasing (negative) and diminishing marginal profit to enhancements in own-product quality.

The final derivative in (10) implies that quality enhancements by the low quality firm will make further improvements by the high quality firm more profitable, *ceteris paribus*. This result has important implications for minimum quality standards. Namely, raising minimum quality standards has both direct and strategic effects. The direct effect is a one for one increase in q_l while the strategic effect is an increase in q_h. I show below that this latter increase is less than one for one.

Proposition 1. *The unique subgame perfect equilibrium quality choices are* $q_l = 0$ *and* $q_h = q_h^\pi$ *where* q_h^π *satisfies*

$$G^h_{q_h}(0, q_h^\pi) - F_{q_h}(q_h^\pi) = 0. \tag{11}$$

Proof. Note first that the subscripts q_l and q_h in (11) denote partial derivatives, I will adopt this convention in the remainder of the chapter. The proof follows directly form the profit functions (9) and the properties of the quality function F. It is never profitable for the low quality firm to adopt a positive level of quality. Knowing this, the optimal choice of the high quality firm is described by (11). Q.E.D.

Note that I assume that (11) holds for $q_h < 1$. This assumption implies that it is not profitable to produce the highest quality product. This is quite likely to be the case for environmental attributes because of technological constraints. If I assumed that $q_h^\pi = 1$ it is easy to see that the results derived below still hold.

In this case $dq_h/dq_l = 0$, and most of the results rely on this derivative being small. In this case the model is less interesting, however, since there is no strategic response to the imposed minimum quality standard.

This market equilibrium illustrates a potential problem with reliance on product labeling to generate competition over environmental quality. While the high quality firm has an incentive to raise the quality of its product, hoped-for competition over product quality will not result. To see this, imagine that some technical innovation reduces the marginal costs of quality development. In this case it is the leading firm that has the desire to improve the quality of its product. The lagging firm has no desire to increase the quality of its product. The same can be said of incentives to engage in R&D to reduce the marginal costs of environmental quality.

The goal of many product labeling schemes is to increase the quality of products consumed, especially if consumption of the products has associated negative externalities. While I do not model consumption externalities here, I can say that labeling that highlights product heterogeneity does make a positive contribution. The high quality firm has an incentive to raise its quality from zero.[4] However, these schemes provide no incentives for the low quality firm to raise its product's quality. In fact such moves would be resisted.

II. ANALYSIS OF A MINIMUM QUALITY STANDARD

As mentioned above, it is possible that a minimum quality standard may be a barrier to entry if set too high. Since I am interested in investigating whether such a standard can improve social welfare, I shall assume that the standard is set low enough so as not to deter entry. One can easily verify that such a standard exists by examining $\pi^l(0, q_h)$, and noting that $F_{q_l}(0) = 0$.

Proposition 2. *There exists a positive level of quality q_s such that if the regulator sets the minimum quality at q_s, both firms will offer a higher level of environmental quality than they would if the standard was not imposed.*

Proof. The fact that the lower quality firm offers a higher level of quality is obvious, $q_l = q_s > 0$. To see that the high quality firm also raises quality totally differentiate (11) to see that

$$\frac{dq_h}{dq_l} = \frac{-G^h_{q_h q_l}}{G^h_{q_h q_l} - F_{q_l q_h}} > 0. \tag{12}$$

The inequality (12) arises from (9) and the fact that quality adoption costs are strictly convex. Q.E.D.

In contrast to models of vertical quality differentiation, the quality adjustments that take place here as a result of the standard, are not welcomed by all consumers. However the quality adjustments lead to enhanced price competition which does benefit all consumers. On balance, every consumer can be made better off at the expense of firm profits.

Proposition 3. *There exists a positive level of quality q_s such that if the regulator sets the minimum quality at q_s (a) all consumers will obtain a greater indirect utility and (b) both firms will see their profits fall.*

Proof. To establish parts (a) and (b) of the proposition note first that the expression (12) attains it maximum at less than 1/5. To see this, suppose that $F_{q_h q_h}$ is zero (we know that it is positive). Then the ratio in (11) becomes $(q_h - q_l)/(8 - 3q_h - q_l)$ which is equivalent to 1/5 at $q_h = 1$ and $q_l = 0$ and is less than 1/5 for all other values.

The proof of part (a) then proceeds as follows. Since $dq_h/dq_l < 1/5$ and $dq_l/dq_l = 1$, both firms will charge lower prices. This can be seen by totally differentiating the price functions (5). Since prices are lower, the only consumers that can be worse off are those that prefer the original qualities $(q_l = 0, q_h = q_h)$. That is, the consumer with preference parameter $\theta = 0$, and consumers with preference parameters in the interval $(x^*, q_h]$. Of these consumers, it is the indifferent consumer that will bear the greatest cost if he continued to purchase the high quality product (in fact we know he does not). Thus, in order to show that all consumers are better off it is sufficient to show that the consumer with preference parameter $\theta = 0$ is better off, and that the indifferent consumers would be better off *even if he continued to purchase the high quality good.*

$$\left.\frac{\partial U(0, p_l, q_l)}{\partial q_l}\right|_{q_l=0} = \left.\left[\frac{\partial p_l}{\partial q_h} + \frac{\partial p_l}{\partial q_l}\frac{\partial q_h}{\partial q_l} - 2t(q_l)\right]\right|_{q_l=0} = \frac{2t}{3}\left(1 - (1 + q_h)\frac{dq_h}{dq_l}\right) > 0,$$

and

$$\left.\frac{\partial U(x^*, p_l, q_l)}{\partial q_l}\right|_{q_l=0} = \left[\frac{\partial p_l}{\partial q_h} + \frac{\partial p_l}{\partial q_l}\frac{\partial q_h}{\partial q_l} - 2t(q_h - x^*)\right]\left.\frac{\partial q_h}{\partial q_l}\right|_{q_l=0}$$

$$= \frac{4t}{3}\left(1 - (1 + q_h)\frac{dq_h}{dq_l}\right) + 2tx^*\frac{\partial q_h}{\partial q_l} > 0.$$

To establish part (b) of the proposition note that:

$$\frac{dG^h}{dq_l} = G^h_{q_l} + G^h_{q_l}\left(\frac{\partial q_h}{\partial q_l}\right) - F_{q_l}\frac{\partial q_h}{\partial q_l} < 0.$$

The inequality above arises from the fact that $|G^h_{q_l}| > G^h_{q_h}$. Since $q_h > q_l$, an increase in level of quality of the low quality product has a larger marginal impact on the profit of the high quality firm than an increase in its own product's quality. The opposite is true for the low quality firm. That is $|G^l_{q_l}| < G^l_{q_h}$. Thus the impact of rasing the minimum quality standard on the low quality firm's profits,

$$\frac{dG^l}{dq_l} = G^l_{q_l} + G^l_{q_h}\left(\frac{\partial q_h}{\partial q_l}\right) - F_{q_l},$$

is not immediately clear. However, after some tedious algebra one can show that the sum of the first two terms is zero only if $q_h/q_l = 1/5$ and is negative for values less than this. Thus, profits are always declining for the low quality firm. Q.E.D.

The fact that the low quality firm suffers from the imposition of a minimum quality standard is worth some comment. The low quality firm suffers due to increased price competition. In Ronnen the low quality firm benefits from an increase in the minimum quality standard because its lower price (combined with the increased quality) causes consumer entry at the low end of the market. Such entry is not possible here. If minimum quality is unregulated, it is the consumer at the low end who obtains the greatest surplus from the low quality product (recall $q_l = 0$). If they do not buy the product no one will.

In order to determine the overall impact of an increase in the minimum quality standard on welfare I calculate the following measures of social welfare:

$$\int_0^{x^*} (s - p_l - t(z - q_l)^2) + (p_l - c)dz = (s - c)x^* - \frac{t}{3}((x^* - q_l)^3 - (q_l)^3), \quad (13)$$

represents welfare arising from sales by the low quality firm and

$$\int_{x^*}^{1} (s - p_h - t(z - q_h)^2) + (p_h - c)dz = (s - c)(1 - x^*)$$

$$- \frac{t}{3}((1 - q_h)^3 - (x^* - q_h)^3), \quad (14)$$

is welfare arising from sales by the high quality firm. Summing (13) and (14), and using equation (7), it is easy to show that the total surplus generated from the sales of the two firms is:

$$V(q_l, q_h) = s - c - \frac{t}{36}(5q_l^3 + 8q_l^2 + 5q_l^2 q_h - 4q_l - 5q_l q_h^2 + 12 - 32q_h + 28q_h^2 - 5q_h^3). \tag{15}$$

Thus, accounting for quality adoption costs, a full measure of social welfare generated in the market is:

$$W(q_l, q_h) = V(q_l, q_h) - F(q_l) - F(q_h). \tag{16}$$

Proposition 4. *There exists a positive level of quality, q_s, such that if the minimum quality standard is set at q_s, social welfare will rise from its level under a labeling-only regulatory policy.*

Proof. Differentiating (16) and evaluating the derivative at market equilibrium ($q_l = 0$, $q_h = q_h^{\pi}$) I obtain:

$$V_{q_l} + (V_{q_h} - F_{q_h})\frac{dq_h}{dq_l} = V_{q_l} + (V_{q_h} - G_{q_h}{}^{h})\frac{dq_h}{dq_l} > 0. \tag{17}$$

The equality in (17) arises from first order condition (11). Thus, from Proposition 5 below, the social surplus arising from sales of the high quality product will fall as a result of the minimum quality standard. However, using the fact that $q_h/q_l < 1/5$ straightforward calculations establish the inequality. Q.E.D.

Proposition 4 shows that the gains to welfare from an increase in the minimum quality standard are more than the simple decline in price (which is simply a transfer). In the market equilibrium the low quality firm provides too little quality and the high quality firm too much. However, since profit considerations prevent the high quality firm from matching the quality increase imposed by the regulator, the surplus generated at the low end of the quality spectrum outweighs the lost surplus incurred at the high end of the quality spectrum.

The reader should note that in Proposition 4 I have considered only a small increase in the minimum quality standard (from zero). It is not true in general that increasing the minimum quality standard will raise welfare (the associated quality costs might outweigh the benefits). It is also the case that the imposition of a minimum quality standard cannot be used to achieve the first best choices of quality. This result arises from the fact that the high quality seller always has an incentive to produce too much quality, as the proof of the following proposition illustrates.

Proposition 5. *A policy of minimum quality standards will not result in first best levels of quality in the market.*

Proof. The first best levels of low and high quality (q_l^* and q_h^* respectively) are defined by the following first order conditions:

$$V_{q_l}(q_l^*, q_h^*) - F_{q_l}(q_l^*) = 0,\ V_{q_h}(q_l^*, q_h^*) - F_{q_h}(q_h^*) = 0. \tag{18}$$

Straightforward calculation shows, however, that $G_{q_h}(q_l, q_h) > Vq_l(q_l, q_h)$ for all feasible values of q_l and q_h. Thus, social welfare will be falling in the level of the minimum quality standard at a level below its socially optimal value. Q.E.D.

Proposition 5 is not surprising. The regulator would like to control both levels of quality in order to achieve the first best result. The minimum quality standard however gives the regulator control over only one quality level. It is worth noting, however, that the difference between q_h and q_h^* is falling in $F_{qh}q_h$. Thus, minimum quality standards will be able to achieve the an outcome that is closer to the first best outcome the more technologically constrained the high quality firm is.

Before concluding this chapter it is worth noting how the results of the model carry over to a multi-firm setting. As discussed at the outset of Section II, it can never be an equilibrium for firms to choose the same level of product quality (profits inclusive of quality costs would be negative). Consequently, in any n-firm equilibrium that exists firms could be ranked according to their quality level with one firm choosing a zero quality level. Given that $F_q(0) = 0$, it is straightforward to show that the results above hold. That is there will exist a small increase in the minimum quality of the lowest firm that will benefit all consumers and enhance social welfare.

III. CONCLUSION

This chapter has demonstrated the impact of product labeling and minimum quality standards in a market characterized by horizontal preferences over environmental quality. Since product labels often focus on one attribute (such as fuel economy) it seems natural to assume that advances in that attribute may come at the expense of other product attributes. Within the context of such a model I have shown that hoped-for competition resulting from product labeling may not arise. Instead firms may exploit product differences through relaxed price competition. In this case the market will provide too much quality at the top end and too little quality at the bottom.

I have demonstrated that the imposition of a minimum quality standard may be in order. If properly set, the minimum quality standard may improve the value of a purchase for all consumers. Any consumers that incur losses from the resulting product repositioning are more than compensated by declines in prices.

The more constrained is the high quality firm in making adoptions, the more beneficial to consumers is the imposition of the standard. Price competition will increase, and welfare losses resulting from the high quality firm's strategic response will be diminished. I have shown that, while firms lose from the imposition of a minimum quality standard, such a standard can increase social welfare, if set properly. Furthermore, the losses incurred by firms as a result of the minimum quality standard are less troubling when one considers a dual policy of minimum quality standard and product labels. In this case, the pricing effects of a standard work to reallocate some of the surplus firms enjoy as a result of the labeling regulation.

The results of this chapter generally favor the imposition of a minimum quality standard. However, I should add some words of caution. First the model I have considered is not dynamic. In a dynamic setting I have shown elsewhere that minimum quality standards may harm innovation, leading to foregone welfare gains (see Maxwell (1998)). Second, the imposition of a minimum quality standard will cause some consumers to switch consumption of the high quality product to the (now higher) low quality product. It is easy to posit a consumption externality function that would generate greater negative externalities after the imposition of a minimum quality standard (whether such a function is realistic is another matter). Such increases would diminish the welfare enhancing results I have derived.

NOTES

1. The basic set up of my model is well known in the literature, see Tirole (1988).
2. This quadratic specification of the indirect utility function leads to demand functions that are continuous in both price and location. I assume that s is sufficiently large so that each consumer in the market will buy one unit from one of the two firms.
3. Allowing marginal production costs to rise with quality will not alter the qualitative results of the model.
4. In fact, as I show below, the high quality firm adopts too much quality, however, the model does not include externalities. If these drop with quality, the profit maximizing level of product quality and the socially optimal level will ten towards each other.

REFERENCES

Beath, J., & Katsoulacos, Y. (1991). *The Economic Theory of Product Differentiation*. New York, New York: Cambridge University Press.

Besanko, D., Donnefeld, S., & White, L. J. (1988). The Multiproduct Firm, Quality Choice and Regulation. *The Journal of Industrial Economics*, *36*, 411–429.

Leland, H. E. (1979). Quacks, Lemons, and Licencing: A Theory of Minimum Quality Standards. *Journal of Political Economy*, *87*, 1328–1346.

Maxwell, J. W. (1998). Minimum Quality Standards as a Barrier to Innovation. *Economics Letters*, *58*, 355–360.

Ronnen, U. (1991). Minimum Quality Standards, Fixed Costs, and Competition. *RAND Journal of Economics*, *22*(4), 490–504.

Shaprio, C. (1983). Premiums for High Quality Products as Returns to Reputations. *Quarterly Journal of Economics*, *98*, 3–314.

Tirole, J. (1988). *The Theory of Industrial Organization*. Boston, Massachusetts: MIT Press.

STRATEGIC DELEGATION IN COURNOT OLIGOPOLY WITH INCOMPLETE INFORMATION

Guido Merzoni

ABSTRACT

This chapter contains a model of strategic delegation from owners to managers in a Cournot duopoly where firms compete under incomplete information on the rival marginal costs and the relations between owners and managers are characterised by moral hazard. It is shown that despite incomplete information the owners are able to use strategically the delegation of the output decision by making the managers' incentive schemes observable. Strategic delegation enhances the equilibrium level of the managers' effort, decreasing all firms' marginal costs and so increasing expected output. When moral hazard results in under-provision of effort, strategic delegation has a counter-balancing effect on the loss of productivity due to agency costs. On the other hand, collusive agreements are shown to weaken the disciplinary role on managers of product market competition. In the linear demand case is also shown that the equilibrium industry output in each state of nature is lower with strategic delegation than otherwise, so that the equilibrium price is distorted toward the monopoly price. Therefore, the expected output increase is all due to the better states of nature becoming more likely. However, at difference from what happens with strategic delegation under complete information in the

Industrial Organization, Volume 9, pages 279–305.

ISBN: 0-7623-0687-4

output setting game, in this model consumers may not benefit from strategic delegation, since consumer's surplus is convex in output.

I. INTRODUCTION

Since the seminal paper by Coase (1937) the firm and the market have been recognised as alternative institutions through which to organise economic activity. However, there is no such thing as 'the firm'. Firms are different from each other and their internal organisation should not be taken as an exogenous variable. Furthermore, firms do not exist in isolation, they interact with each other within markets. The structure of the market the firm is operating in influences its internal organisation, and in particular the way in which the incentives for the different parties contributing to the firm activity are designed. This is, of course, especially true when the actions taken by one firm have an influence on the behaviour of its rivals, i.e. in oligopolistic markets, where firms interact strategically. In those circumstances firms can take decisions allowing them to commit to a particular course of action and gain a strategic advantage upon rivals.

Starting with von Stackelberg's (1934), examples of strategic commitment in oligopoly abound in the literature: Spence (1977) and Dixit (1980) on capital investment, Brander-Spencer (1983) on R&D, Schmalensee (1983) on advertising and Fudenberg-Tirole (1983) on learning by doing are just a few, by now almost classic, examples. Fudenberg-Tirole (1984) and Bulow-Geanakoplos-Klemperer (1985) provide a general framework to build a taxonomy of strategic investment. In all these cases the oligopolistic game described is played in two stages. In the second stage each firm makes a choice about either quantity or price. In the first stage at least one of the firms takes an irreversible decision, considering also the effect of that decision on the rivals' behaviour and, as a consequence, on the outcome of the second stage of the game. The irreversibility of the decision taken in the first stage allows the firms to credibly pre-commit to a certain course of action, which will then be taken as given by their rivals. However, all these papers consider 'the firm' as the subject taking the strategic decisions and do not account for variety in internal organisation.

Developing on the seminal idea by Schelling (1960), in recent years the so-called 'strategic delegation' literature consider the appointment of agents and the provision of incentives for them as pre-commitment devices to be used strategically. A part of this literature focuses on the strategic value of the top managers' incentive schemes in an oligopolistic market (Vickers, 1985; Fershtman-Judd, 1987; Sklivas, 1987; Macho-Stadler-Verdier, 1991; Basu, 1995).[1] These papers provide more examples of strategic commitment in oligopoly, recognising that firms are the result of the interactions of different subjects related to each other through a set of contracts that may be selected strategically. In these papers the remuneration in the optimum incentive scheme for the manager of an oligopolistic firm competing *à la Cournot* is positively correlated to the firm's total revenues as well as to profits. Through such an incentive scheme, the owner of the firm manipulates the objectives of his manager, committing the firm to aggressive behaviour in the product market, discouraging rivals and gaining a strategic advantage. This approach emphasises the 'external' role of the manager. Managers are described as being representatives appointed just to monitor demand and play the market game. Their objectives can be arbitrarily chosen by the owners. No reference is made to the 'internal' role of the manager as the person in charge of the organisation and control of the productive activity. The problem of monitoring their level of effort is not a relevant issue in that framework. The structure of the market does not influence the internal operations and so the efficiency of the firms, but only conditions their selling strategies.

The effect of product market competition on x-inefficiencies caused by asymmetric information in the owner-manager relation has been the main focus of another stream of the literature. These authors investigate the validity of the conventional wisdom view, according to which product market competition disciplines firms, reducing the incidence of agency costs. Hart (1983) shows that the competition of owner-managed firms reduces managerial slack in firms where owners delegate some discretionary power. Scharfstein (1988) points out that Hart's result is sensitive to the assumption he makes of managers' being infinitely risk averse, while different results can be obtained from different assumptions on managerial preferences. Hermalin (1992) shows that the effect of competition on executives' behaviour may be strongly affected by income effects, while the sign of this effect is ambiguous. Martin (1993) gives an example with Cournot competition, which shows that firms efficiency is inversely related to the number of firms in the market. Recently Bertoletti-Poletti (1996) have pointed out that Martin's result is due to a common feature of many of these models, i.e. increasing returns, so that a reduction in the scale of production always reduces the incentives to invest in cost-reducing activities. One of the aims of this paper is to address the problem of the relations between product market competition and x-inefficiency, and, in particular, to see the effect of strategic delegation on managerial inefficiency.

We present a model of strategic delegation from owners to managers in a Cournot duopoly where firms compete under incomplete information on the rival marginal costs and the relations between owners and managers are characterised by moral hazard. The game has two stages: in the first stage managers' incentive schemes are chosen, while in the second stage the output setting game is played by the managers under incomplete information on the rival's marginal cost. The managers do not only play the output-setting game, but they also contribute to the productive activity of the firms through their organisational and monitoring duties, so having a marginal-cost-reducing effect on their firms' activity.

The analysis is meant to verify the validity of the main results of the strategic delegation models, where managers are merely playing the output setting game under perfect information, for this setting where managers, who also contribute to production, select output under incomplete information. With complete information strategic contracting would achieve its commitment effect through a reduction in costs observable to rival firms; with incomplete information this cannot be the case. However, we show that the observability of contracts is enough to accomplish the aim of strategic commitment.

We move, then, to the issues of the effects of strategic contracting and of changes in the degree of product market competition on x-inefficiency. The basic model with moral hazard and strategic delegation is also compared with alternative settings where the distribution of information and decisional power between owners and managers are different or it is not possible to use incentive contracts as a commitment device. Finally, the effect on consumers' welfare of strategic delegation under incomplete information is evaluated for the linear demand case and compared to what happens in this respect under complete information.

In Section II the basic structure of the model is presented; in Section III the equilibrium choices of managers' effort and output are described; in Section IV the equilibrium of the game with strategic delegation is compared with a few alternative settings; Section V contains the analysis of the linear demand case. Section VI concludes.

II. THE MODEL

We consider a duopolistic market with identical firms producing a homogeneous good. Separation between ownership and control prevails. Each firm is owned by a representative individual, the owner, who wants to maximise profits. He delegates to another individual, the manager, the task of running the firm. The activity of the manager usually has an impact on the productivity of the firm. This is taken account of in our model where the manager provides a level of effort which decreases the marginal cost in a way to be made more precise below. The manager exercises his influence in many ways: by monitoring the productive activity, by improving the organisational structure of the firm, by keeping up to date the technology used.

The owner does not observe the manager's level of effort, so that the manager's compensation cannot be made contingent on that. Before the contract is signed, neither player observes the actual level of the marginal cost since some productivity parameter (state of the world) is unknown at this stage (e.g. the quality of inputs, the conditions of the machinery). The manager observes the level of the marginal cost before the output decision is taken. For that reason, each owner delegates the output decision to his manager.

After the decision about output has been made, the owner, as well as any third party, does observe the level of the marginal cost, which provides a signal of the manager's effort. However, the owner never becomes aware of the uncertain productivity parameter, so that the observation of the marginal cost does not reveal to him the level of effort provided by the manager. Nevertheless, a contract in which the manager's compensation is made contingent on the firm's marginal cost is legally enforceable and commits the manager and the owner to its fulfilment, unless they mutually agree otherwise.

Furthermore we assume that the contract is observable. We say that a contract is observable when its contents are known to a third party, who can be sure that it will not be (even secretly) re-negotiated.[2] Observability changes the structure of the game being played, by allowing the player to use the contract for credibly pre-committing to a particular course of action. The assumption of contract observability, while not being by any means innocuous, is widely used in the literature on strategic delegation. For the present case a possible rationale for using it comes from the small number of parties involved in the game.[3] The described situation can be analysed through a multi-stage game, which results from combining a moral hazard principal-agent game with hidden action[4] and a Cournot duopoly game. The timing of the game is the following.

Game 1 – Strategic Delegation

Stage 1. The owner of each firm offers to his manager a contract, observable to their rivals, which establishes a set of levels of remuneration contingent on the marginal cost that will be observed later: $\{(w_b, c_b), (w_g, c_g)\}$, where the subscript b denotes a bad state of the world (high cost) and g denotes a good state of the world (high cost).

Stage 2. Each manager decides whether to accept the offer and, if he does accept, selects his level of effort.

Stage 3. Each manager observes his own firm's marginal cost and, having formed his expectations about the rival firm's marginal cost on the basis of the observed contract, chooses the level of output.

Stage 4. Each owner *observes* his own firm's marginal cost and pays the remuneration to his manager.

At both stage 1 and stage 3, the decision-makers have Cournot-Nash conjectures about the rival firm's behaviour.

It may be useful to stress, that the effort level, despite being chosen by the managers, are actually decided by the owners through the choice of the incentive scheme. By representing the principal-agent game as a mere choice of the manager's effort made by the owner, the model may be treated, for simplicity, as a two-stage game with effort determination in the first stage and choice of output in the second. That is what we are going to do in the following sections, after solving for the incentive scheme which implements any given effort level at minimum cost in the present section.

We now consider the principal-agent game.

We assume that the cost structure is stochastic. For simplicity, the marginal cost is either low, at level c_g, with probability $Prob(g)$ or high, c_b, with probability $[1 - Prob(g)]$.[5] The probability distribution of the marginal cost depends on the managers' effort levels e_i and on his ability, described by the parameter α:[6]

$$Prob(g) = \alpha p(e_i) \qquad i = 1, 2 \tag{1}$$

where i stands for the ith firm and $p(e_i)$ is a twice differentiable and strictly concave function. For the problem to make sense $\alpha p(e_i)$ must satisfy the following condition:

$$0 < \alpha p(e_i) < 1 \qquad \forall e_i \tag{1a}$$

We assume that, within the range of interest for our problem, the probability of having low costs is non-decreasing in the manager's effort. Both the owner and the manager observe this probability distribution over the marginal costs.

The manager's utility function is

$$U = f(w) - xe \tag{2}$$

where f is a strictly increasing and twice continuously differentiable function, w is the manager's remuneration and x is a coefficient representing a measure of his 'laziness' or disutility of effort, which we will take to be positive. As it is immediately clear by looking at (2), the utility function is assumed to be separable in its two arguments. Its first derivatives are

$$\frac{\partial U}{\partial w} > 0$$

$$\frac{\partial U}{\partial e} < 0.$$ [7]

In our analysis we will consider both risk neutral and risk averse managers.[8]

As mentioned above, the objective of the owner is to maximise profit. He is assumed to be risk neutral.

The owner also knows the manager's utility function and his reservation utility $u°$.

In order to solve the principal-agent game we split the principal's problem in two parts as suggested by Grossman-Hart (1983). First we find the 'best', i.e. the least costly, way to induce the manager to choose a certain level of effort to be determined later; then we calculate the level of the manager's effort preferred by the owner.

The solution of the first part of the owner's problem is the solution of the following constrained cost minimisation problem. For a given e_i

$$\min_{w_g,\, w_b} \alpha p(e_i)w_g + [1 - \alpha p(e_i)]w_b \tag{3}$$

s.t.

$$\alpha p(e_i f(w_g) + [1 - \alpha p(e_i)]f(w_b) - xe_i \geq u° \tag{4}$$

$$\alpha p'(e_i)[f(w_g) - f(w_b)] = x.^{9} \tag{5}$$

As shown by Grossman-Hart (1983), for the preference ordering assumed here, at the optimum both constraints bind. The following conditions for a minimum are found:

$$f(w_g) - u° - xe_i - \frac{[1 - \alpha p(e_i)]x}{\alpha p'(e_i)} = 0 \tag{6}$$

$$f(w_b) - u° - xe_i + \frac{p(e_i)x}{p'(e_i)} = 0 \tag{7}$$

Since $f(w)$ is monotonically increasing in w, it has an inverse function $f^{-1}(\cdot)$. Therefore the solution of the minimisation problem is the following:

$$w_g^* = f^{-1}\left\{ u° + xe_i + \frac{[1 - \alpha p(e_i)]x}{\alpha p'(e_i)} \right\} \tag{8}$$

$$w_b^* = f^{-1}\left\{ u° + xe_i - \frac{p(e_i)x}{p'(e_i)} \right\}^{10} \tag{9}$$

The expected value of the remuneration is:

$$E[w^*(e_i, \alpha, x, u^\circ)] = \alpha p(e_i) f^{-1}\left\{ u^\circ + xe_i + \frac{[1-\alpha p(e_i)]x}{\alpha p'(e_i)} \right\}$$

$$+ [1-\alpha p(e_i)] f^{-1}\left\{ u^\circ + xe_i - \frac{p(e_i x)}{p'(e_i)} \right\} \tag{10}$$

The manager will get w_g^* if costs are low and w_b^* if costs are high. The incentive scheme in (10), which implements any given effort level at minimum cost, is a function of the manager's ability, α, his disutility of effort, x, and his reservation utility, u°. The structure of the game implies that the manager chooses between a set of lotteries. These lotteries differ between each other just for their probability distributions, determined by the level of effort chosen, and not for the size of the prizes (i.e. the levels of remuneration), because w_g^* and w_b^* have already been fixed by the owner when the manager's choice actually occurs.

From (10) it is easy to see that $E(w^*) > f^{-1}(u^\circ + xe_i)$ if $f(w)$ is strictly concave. Therefore, the cost of effort will be larger with asymmetric information than in the case when the owner observes effort and just provides the manager his reservation utility through a fixed remuneration. $E(w^*)$ is monotonically increasing in e_i, since

$$\frac{\partial E(w^*)}{\partial e_i} = \alpha p'(e_i(w_g^* - w_g^*) + \alpha p(e_i[1-\alpha p(e_i)] \frac{xp''(e_i)}{\alpha [p'(e_i)]^2}$$

$$\times \left\{ f^{-1\prime}\left\{ u^\circ + xe_i - \frac{p(e_i)x}{p'(e_i)} \right\} \right.$$

$$\left. - f^{-1\prime}\left\{ u^\circ + xe_i + \frac{[1-\alpha p(e_i)]x}{\alpha p'(e_i)} \right\} \right\} > 0 \tag{10a}$$

because $p(e_i$ is concave and $f^{-1}(\cdot)$ is monotonically increasing. It is also easy to see that in order to increase the effort provided by the manager it is necessary to increase the spread between the payments contingent on the two levels of marginal cost, since

$$\frac{\partial w_g^*}{\partial e_i} = -[1-\alpha p(e_i)] \frac{xp''(e_i)}{\alpha [p'(e_i)]^2} f^{-1\prime}\left\{ u^\circ + xe_i + \frac{[1-ap(e_i)]x}{\alpha p'(e_i)} \right\} > 0 \tag{10b}$$

while

$$\frac{\partial w_b^*}{\partial e_i} = \alpha p(e_i) \frac{xp''(e_i)}{\alpha[p'(e_i)]^2} f^{-1\prime} \left\{ u^\circ + xe_i - \frac{\alpha p(e_i)x}{\alpha p'(e_i)} \right\} < 0 \tag{10c}$$

However, we cannot generally establish whether the marginal cost of effort is everywhere larger under asymmetric information, so that in equilibrium we might have either under- or over-provision of effort. Furthermore, we cannot generally say anything about the convexity in e_i *of* $E(w^*)$; in the analysis that follows for simplicity we assume that $E(w^*)$ is well-behaved.

III. EQUILIBRIUM OUTPUT AND EFFORT PROVISION WITH STRATEGIC DELEGATION

So far, only the shape of the incentive scheme which implements any given effort level at minimum cost has been calculated. However, in order to determine the actual value for w_g^* and w_b^*, the owners have to choose the level of the manager's effort they want to implement. This choice completes the effort stage of the game.

We have assumed that each manager's preferences and remuneration scheme are observable. Each manager forms his expectations about the rival's marginal cost on the basis of that observation.[11] Since the managers' effort decreases the probability of their respective firms having a high marginal cost, the choice of the effort level has a strategic value, influencing the outcome of the output-setting game. The owner takes into account that strategic effect of his decision about his manager's effort. The game is solved starting from the output stage and moving backwards. The appropriate solution concept to be used is sub-game perfect equilibrium.

The decision maker in the second stage is the manager of the firm. He is indifferent between the levels of output: hence we assume, as it is often done in the literature, that he will choose in accordance with the owner's will and maximise profits. When he chooses the level of output, the manager observes his own firm's marginal cost and thereby knows what his remuneration is going to be. In his choice, uncertainty enters only with respect to the rival firm's costs structure. Each manager does not know whether he is playing a high or a low costs firm, i.e. he does not know his opponent's type.

Hence, in this stage we have a game of incomplete information, in which the beliefs depends on the decisions about effort taken in the previous stage. The game is solved as if there were four players, one for each type. The appropriate equilibrium concept to apply is the Bayesian equilibrium. Each manager faces either of two different objective functions according to the level of his firm's marginal cost. Both are weighted sum of the objective functions he would face if he knew the actual value of the rival's marginal cost. The maximisation problem for manager i is:

$$\underset{q_i}{\text{Max}}\ \pi_{ig} = \alpha_j p(e_j)[Y(q_i, q_{jg})q_i] + [1 - \alpha_j p(e_j)][Y(q_i, q_{jb})q_i] - w^*_{ig} - c_{ig}q_i \qquad i, j = 1, 2 \quad i \neq j \quad (11)$$

if the marginal cost is c_{ig} and

$$\underset{q_i}{\text{Max}}\ \pi_{ib} = \alpha_j p(e_j)[Y(q_i, q_{jg})q_i] + [1 - \alpha_j p(e_j)][Y(q_i, q_{jb})q_i] - w^*_{ib} - c_{ib}q_i \qquad i, j = 1, 2 \quad i \neq j \quad (11a)$$

if the marginal cost is c_{ib}.

Here π is the expected profit and $Y(q_i, q_j)$ is the inverse demand function,

$$\frac{\partial Y}{\partial q_i} = \frac{\partial Y}{\partial q_j} < 0.$$

The simultaneous solutions of the four maximisation problems, two for each player, gives us the Bayesian equilibrium of this stage game. Four first-order conditions like the following are obtained:

$$\frac{\partial \pi_{ig}}{\partial q_i} = \alpha_j p(e_j)[Y'_{q_i}(q_i, q_{jg})q_i + Y_{q_i}(q_i, q_{jg}) - c_{ig}] + [1 - \alpha_j p(e_j)][Y'_{q_i}(q_i, q_{jb})q_i + Y_{q_i}(q_i, q_{jb}) - c_{ib}] = 0 \qquad (12)$$

The reaction functions for the choice of output made by the four types of firms will be:

$$q_{ig} = \hat{q}_{ig}(q_{jg}, q_{jb}, \alpha_j, e_j) \qquad i, j = 1, 2 \quad i \neq j$$
$$q_{ib} = \hat{q}_{ib}(q_{jg}, q_{jb}, \alpha_j, e_j) \qquad i, j = 1, 2 \quad i \neq j \qquad (13)$$

All the equilibrium values are functions of α_i, e_i, α_j, e_j:

$$q^*_{ig} = \hat{\hat{q}}_{ig}(\alpha_i, e_i, \alpha_j, e_j) \qquad i,j = 1, 2 \quad i \neq j$$
$$q^*_{ib} = \hat{\hat{q}}_{ib}(\alpha_i, e_i, \alpha_i, e_j)^{12} \qquad i, j = 1, 2 \quad i \neq j \qquad (14)$$

Therefore the output level of a firm will depend on the effort levels and the ability of its own and the other firm's managers.

Now we move back to the first stage of the game in order to analyse how the managers' effort levels are chosen. When they choose the incentive schemes, the owners do not know their own firms' marginal cost and, as a consequence, which game their managers are playing in the second stage. Therefore, they maximise the following expected profit function, including the two alternative continuations weighted by the objective probability distribution on them:

$$\begin{aligned}\max_{e_i} \pi_i = {} & \alpha_i p(e_i)\{\alpha_j p(e_j)[Y(q^*_{ig_i}, q^*_{jg})q^*_{ig}] \\ & + [1-\alpha_j p(e_j)][Y(q^*_{ig_i}, q^*_{jb})q^*_{ig}] - w^*_{ig} - c_{ig}q^*_{ig}\} \\ & + [1-\alpha_i p(e_i)]\{\alpha_j p(e_j)[Y(q^*_{ib_i}, q^*_{jg})q^*_{ib}] \\ & + [1-\alpha_j p(e_j)][Y(q^*_{ib_i}, q^*_{jb})q^*_{ib}] - w^*_{ib} - c_{ib}q^*_{ib}\} \quad i,j=1,2 \quad i\neq j \end{aligned} \tag{15}$$

The simultaneous solution of two maximisation problems like this (one for each firm) gives us the equilibrium of the effort-setting game.

The reaction functions for the choice of the manager's effort are:

$$e_i = e_i^{\wedge}(e_j, \alpha_i, \alpha_j) \qquad i,j=1,2 \quad i\neq j \tag{16}$$

The equilibrium value for the effort level will be a function of both managers' ability:

$$e_i^* = e_i^{\wedge\wedge}(\alpha_i, \alpha_j) \qquad i,j=1,2 \quad i\neq j \tag{17}$$

Through a simple inspection, we note that equation (15) is just the sum of the maximum value functions corresponding to the output-setting game in equations (11) and (11a), weighted by the probability distribution over the two states of the world. Hence, we can write (15) as follows:

$$\max_{e_i} \pi_i = \alpha_i p(e_i)\pi^*_{ig} + [1-\alpha_i p(e_i)]\pi^*_{ib} \qquad i,j=1,2 \quad i\neq j \tag{18}$$

Differentiating with respect to e_i we get:

$$\frac{\partial \pi_i}{\partial e_i} = \alpha_i p'(e_i)(\pi^*_{ig} - \pi^*_{ib}) + \alpha_i p(e_i)\left(\frac{\partial \pi^*_{ig}}{\partial e_i} - \frac{\partial \pi^*{}_{ib}}{\partial e_i}\right) + \frac{\partial \pi^*_{ib}}{\partial e_i} = 0 \tag{19}$$

where

$$\frac{\partial \pi^*_{ig}}{\partial e_i} = \frac{\partial \pi^*_{ig}}{\partial q^*_{ig}}\frac{\partial q^*_{ig}}{\partial e_i} + \frac{\partial \pi^*_{ig}}{\partial q^*_{jg}}\frac{\partial q^*_{jg}}{\partial e_i} + \frac{\partial \pi^*_{ig}}{\partial q^*_{jb}}\frac{\partial q^*_{jb}}{\partial q_i} - \frac{\partial w^*_{ig}}{\partial e_i} \tag{20a}$$

and

$$\frac{\partial \pi^*_{ib}}{\partial e_i} = \frac{\partial \pi^*_{ib}}{\partial q^*_{ib}}\frac{\partial q^*_{ib}}{\partial e_i} + \frac{\partial \pi^*_{ib}}{\partial q^*_{jg}}\frac{\partial q^*_{jg}}{\partial e_i} + \frac{\partial \pi^*_{ib}}{\partial q^*_{jb}}\frac{\partial q^*_{jb}}{\partial e_i} - \frac{\partial w^*_{ib}}{\partial e_i} \tag{20b}$$

Taking into account that both $\frac{\partial \pi^*_{ig}}{\partial q^*_{ig}}$ and $\frac{\partial \pi^*_{ib}}{\partial q^*_{ib}}$ are equal to zero for the envelope theorem and substituting (20a) and (20b) into (19) we get:

$$\frac{\partial \pi_i}{\partial e_i} = \alpha_i p'(e_i(\pi^*_{ig} - \pi^*_{ib}) + \alpha_i p(e_i)\left(\frac{\partial w^*_{ib}}{\partial e_i} - \frac{\partial w^*_{ig}}{\partial e_i}\right) - \frac{\partial w^*_{ib}}{\partial e_i}$$

$$+ \alpha_i p(e_i)\left(\frac{\partial \pi^*_{ig}}{\partial q^*_{jg}}\frac{\partial q^*_{jg}}{\partial e_i} + \frac{\partial \pi^*_{ig}}{\partial q^*_{jb}}\frac{\partial \pi^*_{jb}}{\partial e_i}\right)$$

$$+ [1 - \alpha_i p(e_i)]\left(\frac{\partial \pi^*_{ib}}{\partial q_{jg}}\frac{\partial q^*_{jg}}{\partial e_i} + \frac{\partial \pi^*_{ib}}{\partial q^*_{jb}}\frac{\partial q^*_{jb}}{\partial e_i}\right) = 0 \qquad (21)$$

Equation (21) characterises the solution of the effort-setting game when the incentive scheme and, as a consequence, the level of effort can be chosen strategically.

IV. THE EFFORT PROVISION UNDER ALTERNATIVE SETTINGS

In this section we compare the equilibrium of Game 1, with a few alternative settings, where the original model is slightly modified. This allows us to consider the effect on the equilibrium of some of the main features of the model, like the presence of moral hazard, the possibility of delegation and of a strategic use of it, the disciplinary role of product market competition.

IV.1. No Moral Hazard

First, we compare the equilibrium of the previous section with the first best with strategic delegation, i.e. with the case when observable managerial contracts can be made contingent on *e*, since managerial effort is verifiable and so there is no moral hazard. The optimal contract contingent on *e* provides the manager with just his reservation utility, so that $w^*_i = u^\circ + xe_i$ and the first order condition for the effort setting game is

$$\frac{\partial \pi_i}{\partial e_i} = \alpha_i p'(e_i)(\pi^*_{ig} - \pi^*_{ib}) + \alpha_i p(e_i)\left(\frac{\partial \pi^*_{ig}}{\partial q^*_{jg}}\frac{\partial q^*_{jg}}{\partial e_i} + \frac{\partial \pi^*_{ig}}{\partial q^*_{jb}}\frac{\partial q^*_{jb}}{\partial e_i}\right)$$

$$+ [1 - \alpha_i p(e_i)]\left(\frac{\partial \pi^*_{ib}}{\partial q^*_{jg}}\frac{\partial q^*_{jg}}{\partial e_i} + \frac{\partial \pi^*_{ib}}{\partial q^*_{jb}}\frac{\partial q^*_{jb}}{\partial e_i}\right) - x = 0 \tag{22}$$

The effort levels that satisfy (21) and (22) are generally different, so that the equilibrium effort of Game 1 is not first best; however, as noted above, we cannot generally establish whether (21) implies under- or over-provision of effort.

IV.2. Unobservable Contracts and Non-Strategic Delegation

Then we move to consider a different setting, where strategic delegation is not an available option, because the contract selected in one firm is not observable by the rival.

Game 2 – Unobservable Contracts and Non-Strategic Delegation
Stage 1. The owner of each firm offers to his *manager* a contract, not observable to their rivals, which establishes a set of levels of remuneration contingent on the marginal cost that will be observed later: $\{(w_b, c_b), (w_g, c_g)\}$.
Stage 2. Each manager decides whether to accept the offer and, if he does accept, selects his level of effort.
Stage 3. Each manager observes his own firm's marginal cost and chooses the level of output.
Stage 4. Each owner observes his own firm's *marginal* cost and pays the remuneration to his manager.

As a consequence of contracts being unobservable, in the output setting game each manager rationally foresees that the rival's effort will be the equilibrium effort of the principal-agent game played without taking strategic consideration into account,[13] i.e. solving the following equilibrium condition:

$$\frac{\partial \pi_i}{\partial e_i} = \alpha_i p'(e_i)(\pi^*_{ig} - \pi^*_{ib}) + \alpha_i p(e_i)\left(\frac{\partial w^*_{ib}}{\partial e_i} - \frac{\partial w^*_{ig}}{\partial e_i}\right) - \frac{\partial w^*_{ib}}{\partial e_i}$$

$$= \alpha_i p'(e_i)[(R^*_{ig} - R^*_{ib}) - (q^*_{ig}c_{ig} - q^*_{ib}c_{ib}) - (w^*_{ig} - w^*_{ib})] - \alpha_i p(e_i)$$

$$\times \left(\frac{\partial w^*_{ig}}{\partial e_i} - \frac{\partial w^*_{ib}}{\partial e_i} \right) - \frac{\partial w^*_{ib}}{\partial e_i} = 0 \tag{23}$$

where R^*_{ig} and R^*_{ib} are the expected revenues contingent to respectively the good and the bad state of the world.

Now we are ready to prove the following proposition.

Proposition 1. If the products of the two firms are strategic substitutes and the equilibrium level of output decreases in the rival manager's effort, strategic delegation implies an equilibrium level of effort larger than in the case of non-strategic delegation.

Proof. By comparing equations (21) and (23) we notice that (21) contains the following extra term:

$$\alpha_i p(e_i) \left(\frac{\partial \pi^*_{ig}}{\partial q^*_{jg}} \frac{\partial q^*_{jg}}{\partial e_i} + \frac{\partial \pi^*_{ig}}{\partial q^*_{jb}} \frac{\partial q^*_{jb}}{\partial e_i} \right) + [1 - \alpha_i p(e_i)] \left(\frac{\partial \pi^*_{ib}}{\partial q^*_{jg}} \frac{\partial q^*_{jg}}{\partial e_i} + \frac{\partial \pi^*_{ib}}{\partial q^*_{jb}} \frac{\partial q^*_{jb}}{\partial e_i} \right).$$

This is positive because the products are strategic substitutes and the levels of output decrease in the rival's effort. Therefore, for concavity of π_i *in* e_i, e^*_i is greater when delegation is strategic than otherwise. (Q.E.D.)

Therefore, when the owner of one firm is able to use strategically his manager's incentive contract, he will select, caeteris paribus, an equilibrium level of effort higher than otherwise, decreasing the expected level of marginal cost. However, it is important to notice that strategic delegation equilibrium implies a level of the manager's effort that does not minimise total costs, as it would be the case were the owners playing.

IV.3. No Delegation

By delegating the output decision to his manager, the owner takes advantage of the manager's capability to distinguish between different states of nature in order to increase revenues. When the choice of output cannot be delegated, so that the firm is owner-managed and the agent we have called 'manager' so far only acts as a supervisor of production, the game becomes the following.

Game 3 – Owner-Managed Firm (No Delegation)
Stage 1. The owner of each firm offers to the supervisor of production a contract, not observable to their rivals, which establishes a set of levels of remuneration contingent on the marginal cost that will be observed later: $\{(w_b, c_b), (w_g, c_g)\}$.

Stage 2. Each supervisor decides whether to *accept* the offer and, if he does accept, selects his level of effort.
Stage 3. Each owner selects output, before *observing* the realisation of marginal cost.
Stage 4. Each owner observes his own firm's marginal cost and pays the remuneration to his supervisor.

In this case the f.o.c. for the effort-setting game only implies total expected costs minimisation, since the level of effort does not have any effect on the choice of output, apart from that going through the expected value of marginal cost:

$$\frac{\partial \pi_i}{\partial e_i} = -\alpha_i p'(e_i)[(w^*_{ig} - w^*_{ib})] - \alpha_i p(e_i)\left(\frac{\partial w^*_{ig}}{\partial e_i} - \frac{\partial w^*_{ib}}{\partial e_i}\right)$$

$$-\frac{\partial w^*_{ib}}{\partial e_i} - \alpha_i p'(e_i) q^*_i (c_{ig} - c_{ib}) = 0 \tag{24}$$

As a term of comparison we consider the game with non-strategic delegation and re-write for convenience the f.o.c. for that case:

$$\frac{\partial \pi_i}{\partial e_i} = \alpha_i p'(e_i)[(R^*_{ig} - R^*_{ib}) - (q^*_{ig} c_{ig} - q^*_{ib} c_{ib})] - \alpha_i p'(e_i)[(w^*_{igh} - w^*_{ib})]$$

$$-\alpha_i p(e_i)\left(\frac{\partial w^*_{ig}}{\partial e_i} - \frac{\partial w^*_{ib}}{\partial e_i}\right) - \frac{\partial w^*_{ib}}{\partial e_i} = 0 \tag{23}$$

Now we are ready to prove the following proposition.

Proposition 2. The delegation of the output decision to an informed manager implies a higher equilibrium level of effort than the corresponding case without delegation.

Proof. By comparing equation (23) and (24), we notice that the left hand side of (23) is larger than the left hand side of (24), since the difference in profit gross of the manager's remuneration between the good and the bad state of nature when an informed manager chooses q_i exceed the difference in production costs between the two states when a uninformed owner makes the decision. The result follows from concavity of the expected profit function. (Q.E.D.)

Therefore, the use of delegation enhances the profitability of decreasing the firm's marginal cost. As a consequence the equilibrium level of marginal cost decreases.

The fact that the level of effort of one firm's manager increases when his incentive scheme is used strategically or delegation takes place, does not necessarily mean that the same happens for all firms in the industry, when all the owners behave in the same way. When the levels of effort are strategic substitutes, as we have assumed for Proposition 1, the outcome will depend on which between the own and the cross effect prevails. Proposition 3, which follows the line of reasoning of the paper by Brander & Spencer (1983) on R&D rivalry, states the conditions for the net effect to be positive.[14]

Proposition 3. If the firms are symmetric and the equilibrium in the effort setting game is stable:

- the equilibrium level of effort of each firm's manager in the strategic delegation game is greater than in the non-strategic one;
- the equilibrium level of effort of each firm's manager in the game with delegation is greater than without delegation.

Proof. (see the Appendix).

Therefore delegation, strategic and non-strategic decrease, the expected level of marginal cost of all the firms in the industry.

IV.4. Product Market Competition and Agency Costs

Before closing this section we add a final remark on the relations between product market competition, strategic delegation and agency costs. The literature does not always provide support to the conventional wisdom argument, according to which product markets discipline firms, so reducing agency costs.[15] In our framework the second best equilibrium of the moral hazard game described in Section II implies a distortion in the level of effort, whose sign cannot be unequivocally established, so that we might have either under – or over-provision of effort. When moral hazard implies under-provision of effort, Proposition 3 has an important corollary: strategic delegation tends to counterbalance the x-inefficiency due to agency costs.

Corollary 4. In a duopolistic market where the products are strategic substitutes, strategic delegation of the output decision to a manager reduces the under-provision of managerial effort due to moral hazard, if that occurs.

Finally, we investigate how the equilibrium effort is affected by whether or not the firms relax product market competition by colluding to fully extract the potential joint surplus of the market. The f.o.c. for the choice of effort of a colluding firm would be

$$\frac{\partial \pi_i}{\partial e_i} = \alpha_i p'(e_i)[(\pi_{ig}^c - \pi_{ib}^c)] - \alpha_i p(e_i)\left(\frac{\partial w_{ig}^*}{\partial e_i} - \frac{\partial w_{ib}^*}{\partial e_i}\right) - \frac{\partial w_{ib}^*}{\partial e_i} = 0 \quad (25)$$

where π_{ig}^c and π_{ib}^c are the profit of collusion when the marginal costs are respectively low and high. If we compare (25) with the f.o.c. of the game with strategic delegation (21), we notice that (21) contains the same extra-term as for the case of non-strategic delegation. In fact, colluding firms have no incentive to over-invest to discourage rivals. Furthermore, in (25) we have $\alpha_i p'(e_i)[(\pi_{ig}^c - \pi_{ib}^c)]$ in place of $\alpha_i p'(e_i)[(\pi_{ig}^* - \pi_{ib}^*)]$ in (21). Then, we can establish the following lemma.

Lemma 5. The equilibrium level of effort of the managers is lower when firms collude than without collusion if $\alpha_i p'(e_i)[(\pi_{ig}^c - \pi_{ib}^c)] < \alpha_i p'(e_i)[(\pi_{ig}^* - \pi_{ib}^*)]$.

Proof. If $\alpha_i p'(e_i)[(\pi_{ig}^c - \pi_{ib}^c)] < \alpha_i p'(e_i)[(\pi_{ig}^* - \pi_{ib}^*)]$ the left hand side of (25) is always smaller than the left hand side of (21). The result follows from concavity of the profit function. (Q.E.D.)

A simple setting where the condition in Lemma 5 is satisfied is the case of linear demand with symmetric costs, as shown in the following section. Lemma 5 shows that when agency costs imply under-provision of effort, an increase of competitiveness of the product market, represented by the removal of collusion, disciplines firms and reduces x-inefficiencies.

V. THE LINEAR DEMAND CASE

In order to get some further insights, we consider the simple case of linear demand:

$$Y(q_i, q_j) = a - q_i - q_j. \quad (26)$$

The first order conditions for firm i in the two states of nature are

$$\frac{\partial \pi_{ig}}{\partial q_i} = \alpha_j p(e_j)(a - 2q_{ig} - q_{jg} - c_{ig}) + [1 - \alpha_j p(e_j)](a - 2q_{ig} - q_{jb} - c_{ig}) = 0 \quad (27a)$$

$$\frac{\partial \pi_{ib}}{\partial q_i} = \alpha_j p(e_j)(a - 2q_{ib} - q_{jg} - c_{ib}) + [1 - \alpha_j p(e_j)](a - 2q_{ib} - q_{jb} - c_{ib}) = 0 \quad (27b)$$

Hence, the reaction functions are:

$$q_{ig} = \frac{a - c_{ig} - \alpha_j p(e_j)q_{jg} + [1 - \alpha_j p(e_j)]q_{jb}}{2} \quad (28a)$$

$$q_{ib} = \frac{a - c_{ib} - \alpha_j p(e_j)q_{jg} + [1 - \alpha_j p(e_j)]q_{jb}}{2} \quad (28b)$$

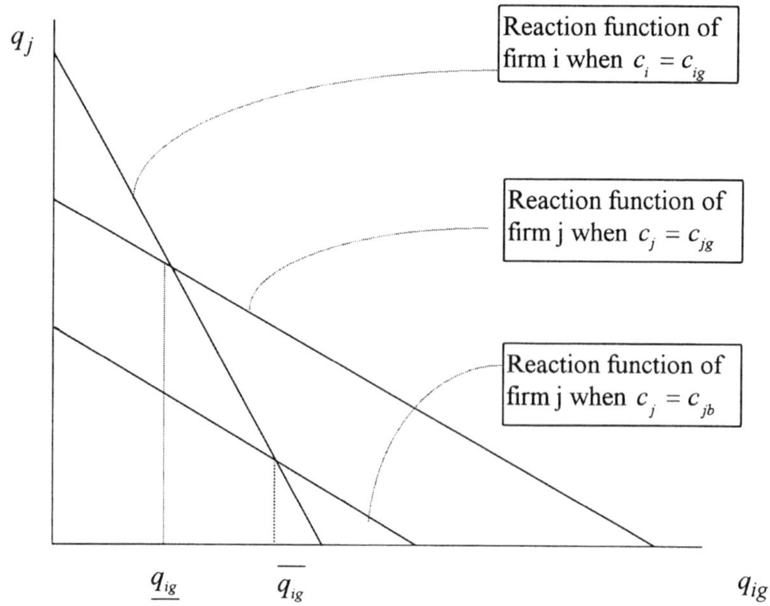

Fig. 1. Reaction functions for the output setting game.

which are linear in the expected value of the rival's output.

Figure 1 below illustrates the situation facing the manager of firm i's when he plays the output setting game, $c_i = c_{ig}$ and he is uncertain on the actual value of firm j's cost between c_{jg} and c_{jb}.

The equilibrium output of firm *i* depends on its manager's expectations on the value of q_j and will be in the interval between $\underline{q_{ig}}$ and $\overline{q_{ig}}$. Straightforward comparative statics on the reaction functions shows that the output of both firms in all states of nature are inversely related to the rival manager's ability and effort.

The equilibrium values for the output-setting game are the following:

$$q^*_{ig} = \frac{a + \alpha_i p(e_i)\left(\dfrac{c_{ib} - c_{ig}}{2}\right) - \dfrac{1}{2}c_{ib} - \dfrac{3}{2}c_{ig} - 2\alpha_j p(e_j)\left(\dfrac{c_{jb} - c_{jg}}{2}\right) + c_{jb}}{3} \tag{29a}$$

$$q^*_{ib} = \frac{a + \alpha_i p(e_i)\left(\dfrac{c_{ib} - c_{ig}}{2}\right) - 2c_{ib} - 2\alpha_j p(e_j)\left(\dfrac{c_{jb} - c_{jg}}{2}\right) + c_{jb}}{3} \tag{29b}$$

The explicit solution of the output-setting game allows us to prove the three following Lemmas.

Lemma 6. If the demand is linear, the managers' effort levels are strategic substitutes.

Proof. (see the Appendix).

Lemma 7. If the demand is linear, a rise in the manager's effort increases the equilibrium levels of output of his own firm and decreases the levels of output of the rival.

Proof. Direct from (29a) and (29b).

Lemmas 6 and 7 show that the linear demand case satisfies the conditions for Proposition 1 to hold.

Lemma 8. If the demand is linear, a rise in the manager's ability increases the equilibrium levels of output of his own firm and decreases the levels of output of the rival.

Proof. Direct from (29a) and (29b).

Lemma 8 shows the consequences of the choice of the manager's ability on his own and the rival firm's behaviour. In particular, given the existence of a negative cross effect, we can see that also the choice of the manager's ability can be used strategically.

Then, we consider the total effect on expected output of the strategic behaviour of both firms. For Proposition 3 we know that the equilibrium level of effort of both firms is greater when they behave strategically than otherwise. Hence, in order to evaluate the whole effect of the strategic behaviour on total industry expected output is enough to inspect the effect of a change in just one firm's manager effort level.

Proposition 9. If the demand is linear, when all firms choose their manager's effort strategically, the expected total industry output increases.

Proof. From (29a) and (29b) it is easy to see that

$$\frac{\partial E(q_i^*)}{\partial e_i}+\frac{\partial E(q_j^*)}{\partial e_i}=\alpha_i p'(e_i)\left[\frac{2(c_{ib}-c_{ig})}{3}-\frac{(c_{ib}-c_{ig})}{3}\right]=\alpha_i p'(e_i)\left(\frac{c_{ib}-c_{ig}}{3}\right)>0. \tag{30}$$

The same holds for the change in e_j.

Since the net result of the strategic behaviour of both firms is an increase in both e_i and e_j, the total variation in the industry output will be positive (Q.E.D.).

It is interesting to note that in the symmetric case, when $c_{ig} = c_{jg}$, $c_{ib} = c_{jb}$ and $\alpha_j = \alpha_i$, the expected output increase due to strategic over-investment in effort depends on the good state of nature becoming more likely, while in each state the equilibrium output is lower for both firms. This is implied by the following lemma.

Lemma 10. In the symmetric case, if the demand is linear, the equilibrium output in each state of nature is smaller when the managers' effort levels are chosen strategically than otherwise.

Proof. From (29a) and (29b) we know that

$$\frac{\partial q^*_{ig}}{\partial e_i} + \frac{\partial q^*_{ig}}{\partial e_j} = \frac{\alpha_i p'(e_i)\dfrac{(c_{ib} - c_{ig})}{2} - 2\alpha_j p'(e_j)\left(\dfrac{c_{jb} - c_{jg}}{2}\right)}{3} \tag{31}$$

In the symmetric case this simplifies to

$$\frac{\partial q^*_{ig}}{\partial e_i} + \frac{\partial q^*_{ig}}{\partial e_j} = -\frac{\alpha p'(e)\dfrac{(c_b - c_g)}{2}}{3} < 0. \tag{32}$$

We have already shown that equilibrium effort levels are larger in the strategic equilibrium than otherwise. Therefore, q^*_{ig} is smaller in the strategic equilibrium than otherwise. The same holds for q^*_{ib}, q^*_{jg} and q^*_{jib}. (Q.E.D.)

In the strategic delegation case the equilibrium output in each state of nature is lowered by the increase in the expectations that the rival is in the good state due to the observation of a contract that implies over-investment in effort. This forces the managers to exercise restraint and reduce the output their firms produce. Given the reduction in output in each state of nature, the increase in the expected level of output in the strategic delegation case entirely depends on the increased probability of being in the good states.

An immediate corollary of Lemma 10 is the following.

Corollary 11. In a Cournot duopoly with incomplete information the equilibrium price in each state of nature will be higher in the strategic delegation equilibrium than otherwise, i.e. it will be pushed towards monopoly price.

When the realisation of the shocks on costs in the two firms is the same, this directly implies that total revenues gross of managerial compensation are higher in both firms; while the effect on profit is ambiguous.

The effect of strategic delegation on consumer welfare is also ambiguous. We have shown that total industry expected output is higher in the strategic delegation equilibrium than otherwise. However, total output in each state of nature is lower, while with linear demand consumer surplus is convex in output.

The present framework of incomplete information in the output setting game and related results are to be compared with models of strategic delegation under complete information like those in Fershtman-Judd (1987) and Sklivas (1987). In those models strategic delegation with Cournot competition increases the equilibrium output and so decreases the equilibrium price and consumer surplus. We have seen that under incomplete information the reverse may hold.

Finally, we can easily show that with linear demand the condition in Lemma 5 is satisfied in the symmetric case, and so collusion has an effort-reducing effect on managers.

Lemma 12. If the demand is linear and costs are symmetric $\alpha_i p'(e_i)[(\pi^c_{ig} - \pi^c_{ib})] < \alpha_i p'(e_i)[(\pi^*_{ig} - \pi^*_{ib})]$.

Proof. When marginal costs are symmetric the most obvious division of market shares between colluding firms would allocate half of the production to each. Each firm's profit would be half of the monopoly profit. When demand is like in (26), the derivative of firm i's profit with respect to c_i in the symmetric case is:

$$\frac{\partial \pi^*_i}{\partial c_i} = -\frac{4}{9}(a - 2c_i + c_j) = -\frac{4}{9}(a - c) \tag{33}$$

without collusion and

$$\frac{\partial \pi^c_i}{\partial c_i} = -\frac{1}{4}(a - c) \tag{34}$$

with collusion. Hence the variation in profit due to a shift from a bad to a good state of nature is smaller with collusion than otherwise. (Q.E.D.)

VI. CONCLUSIONS

In this chapter we have been shown that despite incomplete information in the product market the owners are able to use strategically the delegation of the output decision to managers by merely making the managers' incentive schemes observable. Strategic delegation enhances the equilibrium level of the managers' effort, decreasing all firms' marginal costs. Equilibrium effort with strategic delegation is above that minimising each firm's total cost, the level which would be selected with no delegation, i.e. were the owner playing the product market game.

Moral hazard is another source of distortion of the equilibrium effort, which can generate under- but also over-provision of effort. When the former is the case, strategic delegation has a counter-balancing effect on the loss of productivity due to agency costs. Also, an increase in the degree of competitiveness of the product market, represented by the elimination of collusion, increases equilibrium effort and then have similar effect on agency distortions.

The analysis of the linear demand case has allowed us to show some interesting features of the strategic delegation equilibrium under incomplete information. The expected value of the equilibrium industry output is higher with strategic delegation than otherwise; however, the equilibrium industry output in each state of nature is lower, so that the equilibrium price is distorted toward monopoly price. Therefore, the expected output increase is all due to the better states of nature becoming more likely. However, differently from what happens with strategic delegation under complete information in the output setting game, in this model consumers may not benefit from strategic delegation, since consumer's surplus is convex in output.

NOTES

1. Other applications concern the credibility of monetary policy (for example Rogoff (1985) and Lohmann (1992)) or bargaining through agents (Katz (1991)).
2. For a discussion of the role of observability, see Dewatripont (1988).
3. The fewer are the players, the more they become 'visible' to each others, i.e. the easier it becomes to control what is going on in the rival firms.
4. The classical references for the principal-agent literature are Mirrlees (1975), Holmstrom (1979), Grossman-Hart (1983) and Hart-Holmstrom (1987).
5. This formulation with only two possible levels of the outcome of the game follows Grossman & Hart (1983) and Kreps (1990).
6. Alternatively, α can be interpreted as an industrial sector parameter, which is larger the more intensive is the production technology in manager's effort.
7. The separability assumption implies that the manager's preferences over income are independent of his effort.
8. Risk averse with respect to the w component.
9. (4) and (5) are the usual participation and incentive constraints. Writing the incentive constraint as in (5), we apply the First Order Approach, which can be used since the function $\alpha p(e_i)$ described above satisfies both the Monotone Likelihood Ratio Condition and the Concavity of the Distribution Function Condition on this see also Jewitt (1988)). Through the incentive constraint (5) the solution of the manager's maximisation problem is incorporated in the owner's minimisation problem, whose solution is then a sub-game perfect equilibrium.
10. Note that for given w_b^*, w_g^* equations (8) and (9) can be solved for α and e_i. They are just a rewriting of the two constraints (4) and (5), that bind in equilibrium. (5) has a unique solution for e_i, whenever the two conditions mentioned in the previous footnote

are satisfied and so the first-order approach is valid; given (5) the left-hand side of (4) is monotonically increasing in α, and so has a unique solution for α. Hence the rival manager is able to ascertain the equilibrium level of effort and the ability of firm i's manager only on the basis of the observation of his contract, x, u°, and the knowledge of function $p(e_i)$.

11. When the managers' contracts are not observable, the analysis proceeds as if the choices concerning the incentive scheme and the level of output were taken simultaneously and no strategic consideration applies. I have analysed the case of unobservable agency contracts with symmetric information about costs in Merzoni (1991). The equilibrium level of effort in that case is the value which minimises total costs.

12. Given the linearity in q of the variable cost function, a sufficient condition for the existence of the equilibrium is that the demand function be weakly concave.

13. On game played through agents with unobservable contracts, see Katz (1991).

14. They consider the duopoly case, but the main result extend to oligopolies with more than two firms.

15. As noticed in the introduction, this problem is discussed in different settings by Hart (1983), Scharfstein (1988), Hermalin (1992), Martin (1993), Bertoletti & Poletti (1996).

ACKNOWLEDGMENTS

This chapter is based on Chapter 1 of my Ph.D. thesis at the Department of Economics, University of Warwick, U.K. I would like to thank Gareth Myles, Jonathan Thomas and Norman Ireland for their very useful suggestions and advice, Piergiovanna Natale and Carlo Beretta for their comments to a previous version. The responsibility for all the remaining mistakes is only mine.

REFERENCES

Basu, K. (1995). Stackelberg equilibrium in oligopoly: an explanation based on managerial incentives. *Economics Letters*, *49*, 459–464.

Bertoletti P., & Poletti, C. (1996). A Note on Endogenous Firm Efficiency in Cournot Models of Incomplete Information. *Journal of Economic Theory*, *71*, 303–310.

Brander, J., & Spencer, B. (1983). Strategic Commitment with R&D: The Symmetric Case. *The Bell Journal of Economics*, *14*, 225–235.

Bulow, J., Geanakoplos, J., & Klemperer, P. (1985). Holding Idle Capacity to Deter Entry. *Economic Journal*, *95*, 178–182.

Coase, R. H. (1937). The Nature of the Firm. *Economica*, *4*, 386–405.

Dewatripont, M. (1988). Commitment Through Renegotiation-Proof Contracts with Third Parties. *Review of Economic Studies*, *55*, 377–390.

Dixit, A. (1980). The Role of Investment in Entry Deterrence. *Economic Journal*, *90*, 95–106.

Fershtman, C., & Judd, K. (1987). Equilibrium Incentives in Oligopoly. *American Economic Review*, *77*, 927–940.

Fudenberg, D., & Tirole, J. (1983). Learning by Doing and Market Performance. *Bell Journal of Economics*, *14*, 522–530.

Fudenberg, D., & Tirole, J. (1984). The Fat Cat Effect, the Puppy Dog Ploy and the Lean and Hungry Look. *American Economic Review, Papers and Proceedings, 74*, 361–368.

Grossman, S. J., & Hart, O. D. (1983). An Analysis of the Principal-Agent Problem. *Econometrica*, *51*(1), 7–45.

Hart, O. (1983). The market mechanism as an incentive scheme. *The bell Journal of Economics*, *14*, 366–382.

Hart, O., & Holmstrom, B. (1987). The Theory of Contracts. In: T. Bewley (Ed.), *Advances in Economic Theory*.

Hermalin, B. (1992). The effects of competition on executive behavior. *Rand Journal of Economics*, *23*, 350–365.

Holmstrom, B. (1979). Moral Hazard and Observability. *The Bell Journal of Economics*, *10*, 74–91.

Jewitt, I. (1988). Justifying the First-Order Approach to Principal-Agent Problems. *Econometrica*, *56*(5), 1177–1190.

Katz, M. (1991). Game-playing agents: unobservable contracts as pre-commitments. *Rand Journal of Economics*, *22*(3), 307–328.

Kreps, D. M. (1990). *A Course in Microeconomic Theory*. New York, Harvester Wheatsheaf.

Lohmann, S. (1992). Optimal Commitment in Monetary Policy. *American Economic Review*, *82*, 273–286.

Macho-Stadler, I., & Verdier, T. (1991). Strategic Managerial Incentives and Cross Ownership Structure: A Note. *Journal of Economics*, *53*(3), 285–297.

Martin, S. (1993). Endogenous Firm Efficiency in a Cournot Principal-Agent Model. *Journal of Economic Theory*, *59*, 445–450.

Merzoni, G. (1991). Qualità dei managers e incentivi in un modello di duopolio con separazione tra proprietà e controllo. *Annali della Fondazione Einaudi – Torino*, *25*, 137–159.

Merzoni, G. (1991). Managers quality and incentives in a model of duopoly with separation between ownership and control. *Annals of Einaudi Foundation – Torino*, *25*, 137–159.

Mirrlees, J. A. (1975). The Theory of Moral Hazard and Unobservable Behaviour – Part I, Nuffield College, Oxford, mimeo. Published in 1999 on the *Review of Economic Studies*, *66*(1), 3–21.

Rogoff, K. (1985). The Optimal Degree of Commitment to an Intermediate Monetary Target'. *Quarterly Journal of Economics*, *100*(4), 1169–1190.

Scharfstein, D. (1988). Product-market competition and managerial slack. *Rand Journal of Economics*, *19*, 147–155.

Schelling, T. C. (1960). *The Strategy of Conflict*. Cambridge, Mass. Harvard University Press.

Schmalensee, R. (1983). Advertising and Entry Deterrence: An Exploratory Model. *Journal of Political Economy*, *90*, 636–653.

Sklivas, S. D.(1987). The Strategic Choice of Management Incentives. *The Rand Journal of Economics*, *18*, 452–458.

Spence, A. M. (1977). Entry, Capacity, Investment and Oligopolistic Pricing. *Bell Journal of Economics*, *8*, 534–544.

Vickers, J. (1985). Delegation and the theory of the firm. *Economic Journal, Supplement*, *53*, 138–147.

von Stackelberg, H. (1934) *Marktform und Gleichgewicht*. Vienna, Julius Springer.

APPENDIX

Proof of Proposition 3. As in Brander & Spencer (1983), in order to compare strategic and non-strategic equilibria, we apply the mean value theorem. We start writing the following expression, showing the two components of the total variation in $\frac{\partial \pi_i}{\partial e_i}$ caused by a shift from one equilibrium to another.

$$\Delta \frac{\partial \pi_i}{\partial e_i} = \frac{\partial^2 \pi_i}{\partial e_i^2} \Delta e_i + \frac{\partial^2 \pi_i}{\partial e_i \partial e_j} \Delta e_j \tag{A1}$$

where Δ indicates any difference between two equilibrium. The same kind of expression can be written for $\frac{\partial \pi_i}{\partial e_j}$, so that we obtain a two equations system, which is solved for Δe_i and Δe_j. Adding the two expressions obtained we get

$$\Delta e_i + \Delta e_j = \frac{\left(\frac{\partial^2 \pi_i}{\partial e_i^2} - \frac{\partial^2 \pi_i}{\partial e_i \partial e_j} \right) \Delta \frac{\partial \pi_j}{\partial e_j} + \left(\frac{\partial^2 \pi_j}{\partial e_j^2} - \frac{\partial^2 \pi_j}{\partial e_j \partial e_i} \right) \Delta \frac{\partial \pi_i}{\partial e_i}}{\frac{\partial^2 \pi_i}{\partial e_i^2} \frac{\partial^2 \pi_j}{\partial e_j^2} - \frac{\partial^2 \pi_i}{\partial e_i \partial e_j} \frac{\partial^2 \pi_j}{\partial e_j \partial e_i}} \tag{A2}$$

The denominator in (A2) is positive, being the determinant of the Jacobian matrix for the profit maximisations of the two firms. Furthermore, for local stability and symmetry $\frac{\partial^2 \pi_i}{\partial e_i^2} < \frac{\partial^2 \pi_i}{\partial e_i \partial e_j}$ and $\frac{\partial^2 \pi_j}{\partial e_j^2} < \frac{\partial^2 \pi_j}{\partial e_j \partial e_i}$.

$\Delta \frac{\partial \pi_i}{\partial e_i}$ and $\Delta \frac{\partial \pi_j}{\partial e_j}$ remain to be signed. $\Delta \frac{\partial \pi_i}{\partial e_i}$ can be written as the difference between the value of $\frac{\partial \pi_i}{\partial e_i}$ at the strategic $\frac{\partial \pi_i}{\partial e_i^s}$ and at the non-strategic equilibrium $\frac{\partial \pi_i}{\partial e_i^n}$:

$$\Delta \frac{\partial \pi_i}{\partial e_i} = \frac{\partial \pi_i}{\partial e_i^s} - \frac{\partial \pi_i}{\partial e_i^n} \tag{A3}$$

The first term on the right-hand side is zero for profit maximisation, while $\frac{\partial \pi_i}{\partial e_i^n}$ is positive, since at e_i^n the strategic distortion is not taken into account in the maximising choice. The same holds for $\Delta \frac{\partial \pi_j}{\partial e_j}$. Therefore, (A2) is positive. In the same way (A2) is shown to be positive for a change from an equilibrium without delegation to an equilibrium with delegation. (Q.E.D.).

Proof of Lemma 6. If the managers' effort levels are strategic substitutes, the reaction functions for the effort-setting game must be downward sloping:

$$\frac{\partial e_i}{\partial e_j}=\frac{\dfrac{\partial^2\pi_i}{\partial e_i\partial e_j}}{\dfrac{\partial^2\pi_i}{\partial e_i^2}}<0. \tag{A4}$$

We know that $\frac{\partial^2\pi_i}{\partial e_i^2}$ is negative, being a second-order condition for profit maximisation, while $\frac{\partial^2\pi_i}{\partial e_i\partial e_j}$ remains to be signed. In the linear demand case, taking into account that $\frac{\partial Y}{\partial q_{jb}}=\frac{\partial Y}{\partial q_{jg}}=\frac{\partial Y}{\partial q_j}$ and $\frac{\partial q_{jb}}{\partial e_i}=\frac{\partial q_{jg}}{\partial e_i}=\frac{\partial q_j}{\partial e_i}$, $\frac{\partial \pi_i}{\partial e_i}$ may be written as follows:

$$\frac{\partial \pi_i}{\partial e_i}=\alpha_i p'(e_i)(\pi^*_{ig}-\pi^*_{ib})$$

$$+\left\{\alpha_i p(e_i)\left(q_{ig}\frac{\partial Y}{\partial q_j}\frac{\partial q_j}{\partial e_j}-q_{ib}\frac{\partial Y}{\partial q_j}\frac{\partial q_j}{\partial e_i}-\frac{\partial w^*_{ig}}{\partial e_i}+\frac{\partial w^*_{ib}}{\partial e_i}\right)+q_{ib}\frac{\partial Y}{\partial q_j}\frac{\partial q_j}{\partial e_i}-\frac{\partial w^*_{ib}}{\partial e_i}\right\} \tag{A5}$$

Therefore,

$$\frac{\partial^2\pi_i}{\partial e_i\partial e_j}=\alpha_i p'(e_i)\left(\frac{\partial \pi^*_{ig}}{\partial e_j}-\frac{\partial \pi^*_{ib}}{\partial e_j}\right)$$

$$+\left\{\alpha_i p(e_i)\left(\frac{\partial q_{ig}}{\partial e_j}\frac{\partial Y}{\partial q_j}\frac{\partial q_j}{\partial e_i}-\frac{\partial q_{ib}}{\partial e_j}\frac{\partial Y}{\partial q_j}\frac{\partial q_j}{\partial e_i}-\frac{\partial^2 w^*_{ig}}{\partial e_i\partial e_j}+\frac{\partial^2 w^*_{ib}}{\partial e_i\partial e_j}\right)\right.$$

$$\left.+\frac{\partial q_{ib}}{\partial e_j}\frac{\partial Y}{\partial q_j}\frac{\partial q_j}{\partial e_i}-\frac{\partial^2 w^*_{ib}}{\partial e_i\partial e_j}\right\} \tag{A6}$$

From (8) and (9) we know that $\frac{\partial^2 w^*_{ig}}{\partial e_i\partial e_j}=\frac{\partial^2 w^*_{ib}}{\partial e_i\partial e_j}=0$, while from *(29a)* and *(29b)* we know that $\frac{\partial q_{ig}}{\partial e_j}\frac{\partial Y}{\partial q_j}\frac{\partial q_j}{\partial e_i}=\frac{\partial q_{ib}}{\partial e_j}\frac{\partial Y}{\partial q_j}\frac{\partial q_j}{\partial e_i}<0$. Furthermore, in the linear demand case, given that $\frac{\partial q_{ig}}{\partial e_j}=\frac{\partial q_{jb}}{\partial e_j}$, the following hold

$$\frac{\partial \pi^*_{ig}}{\partial e_j} = \alpha_j p'(e_j)\{[Y(q^*_{jg}, q_{jg})q^*_{ig} - w^*_{ig} - c_{ig}q^*_{ig}] - [Y(q^*_{ig}, q_{jb})q^*_{ig} - w^*_{ig} - c_{ig}q^*_{ig}]\} \tag{A7}$$

$$\frac{\partial \pi^*_{ib}}{\partial e_j} = \alpha_j p'(e_j)\{[Y(q^*_{ib}, q_{jg})q^*_{ib} - w^*_{ib} - c_{ib}q^*_{ib}] - [Y(q^*_{ib}, q_{jb})q^*_{ib} - w^*_{ib} - c_{ib}q^*_{ib}]\} \tag{A8}$$

The difference between $\frac{\partial \pi^*_{ig}}{\partial e_j} - \frac{\partial \pi^*_{ib}}{\partial e_j}$ is negative since $\frac{\partial^2 \pi_i}{\partial q_i \partial q_j} < 0$ for output being strategic substitutes and then the variation in expected profit of firm i due to a change in q_i when firm j is in the bad state of nature is greater than the corresponding change when firm j is in the good state of nature. Therefore, (A6) is negative and (A4) is negative as well (Q.E.D.).